Scaling Urban Environmental Challenges

Scaling Urban Environmental Challenges

From Local to Global and Back

Edited by
Peter J. Marcotullio
and
Gordon McGranahan

London • Sterling, VA

First published by Earthscan in the UK and USA in 2007

Copyright © International Institute for Environment and Development and United Nations University/Institute of Advanced Studies, 2007

All rights reserved

ISBN: 978-1-84407-323-8   paperback
       978-1-84407-322-1   hardback

Typesetting by JS Typesetting Ltd, Porthcawl, Mid Glamorgan
Printed and bound in the UK by Cromwell Press, Trowbridge
Cover design by Philip Peake
Cover photo by Mark Edwards/Still Pictures

For a full list of publications please contact:

Earthscan
8–12 Camden High Street
London, NW1 0JH, UK
Tel: +44 (0)20 7387 8558
Fax: +44 (0)20 7387 8998
Email: earthinfo@earthscan.co.uk
Web: **www.earthscan.co.uk**

22883 Quicksilver Drive, Sterling, VA 20166-2012, USA

Earthscan is an imprint of James and James (Science Publishers) Ltd and publishes in association with the International Institute for Environment and Development

A catalogue record for this book is available from the British Library

Library of Congress Cataloging-in-Publication Data

Scaling urban environmental challenges : from local to global and back / edited by Peter J. Marcotullio and Gordon McGranahan.
    p. cm.
  Includes bibliographical references.
  ISBN-13: 978-1-84407-323-8 (pbk.)
  ISBN: 1-84407-323-8 (pbk.)
  ISBN-13: 978-1-84407-322-1 (hardback)
  ISBN-10: 1-84407-322-X (hardback)
  1. Urbanization–Environmental aspects. 2. Cities and towns–Growth. 3. Urban ecology. 4. Sustainable development. I. Marcotullio, Peter, 1957– II. McGranahan, Gordon.
  HT361.C33 2006
  307.76–dc22
                                                                2006021887

The paper used for the text pages of this book is FSC certified. FSC (the Forest Stewardship Council) is an international network to promote responsible management of the world's forests.

Printed on totally chlorine-free paper

Contents

List of Figures, Tables and Boxes		*vii*
List of Contributors		*xi*
List of Acronyms and Abbreviations		*xv*

1 **Scaling the Urban Environmental Challenge**   1
 Peter J. Marcotullio and Gordon McGranahan

2 **Urban Transitions and the Spatial Displacement of Environmental Burdens**   18
 Gordon McGranahan

3 **Variations of Urban Environmental Transitions: The Experiences of Rapidly Developing Asia-Pacific Cities**   45
 Peter J. Marcotullio

4 **In Pursuit of a Healthy Urban Environment in Low- and Middle-income Nations**   69
 David Satterthwaite

5 **Improving Urban Water and Sanitation Services: Health, Access and Boundaries**   106
 Kristof Bostoen, Pete Kolsky and Caroline Hunt

6 **Poverty and the Environmental Health Agenda in a Low-income City: The Case of the Greater Accra Metropolitan Area (GAMA), Ghana**   132
 Jacob Songsore and Gordon McGranahan

7 **Dynamics of Growth and Process of Degenerated Peripheralization in Delhi: An Analysis of Socio-economic Segmentation and Differentiation in Micro-environments**   156
 Amitabh Kundu

8	Motorization in Rapidly Developing Cities *Yok-shiu F. Lee*	179
9	A Comparative Perspective on Urban Transport and Emerging Environmental Problems in Middle-income Cities *Jeff Kenworthy and Craig Townsend*	206
10	Fixing Environmental Agendas in Mexico *Priscilla Connolly*	235
11	In Pursuit of the Sustainable City *Graham Haughton*	274
12	The Metabolism of Urban Affluence: Notes from the Greater Manchester City-region *Joe Ravetz*	291
13	Locating the 'Local Agenda': Preserving Public Interest in the Evolving Urban World *Jeb Brugmann*	331

Index   *355*

List of Figures, Tables and Boxes

Figures

2.1	A stylized urban environmental transition	22
2.2	Consumption of CFCs and halons pre- and post-Montreal Protocol	32
5.1	The water balance	107
5.2	F-diagram	110
5.3	Variation on access in various surveys	114
5.4	Relation between water consumption and time involved in water collection	116
5.5	Scales of the urban environment, as seen by a householder	117
5.6	Scales of water supply infrastructure matched to the urban environment	119
5.7	Water Supply infrastructure and priorities, as seen by technical professionals	120
5.8	Relationship between urban water access, national water stress and national GDP per capita	125
8.1	Urban air pollution problems and level of economic development	181
8.2	Estimated growth in ownership and use of motor vehicles since 1950	183
9.1	Simple generic model of urban transport and land-use evolution in developing cities	218
12.1	Urban environment framework	294
12.2	Greater Manchester location	295
12.3	From material flow to information flow	297
12.4	Urban transitions and cumulative effects	303
12.5	Transitions in production: The sustainable firm	304
12.6	Transitions in consumption: Shifting boundaries	307
12.7	Resource flow framework	317
12.8	The global urban environmental system	318
12.9	Breakdown of CO_2 emissions and ecological footprint	322

Tables

2.1	Carbon dioxide emissions per capita for Beijing and Tokyo	30
3.1	Comparative dates of similar GDP per capita levels in selected Asia-Pacific economies, the US, UK and France, 2000	55
4.1	Estimates for the proportion of people without 'adequate' provision for water and sanitation in urban areas	80
4.2	Infant and child mortality rates for urban and rural populations in selected nations, estimated mortality rates amongst infants (age less than 1) and children (ages 1–4 years)	84
4.3	Infant and under-five mortality rates and diarrhoea prevalence in Kenya	86
5.1	Health impacts of water- and sanitation-related diseases	109
5.2	Water supply and sanitation technologies considered to be improved and unimproved in the WHO/UNICEF Global Assessment 2000	112
5.3	Scales of the urban environment and water, sewerage and drainage and solid waste-related infrastructure issues	118
6.1	Access to water and sanitary services by wealth quintile of household (%)	137
6.2	Relationship between wealth and pests, and pest-control methods (%)	140
6.3	Relationship between wealth and indoor air and housing problems	141
6.4	Socio-economic characteristics and children's diarrhoea prevalence (%)	143
6.5	Access to environmental services and children's diarrhoea prevalence (%)	144
6.6	Efficiency of environmental service and children's diarrhoea prevalence (%)	145
6.7	Crowding and children's diarrhoea prevalence (%)	145
6.8	Hygiene behaviour and children's diarrhoea prevalence (%)	146
6.9	Number of risk factors a household faces, and children's diarrhoea prevalence	146
6.10	Residential sector and children's diarrhoeal prevalence (%)	147
6.11	Approximate relative risk of environmental factors with respect to diarrhoea among children under age six	147
6.12	Environmental risk factors and children's acute respiratory infection prevalence	148
6.13	Environmental risk factors and prevalence of respiratory problem symptoms among (female) principal homemakers	149
6.14	Approximate relative risk of environmental factors for respiratory disease	150
6.15	Summary age-adjusted mortality differentials between socio-environmental zones in Accra (Ghana) 1991; mortality rates (per 10,000) and Relative risks (RR)	151

List of Figures, Tables and Boxes ix

7.1	Demographic profile of the national capital territory of Delhi: 1951–2001	158
7.2	Workers (work participation rates) by principal and principal+subsidiary status in Delhi and India in 15+ age groups in various National Sample Survey rounds (%)	161
7.3	Households with access to basic amenities across towns in Delhi in 1991 (%)	172
8.1	Geographic distribution of motor vehicles, 1980–1996	183
8.2	Population distribution and number of cars in different economic groups, 1990	184
8.3	Estimates of daily vehicular trips made by different modes of travel within different economic groups, 1990	185
8.4	Patterns of private transportation in cities of different levels of development, 1990	185
8.5	Patterns of private transportation in wealthy and developing Asian cities	186
8.6	Modal split, motorization and per capita GNP for selected metropolitan areas	188
8.7	Projected growth of global motor vehicle fleet by national income level, 1995–2050	189
8.8	Emissions from transport: Local, regional, and global effects	190
8.9	Extent of exceedance of air concentrations of transport-related pollutants	192
8.10	Contribution of motor vehicles to conventional pollutant emission in selected cities	193
8.11	Estimates of roadside populations by economic group, 1990	194
9.1	Urban areas in the Millennium Cities Database for Sustainable Transport, grouped according to region and income groups, and populations (millions)	208
9.2	Land-use and transport system characteristics by groupings of cities, 1995	212
9.3	Selected transport factors with a key influence on environmental quality, normalized according to affluence	228
11.1	Dubious dualisms	281
12.1	Vital trends and statistics in Greater Manchester	296
12.2	Key transitions and implications for the UET	310
12.3	Spatial dynamics and environmental metabolism	312
12.4	Urban hierarchy and environmental metabolism	314
12.5	Material flows in the UK regional economy	320
12.6	Global footprint comparisons and trends	324
13.1	Institutional origins of LA21 planning	338
13.2	LA21 activity by national GNP	344
13.3	Substantive focus of LA21 processes, by region and income level	351

Boxes

4.1 Bangalore's water and sanitation problems, despite its
 prosperity and economic success 81

List of Contributors

Kristof Bostoen is a research fellow at the London School of Hygiene and Tropical Medicine (LSHTM). Before joining the school he worked for various organizations such as Médecins Sans Frontières (MSF), International Committee of the Red Cross (ICRC), OXFAM and Save the Children providing water and sanitation services in conflict situations. He undertook consultancies for the World Health Organization (WHO), United Nations Children's Fund (UNICEF), United Nations Human Settlements Programme (UN-Habitat), United Nations Environment Programme (UNEP), United Nations Development Programme (UNDP), ICRC and Save the Children in a wide variety of countries. Currently his main interests are developing household survey methodologies for the water and sanitation sector with a particular interest in alternative representative sampling methods that do not require detailed sample frames. Apart from lecturing on various courses, he is the study unit organizer for the advance distance-learning course Water and Sanitation at the University of London.

Jeb Brugmann has worked with local communities to help them assert their development objectives in a globalizing world for more than 20 years and in more than 30 countries. From 1983–1989 he was a leading actor in the North American 'municipal foreign policy' movement. In 1990 he founded The International Council for Local Environmental Initiatives (ICLEI), the international environmental agency for local governments. As ICLEI Secretary General from 1990–2000, he founded the worldwide Local Agenda 21 (LA21) and Cities for Climate Protection campaigns. Since that time he has worked as a development strategy consultant for major cities, development agencies, non-governmental organizations (NGOs) and global companies. Recognizing the increasing dominance of markets over development, in 2004 he co-founded The Next Practice (TNP), the leading management consultancy specializing in low-income or 'base of pyramid' markets. As a TNP Partner he works with large corporations, NGOs and community-based entrepreneurs to develop scalable businesses that provide product solutions and livelihood opportunities for underserved, low-income populations.

Priscilla Connolly, British born, is a distinguished professor at the Universidad Autónoma Metropolitana, Azcapotzalco, Mexico City, where she has lived and worked since 1972 both in the non-governmental sector and in academic institutions. Her published research embraces a wide range of urban problems including housing policy, irregular settlement and land markets, transport, employment and urban segregation, as well as contemporary and historical studies of the construction industry, public investment, employment and local government finances. After coordinating the setting up of a Geographic Information System for the Mexico City Urban Observatory (OCIM-SIG), she is currently researching into the impact of digital revolution on the way people envisage, plan and experience the city, including its environmental risks and impacts.

Graham Haughton is professor of human geography at the University of Hull, UK.

Caroline Hunt is a freelance consultant in environmental and public health. Her background is in environmental epidemiology. She has undertaken research, consultancy and public health service work in a broad range of countries, as well as the UK. She worked and taught at LSHTM for seven years. During this time one of her roles was as Associate Director of the Water and Environmental Health at London and Loughborough (WELL) Resource Centre, a resource centre for the UK Department for International Development.

Jeff Kenworthy is professor in sustainable cities in the Institute for Sustainability and Technology Policy at Murdoch University in Perth, Western Australia. He has been working for 27 years in the urban transport and planning field and is co-author of *Cities and Automobile Dependence: An International Sourcebook* (1989), *Sustainability and Cities: Overcoming Automobile Dependence* (1999) and *An International Sourcebook of Automobile Dependence in Cities, 1960-1990* (1999) and over 200 other publications in the field.

Pete Kolsky is a senior water and sanitation specialist at the Energy and Water Department of the World Bank. His main current responsibilities include support to Bank clients and staff to increase the quality and quantity of Bank activities in sanitation and hygiene. Pete was formerly a Senior Lecturer at LSHTM, where he worked with several of the models described in this paper.

Amitabh Kundu is professor and dean of the School of Social Sciences at Jawaharlal Nehru University and a member of the National Statistical Commission. He has been visiting professor at University of Amsterdam, Maison des Sciences de L'homme, Paris, University of Kaiserslautern and South Asian Institute Heidelberg, Germany and a member of various advisory committees/research teams set up by UNDP, United Nations Educational, Scientific and Cultural Organization (UNESCO), United Nations Centre for Human Settlements (UNCHS),

International Labour Organization (ILO), Government of Netherlands, University of Toronto, Sasakawa Foundation, etc. He has been Director at the National Institute of Urban Affairs, Indian Council of Social Science research and Gujarat Institute of Development Research. Currently he is on the Editorial Board of *Manpower Journal*, *Urban India*, *Journal of Educational Planning and Administration* and *Indian Journal of Labour Economics*. He has about 20 books and 200 research articles, published in India and abroad, to his credit.

Yok-shiu F. Lee is associate professor in the Department of Geography at the University of Hong Kong. His current research interests include automobiles in China's cities, urban water resources management in southern China, and urban cultural heritage management in China.

Peter J. Marcotullio is research fellow at the United Nations University (UNU) Institute of Advanced Studies, UNU Office in New York. From 1999–2006 he taught in the Department of Urban Engineering at the University of Tokyo.

Gordon McGranahan is Director of the Human Settlements Group at the International Institute for Environment and Development (IIED). He spent the 1990s at the Stockholm Environment Institute, where he directed their Urban Environment Programme.

Joe Ravetz is Deputy Director at the Centre for Urban and Regional Ecology (CURE) at the University of Manchester. His landmark study *City-Region 2020 – Integrated Planning for a Sustainable Environment* set up a new agenda in planning for sustainable urban and regional development. His current research focuses on environment–development modelling, spatial policy, future studies and evaluation/assessment. In each of these there is both an analytic agenda and great potential in using new technology for interactive tools.

David Satterthwaite is a senior fellow at IIED, and a member of the teaching staff at the University College London and the London School of Economics. He is editor of the journal, *Environment and Urbanization*.

Jacob Songsore is a professor in the Department of Geography and Resource Development, University of Ghana, Legon, former head of department (1996–1999), and since 2003, he has also been the dean of the School of Research and Graduate Studies. He is a member of the New York Academy of Sciences (1995) and fellow, Ghana Academy of Arts and Sciences (2006). His specialties include urban studies and regional development planning with special research interest in urban environmental health.

Craig Townsend is an assistant professor in the Department of Geography, Planning and Environment at Concordia University in Montreal, Canada, where he teaches courses on urban transportation and design, and planning in the developing world. He is currently researching the development of rail rapid transit systems and their impact on built environments in Bangkok. Previously he worked as a research associate at Western Australia's Planning and Transport Research Centre, and as a planner in Thailand. He holds a PhD from Murdoch University's Institute for Sustainability and Technology Policy in Perth, Australia.

Acronyms and Abbreviations

AMA	Accra Metropolitan Area
APEC	Asia Pacific Co-operation
APHRC	African Population and Health Research Center
ARI	acute respiratory infection
B2B	business to business
B2C	business to consumer
BAPEDAL	Badan Pengendalian Dampak Lingkungan (Indonesia Environment Impact Management Agency)
BMR	Bangkok Metropolitan Region
BPEO	Best Practical Environmental Option
C2C	consumer to consumer
C2B	consumer to business
CBD	central business district
CBO	community-based organization
CCP	Cities for Climate Protection Campaign
CDS	City Development Strategies
CFCs	chlorofluorocarbons
CTT	compressed and telescoped transition framework
CURE	Centre for Urban and Regional Ecology
DAC	Development Assistance Committee (OECD)
DALY	disability adjusted life year
DDA	Delhi Development Authority
DEFRA	Department of Environment, Food and Rural Affairs (UK)
DHS	Demographic and Health Survey
DFT	Department for Transport (UK)
DMC	Delhi Municipal Corporation
DMI	Direct Material Input
DOE	US Department of Energy
DPCB	Delhi Pollution Control Board
DPSIR	drivers, pressures, state, impact, response
DTI	Department of Trade and Industry (UK)
EF	ecological footprint
EKC	Environmental Kuznets Curve hypothesis
EPA	US Environmental Protection Agency

EU	European Union
FDI	foreign direct investment
g	gram
GA2000	Global Assessment 2000 Report
GAMA	Greater Accra Metropolitan Area, Ghana
GDP	gross domestic product
GFCF	Gross Fixed Capital Function
gha/cap	global hectares per person
GM	Greater Manchester
GNP	gross national product
GRP	gross regional product
ha	hectare
ICLEI	International Council for Local Environmental Initiatives
ICRC	International Committee of the Red Cross
ICT	information and communications technology
IIED	International Institute for Environment and Development
ILO	International Labour Organization
IMECA	metropolitan air quality index (Mexico)
IMF	International Monetary Fund
INGO	international non-governmental organization
IPAT	Impact = Population times Affluence times Technology
IPC	integrated pollution control
IPPC	Integrated Pollution Prevention and Control
IULA	International Union of Local Authorities
JICA	Japan International Cooperation Agency
JMP	Joint Monitoring Programme
LA21	Local Agenda 21
LCA	lifecycle analysis
LGO	local government organization
LPG	liquefied petroleum gas
LRT	Light Rail Transit
LSHTM	London School of Hygiene and Tropical Medicine
MFP	municipal foreign policy
MICS	Multi Indicator Cluster Survey (UNICEF)
MILAGRO	Megacity Initiative: Local and Global Research Observations
MSF	Médecins Sans Frontières
MWA	Metropolitan Waterworks Authority, Bangkok
NAAQS	National Ambient Air Quality Standards
NASA	National Aeronautics and Space Administration
NCR	national capital region
NCT	national capital territory
NSF	US National Science Foundation
NSS	National Sample Survey
NDMC	New Delhi Municipal Corporation
NGO	non-governmental organization

OCIM-SIG	Mexico City Urban Observatory
ODA	overseas development assistance
OECD	Organisation for Economic Co-operation and Development
PAN	National Action Party
PEMEX	Petróleos Mexicanos
PICCA	Integral Programme against Atmosphere Pollution in the Mexico City Metropolitan Zone
PPP	purchasing power parity
PRD	Democratic Revolutionary Party
PROAIRE	Programme for Improving Air Quality in the Valley of Mexico
R&D	research and development
RR	Relative risks
RUD	Rural and Urban Programme (WHO)
SDP	State Domestic Product
Sida	Swedish International Development Cooperation Agency
SJJ	Slum and Jhuggi Jhompri Department (of the DMC)
SME	small- and medium-sized enterprise
SPARC	Society for the Promotion of Area Resource Centres
STEE	staged-type environmental evolution model
TNP	The Next Practice
TSP	total suspended particles
UA	urban agglomeration
UET	Urban Environmental Transition hypothesis
UITP	International Association of Public Transport
UN	United Nations
UNCED	United Nations Conference on Environment and Development
UNCHS	United Nations Centre for Human Settlements (now UN-Habitat)
UNDESA	United Nations Department of Economic and Social Affairs
UNDP	United Nations Development Programme
UNEP	United Nations Environment Programme
UNESCAP	United Nations Economic and Social Commission for Asia and the Pacific
UNESCO	United Nations Educational Scientific and Cultural Organization
UN-Habitat	United Nations Human Settlements Programme (previously UNCHS)
UNICEF	United Nations Children's Fund (formerly United Nations International Children's Emergency Fund)
UNU	United Nations University
USAID	United States Agency for International Development
WHO	World Health Organization

WTO	World Trade Organization
WELL Resource Centre	Water and Environmental Health at London and Loughborough Resource Centre
WSSD	World Summit on Sustainable Development
WWF	World Wide Fund for Nature

1
Scaling the Urban Environmental Challenge

Peter J. Marcotullio and Gordon McGranahan

Introduction

Urbanization and economic growth are the two quantifiable trends of recent world history most closely associated with conventional 'development'. Like development itself, they represent disputed concepts. The use of 'rural' and 'urban' to describe different forms of human settlement is now being called into question by new indicators and settlement patterns (Champion and Hugo, 2004). Both the measures and meanings of economic growth are also being increasingly challenged. Yet, poorly understood and quantified though they may be, urbanization and economic growth relate to phenomena that are undoubtedly changing humanity and the world we live in – and not just its urban or affluent parts. Moreover, both trends are intimately linked to radical changes in governance. Shifts in governance are even less amenable to quantification, but equally central to the challenges we now face.

Over the course of the 20th century, the world's urban population increased 13-fold, expanding from about 13 per cent of the total population in 1900 to almost half by 2000 (UN, 1980, 2004). Over the same period, economic production grew 19-fold, with gross domestic product (GDP) per capita increasing from about US$1300 (1990 prices) to US$6100 (IMF, 2000; Van den Berg, 2002). These two trends were interdependent. The economic growth characteristic of the industrial revolution was heavily dependent on urbanization, and vice versa.

Both urbanization and economic growth have been experienced unevenly. Urbanization is uneven by definition: if half the world's population becomes urban, the other half remains rural. In practice, economic growth has also been experienced unevenly: at the end of the 20th century, almost half of the world's population was still living on less than US$2 per day (Chen and Ravallion, 2004) – roughly half of average earnings in 1900. And while the economic growth of the last two centuries has been concentrated in urban centres, both urbanization and economic growth have been concentrated in certain regions and countries.

Alongside the challenge of inequality, some of the greatest challenges posed by urbanization and economic growth have been environmental. These challenges range in scale from the local environmental health problems that result in the deaths of millions of people every year, to the global climate change that threatens to disrupt our life-support systems in the coming decades.

The combination of urbanization and economic growth is often associated with the classical city-scale environmental problems of urban smog, polluted urban waterways and peri-urban resource degradation, which have arisen partly in response to the location of populations and economic production in urban centres. From this perspective, it might seem that the environmental impact of urbanization has been to concentrate environmental impacts spatially.

The dominant trend in recent decades, however, has been a globalization of the environmental burdens, particularly for societies with the highest levels of urbanization and economic output per capita. The effects of these global environmental burdens are delayed, but they are beginning to make themselves felt. Just in the past year, for example, new evidence has been found on climate change, and its impacts on the physical climate, the hydrological cycle and the functioning of ecosystems (Levin and Pershing, 2006). A disproportionate share of the fuel combustion driving greenhouse gas emissions takes place in affluent urban settlements. An even greater share of the final consumption of the goods that greenhouse gas-emitting processes produce is located in these same settlements.

At least superficially going against this globalizing trend, the impacts of many of the most life-threatening urban environmental burdens have actually become increasingly localized over the past two centuries. Urban middle classes and élites in the industrializing cities once lived in fear of epidemics – or what many believed to be gaseous miasmas – rising out of the overcrowded and unsanitary slums. These epidemics also spread from city to city, and country to country. Now, most sanitation-related diseases are endemic to deprived settlements and neighbourhoods. In many urban settlements in Africa, Asia and Latin America, a third or more of the population still live in overcrowded and unsanitary slums. However, their diseases are rarely a serious threat to affluent residents within the affected urban settlements, let alone to urban residents in other cities and countries.

Urbanization has undoubtedly contributed to many of the environmental problems we now face, but it also brings many environmental opportunities. From a population perspective, fertility rates are more inclined to fall in urban settlements. From the perspective of living and working environments, urban settlements provide returns to scale in the provision of piped water and sewerage networks and the distribution of clean fuels. By concentrating polluters and resource users, urban settlement makes them easier to manage and regulate effectively and equitably. From the perspective of global environmental burdens, compact urban settlement has many advantages over rural or suburban sprawl.

Governing and regulating urban environmental burdens has, however, been anything but simple. The model of sanitary reform developed in the late 19th and early 20th centuries is clearly outdated, but most contemporary approaches also have their problems. Conventional forms of resource and pollution management developed since the 1950s, consisting of command-and-control techniques and regulations, have not achieved what was hoped for (Speth, 1992; Holling and Meffe, 1996; Davies and Mazurek, 1999; Folke et al, 2005). A recent assessment of global ecosystems warns of further environmental deterioration with greater impacts on human well-being and ecosystems if new methods for protecting the environment are not found (Hassan et al, 2005). Clearly, addressing the increasing complexity of environmental condition and trends requires an appreciation of the interaction between the scales of impact and governance systems.

This volume considers the full range of urban environmental burdens, from the local environmental health burdens typically associated with urban poverty, to the urban-regional pollution and resource depletion burdens typically associated with motorization and industrialization, to the global ecological footprints (EFs) typically associated with urban affluence. For the purposes of this book, the scale of an urban environmental burden is linked to its spatial extent; if the physical cause and consequence are within the same neighbourhood or district it is of local scale; if the effects span the urban settlement or extend into the surrounding region they are urban-regional; if they cross international borders, and especially if their impact is spread across the continents, they are global. The chapters also explore a range of other spatial aspects to urban environmental burdens and how they relate to economic status, and political influence. Three recurrent questions, addressed from several different perspectives in different chapters of the book, are:

1 How are the spatial characteristics of urban environmental burdens changing?
2 What are the socio-economic and political causes and consequences of these changes?
3 What are the implications for urban environmental policy?

Central to our exploration of these questions is the concept of 'urban environmental transition', and the claim that conventional urban growth and economic development is associated with a shift from immediate, local environmental burdens whose primary impact is on human health, towards delayed and dispersed environmental burdens whose primary impact is on life-support systems (McGranahan et al, 2001). This volume challenges the more economistic interpretations of this transition, and elaborates its political and social aspects. There is general support in this book for the thesis that understanding environmental transitions requires an integrated analysis of biophysical trends and socio-economic development and governance systems (see, for example, Holling, 2001). None of the authors presents the transitions as inevitable, however, or treats the transition model as a sound basis for predicting the future. Rather, the urban environmental transition becomes a conceptual tool to be used critically, often in challenging the prevailing environmental agendas.

This introduction is divided into three further sections. The following section considers the physical and spatial aspects of urban environmental transitions, and how these are addressed in the individual chapters. Following the structure of the book, this section looks first at environmental health agendas in the urban centres of comparatively poor countries, then at motorization and urban environmental agendas in middle-income countries, and finally the pursuit of sustainable urban development in more affluent countries. The next section turns to the changing spatial characteristics of governance and politics, and how these relate to the urban environmental challenges. The final section considers what this all implies for our understanding of urban environmental transitions, and the challenge of addressing the multi-scaled urban environmental burdens.

Urban environmental transitions

For many years there has been controversy over the relationship between economic growth and environmental burdens. At one extreme are those who portray economic growth as inherently destructive of the environment, and at the other those who portray it as environmentally beneficial (or, in more sophisticated versions, initially destructive but eventually beneficial). In Chapter 2, this relationship is examined from a spatial perspective, the author arguing that urban affluence has been associated with more extensive environmental burdens, and that economic growth can be made to look more or less environmentally destructive depending upon which scale of burden the focus is on. By picking local environmental health problems, such as bad household sanitation, economic growth can be made to look good (at least if a significant share of the economic benefits reach those facing the environmental health problems). Alternatively, by picking global sustainability burdens such as greenhouse gas emissions, it can be made to look bad. The relationship also looks different depending on whether urban environmental burdens are assessed in terms of health impacts, economic costs or ecological footprints. Overall, partly because of the spatial shifts, the environmental burdens of affluence tend to amplify economic inequalities.

The patterns we see nowadays, however, are quite different than those experienced in the past. In Chapter 3, the author argues that the urban environmental transitions currently underway in Asia have been affected by what he terms time–space telescoping, with the result that their environmental challenges occur at lower levels of income, rise faster and overlap more than were the experience in most Western cities.

Environmental health agendas and urban poverty

While the 'sanitary revolution' is often linked to 19th century Europe, even official figures for 2000 put the number of urban dwellers without access to improved sanitation at over 400 million (WHO and UNICEF, 2000), and this is a conservative estimate (UN-Habitat, 2003). Inadequate sanitation often goes along with insufficient water, crowding and a range of other health-threatening environmental burdens, which affect a large share of the population in most urban centres in low- or middle-income countries. These urban centres are often also plagued by ambient air pollution and industrial water pollution, and in almost every city there is a wealthy élite whose lifestyles impose a large 'ecological footprint'. At least for those who live in the more deprived neighbourhoods, however, it is the environmental health burdens that tend to be the most immediate and the most severe.

In Chapter 4, the environmental health problems that so often afflict the urban poor are assessed. Improving environmental health conditions in deprived urban settlements is among the stated priorities of local governments, national governments and international agencies, but sustained commitments and progress have been disappointing at every level. In the international arena, a concern about 'urban bias' has shifted efforts at poverty alleviation to rural settings, although in practice the bias does not extend to deprived urban communities. Indeed, there is often a local bias against improving conditions in deprived urban neighbourhoods, justified by the claim that improvements will attract rural migrants, undermining the improvements and adding to urban problems.

While data on urban environmental health conditions are often lacking, there is more than enough evidence to demonstrate that the burden they impose on the more deprived residents is very high and that there is much that can be done to reduce it. Chapter 4 also examines how to go about strengthening the capacity of local governments, and of the urban poor groups who need to work with them, or call them to account.

Water, sanitation and hygiene deficiencies are at the centre of the environmental health problems experienced by deprived urban communities. In affluent cities, where the sanitary revolution is associated with the 19th century, it is all too easy to think water, sanitation and hygiene deficiencies are simple problems, whose solutions are known but not always implemented, for reasons of poverty, incompetence or corruption. The situation presented in Chapter 5 is a rather different one. While our understanding of water-related diseases has improved over the years, the diseases involve a range of different pathogens, and a wide variety of different routes of transmission. There are still a number of common misconceptions about these diseases, and a tendency to focus too narrowly on water quality, to the neglect of the many other pathways through which these diseases spread.

A number of the challenges to improving water, sanitation and hygiene have a strong spatial dimension. Particularly where infrastructure is deficient, there are what the authors refer to as 'boundary problems', ranging from conventional economic externalities to administrative boundaries that are poorly aligned with the limits of physical systems. There are also serious challenges in translating large-scale improvements (e.g. better river water quality) into improvements at the household scale, and important distinctions between the public and domestic domains of disease. Recent epidemiology stresses the need to create the right conditions for households to manage the domestic domain, while spatial focus of water and sanitation engineers tends to be elsewhere.

The importance of domestic environments, including water, sanitation and hygiene in Accra, Ghana, is examined in detail in Chapter 6. Ghana is considered a low-income country, and by international standards average incomes in Accra are low. There are, however, appreciable intra-urban differentials in domestic environmental conditions. Whether in terms of indoor air pollution, water, sanitation, pests or solid waste, environmental hazards are more evident the poorer the household. Moreover, the burdens fall differently within the households: women are more exposed to domestic hazards than men because of their roles within the home, while children and infants are especially vulnerable.

Chapter 7 explores the dynamics of growth and environmental change in Delhi, and the extent to which there has been a shifting of environmental burdens from the core to the periphery, for political as much as physical reasons. The author identifies a process of socio-economic segmentation, and an unequal sharing of environmental costs at the micro-level.

Motorization and urban environmental agendas in middle-income countries

When sanitary reforms began in the 19th century, motorization was not an issue: urban air pollution was largely from industrial emissions and domestic fuel combustion. Now motor vehicles pollute the air in even the poorest urban areas, and are among the persistent polluters in affluent cities. Moreover, as described in Chapter 8, for the middle-income Asian countries now experiencing rapid growth, motorization is extremely rapid and reflects what has been described as a 'compressed and telescoped transition' (Marcotullio and Lee, 2003). Globally, the number of motor vehicles has been growing at a rate of about 5 per cent a year – more than twice the rate of population growth – and a disproportionate share of this growth is taking place in middle-income urban centres.

More generally, as demonstrated in Chapter 9, urban transportation provides a particularly revealing example of how urban environmental burdens change over time and with economic growth. Firstly, urban development is itself inextricably linked to transportation, and the spatial configuration of a city's transportation system is critical to the character and functioning of the city. Secondly, and more important from a research perspective, the Millennium Cities Database (Kenworthy and Laube, 2001) for Sustainable Transport provides the sort of information needed to undertake an international comparison of transports systems, policies and environmental implications, particularly for what are, according to the World Bank's classification, middle-

and high-income countries. Chapter 9 employs this database to examine six different dimensions of urban transport systems, ranging from private motor vehicle ownership to the environmental impacts of urban transport.

If you look closely at individual cities, of course, the changes in environmental burdens and agendas are far more complex than even the best statistics can reveal. As illustrated for Mexico City in Chapter 10, environmental issues and agendas arise in a far from linear fashion. By examining the evolving agendas for water and air pollution, and how they are historically fashioned and politically situated, the author provides a robust antidote to the over-interpretation of statistical trends. Indeed, this complex and historically grounded discussion of how agendas are constructed and fought over is presented as revealing that any notion of an urban environmental transition, relevant to urban centres around the world, is simplistic and misleading. We would argue, however, that this sort of analysis complements the work on urban environmental transitions, exposing only how this work should not be misinterpreted.

Environmental sustainability and urban affluence

Although congestion and ambient air pollution and city water problems are quintessentially urban, and in most senses thoroughly modern, as environmental issues they seem somewhat dated. The latest issues tend to be global, long-term, and a threat to the sustainability of the life support systems we depend on (and to the world economy); global climate change represents far better the latest set of issues to rise to the top of the environmental agenda, at least internationally. One could argue that this international shift in focus has occurred because environmental burdens are becoming more global, and environmentalists are quite justifiably concerned about the future of the planet. On the other hand, one could argue that it has occurred because localized environmental burdens are increasingly restricted to low- and middle-income settings, and environmentalists are more concerned with the environmental burdens of affluence. There is probably some truth to both these claims, but as both the chapters addressing high-income cities demonstrate, it is multi-scaled politics and multi-scaled environmental burdens that characterize, and sometimes confuse, the urban environmental scene in the wealthier parts of the world.

In Chapter 11, the environmental challenges facing the affluent city are examined. The author notes at the outset that 'many of the environmental problems and risks which we might associate with cities are actually felt outside the city and are rooted in social and economic processes that operate without regard to urban boundaries'. Like Chapter 10, however, this chapter focuses more on the politics than the physical characteristics of urban environmental transitions. On the one hand, the author provides an account of the political economy of risk displacement, which has helped to drive shifts in the scale of environmental burdens. On the other hand, he also examines the politics of environmental agenda setting, and provides a critique of simple dualisms, and of the notion that there is an apolitical means of influencing environmental agendas with scientific findings.

Chapter 12 reviews 'the metabolism of urban affluence' and makes very clear the city of Manchester has as complex and challenging an urban environmental agenda as any. The author demonstrates the large resource consumption and waste production levels of the Greater Manchester (GM) Region. With approximately 6 million car trips per day and an energy usage of 90 billion kilowatt hours per year, the city region emits approximately 32 million tonnes of CO_2 per year. Its ecological footprint is approximately 125 times the size of the physical land area of the city region, and the total material consumption of the average person in the city is 25 tonnes per annum.

Local Agenda 21 (LA21), examined by Chapter 13, represents an internationally networked attempt to help urban centres, from the poorest to the most affluent, to address their changing and varied environmental burdens. Part of the challenge for LA21, as for most other initiatives to address urban environmental burdens, is that the political context as well as the physical character of urban environmental burdens is changing. Indeed, the re-scaling of politics is as important as the re-scaling of the burdens themselves, and has had a profound effect on the urban environmental challenge.

Re-scaling politics and the urban environmental challenge

Every chapter in this book is, in one way or the other, about urban environmental governance and scale, although approaches are different. Firstly, most chapters point out the various ways in which urban environmental burdens and governance have been re-scaled over the last few decades. The notion of re-scaling has encouraged debate over which level of governance (local, national or international) is most promising for addressing challenges. Secondly, several chapters examine the politics inherent in the way the scale of urban environmental burdens are presented and debated.

Among environmentalists, the treatment of issues of scale and governance depends on the political setting. For example, environmentalists trying to convince the Manchester authorities to take measures to help address climate change are likely to emphasize its local dimensions. They are unlikely to point out that Manchester could reduce its carbon emissions 10-fold only to be submerged by a climate-induced sea-level rise resulting from the carbon emissions of others. Alternatively, environmentalists trying to convince international donors to fund sanitary improvements in Accra are likely to present that city's deficiencies as part of a global problem, rather than a household or neighbourhood issue.

Unfortunately, issues of environmental scale are notoriously complex, and eager environmentalists are among the less important political actors involved in the re-scaling of urban environmental politics. Indeed, this re-scaling is the outcome of long-term and wide-ranging political changes, only some of which are a response to changing environmental conditions.

Chapter 2, for example, explores the long-term political dimensions of the changing spatial scale of urban environmental burdens. The author finds that both the positive and negative aspects of the relationship have been forged politically. The sanitary movement, which helped address the local water and sanitation problems in early industrial cities, was political. So was the broader environmental movement that followed. Equally important, the tendency to displace urban environmental burdens is also political, and is taking on new significance now that the global 'commons' is under threat. From an urban perspective, however, the global environmental challenge is not simply that of addressing new global threats, but of addressing threats at every scale, distributed unevenly but by no means randomly across the globe. Understanding the evolution of burden equity involves identifying shifts in political control over domestic and international production and consumption processes.

Certain powers once held largely at the central level of governments are being distributed to both supra- and sub-levels. The process that had been slowly advancing since the end of World War II started to accelerate in the early 1970s when deregulation of financial markets and the global credit system undermined the utility of state-level demand management and monetary policies. Thereafter, the increasing globalization of production, competition and financial flows diminished the ability of national governments to insulate the activities within their geographic boundaries from the world economy and its influences.

A result of this 're-territorialization' of power has been called the 'glocal' state (Brenner, 1998; Swyngedouw, 1996, 1997). As the world economic system becomes more globally dispersed, it remains increasingly dependent on highly localized producer networks, labour market processes and local physical infrastructure. Changes in national governance systems have been a response to and effect of 'glocalization'. The national level of state regulation has 'hollowed out', meaning that central powers are being displaced upwards towards supra-national regulatory institutions, devolved downwards towards sub-national scales of governance such as regional and local states (cities) and moving horizontally to inter-regional or trans-local organizations (Jessop, 1994).

These shifts have promoted changes in internal workings of the governments and have a direct bearing on managing environmental conditions within and of cities. During the previous Fordist-Keynesian period, governments used a variety of indirect forms of intervention into markets and social processes to promote the welfare of citizens (through redistributive social welfare policies and the provision of basic infrastructure), industrial development (through subsidies and tax concessions) and the promotion of collective consumption (e.g. through housing, education, transportation and urban planning policies).[1] With glocalization and the hollowing-out of the state, it has become increasingly difficult to regulate markets, processes and development activities at all levels. Current governmental practices have therefore shifted to more direct and locally focused economic development policies.

Some have seen the shifts in governance systems as opportunities for the localization of environmental politics and therefore promote solutions to environmental burdens, at all scales, through local and domestic policies (Hines, 2004). In Chapter 13, for example, the author starts with an explanation of the changing role of urban governance within global economic and political transformations. He suggests that local governance and policies can only be understood in the context of state re-structuring and the relationships of the local to the national and supra-national levels of organization. The autonomy within the local state that is evident within the environmental arena is due to the structural relationship of local states. He moves beyond the 'top-down' and 'bottom-up' perspectives by using the concept of 'autopoietic' for local communities, which describes systems that are primarily concerned with self-reproduction. These systems are resistant to top-down internal management and to direct intervention from outside. The result, in his view, is that local government is where local environmental policies are most likely to succeed. His conclusions suggest the need for a strong interventionist local government that can oversee a coherent implementation plan for LA21. It is only under this strong state that localities can engage in the networks necessary for creating more liveable and sustainable cities.

Chapter 7 also focuses on both local and national political trends in exacerbating, if not creating, the spatial segregation of environmental harms in Delhi. The author describes how low-income groups have been able to maintain a hold in the slums of Delhi's city centre, where they reside in proximate distance to employment and other opportunities. Increasingly, these poor people face pressure generated by a local élite that demands rapid and greater mobility, open space and no slums. Through the apparatus of the court system, for example, slum dwellers have been evicted to outer areas of the city, creating the spatial segregation of social classes. Similarly, though for somewhat different reasons, polluting industries have been moved out, compounding the burdens arising from the lack of infrastructure in the periphery. The author argues that this spatial outcome is a political result, and that national regulations, rather than solely local legislation, have been used in the creation of this 'degenerated peripheralization', so national measures are needed to redress it, even if the ultimate goal is a more participatory local control.

Not all authors within the volume, however, are as sanguine about the potential of local government in the resolution of environmental burdens. Several chapters in the volume also point out the importance of international governance. Underpinning this perspective is the understanding that increasingly old forms of environmental regulations have given way to collaborative, voluntary and market-related forms of governance. Contemporary government policies include promoting and engaging in the commodification of public services, cutting back on state-run social welfare programs, promoting public-private partnerships and local re-development projects such as transportation projects, science parks, conference centres and waterfront strategies (Brenner, 1998). These policies have not only been promoted by industrialized states to enhance capital accumulation within their own territories, but have been part of the developing world agenda through loans and international financial aid packages, and the endorsement of the privatization of public services, encouraged bilaterally and through multilateral organizations.

Chapter 4 considers two different political sets of factors inhibiting improvements in urban environmental health conditions, which may be rooted in urban poverty, but are compounded by a range of governance problems, local, national and international. The author examines the avoidance of risk transfers from low- to middle- and upper-income groups. He identifies a range of local political factors that contributes to the high health risks in many low-income communities, including weak local governments and resistance to the sorts of taxes and charges that could provide the capital base for improving housing conditions, infrastructure and service provision. Better governance systems at the local and national level, along with greater political will to create change, therefore promise to provide the basis of solutions to these problems.

Chapter 4 also identifies how international actors contribute – or fail to contribute – to the resolution of urban environmental health challenges. The author points out that during the 1970s and 1980s, international organizations reduced the already small size of programmes addressing local 'brown' issues in cities. Even UNICEF (United Nations Children's Fund), whose interest in cities was exceptional among international agencies, reduced its programme size during this period. This was a time when development assistance was being criticized for urban bias. Unfortunately for most of the urban poor, who did not benefit from the urban bias (which was more a bias towards the urban affluent), the anti-urban response did help justify the continued neglect of urban poverty issues. There was also increasing scepticism internationally towards social welfare programmes and the Keynesian state, as opinion shifted towards the 'Washington Consensus' (Williamson, 1993). By the 1990s, the World Bank had adopted a neoliberal approach to improving urban services, including water and sanitation.

Private-sector participation was the order of the day – with an emphasis on increasing the participation of large multi-nationals, not improving the participation of the many small water and sanitation enterprises already operating in many low-income settlements. The multi-nationals turned out to be far less willing to invest in expanding water and sanitation coverage than their proponents hoped, and this approach has distracted attention from more serious efforts to address the urban environmental health challenge (McGranahan and Budds, 2003).

Chapter 6 argues that in Accra, Ghana, there are international political components to the creation and maintenance of local environmental health burdens. The authors suggest that the environmental health burdens in Accra result from the class character of the structural adjustment process, which places the greatest burden of adjustment on the urban poor, and precludes the expansion of public services, which for the most part only reach the middle classes. Structural adjustment has also helped to shift gender relations that are so central to household environmental management (Songsore and McGranahan 1998). More recently, and far more in line with the neoliberal designs of structural adjustment, private-sector participation has also become a big issue in Accra. It remains to be seen whether this is having a major impact on local environmental management, or whether it is simply distracting attention from the sort of political changes that really could help deliver better environmental conditions to the urban poor.

Chapter 8 points out both national and international political dimensions to China's current motorization trend. The author notes that the expanding middle class in China has been encouraged to buy cars by international agreements and national policies. China's entry into the World Trade Organization (WTO) has required increased free trade in motor vehicles and parts. National industrial policy, with its emphasis on automobile production, has also promoted automobile consumption. At all levels, policies are converging to increase automobile consumption within the country – which, if the experience of other countries is anything to go by, is likely to have profound implications for urban form, and threatens to lock China into long-term automobile dependence.

Chapter 9 also notes that international agencies, such as the World Bank, through financing infrastructure investments promoting auto use as opposed to transit development, play an important role in transportation governance. The result has been unfortunate, as many high-density cities fail to exploit the advantages that their urban forms allow, and emphasize motorized private transport. In many middle-income cities, while their overall emissions are lower than in high-income cities, the pollution is more concentrated and the congestion more severe.

In addition to the re-scaling of governance, several chapters of this book also examine the politics inherent in the construction of urban environmental burdens, agendas and policies. Increasingly, environmental scholars are interested in how scientific work has been influenced by politics (Glantz, 1979; Ludwig, 2001). This emphasis has not only questioned the origin and purpose of scientific concepts and ideas, but also places doubt upon the certainty of our knowledge of environmental conditions (e.g. pollution levels and soil degradation) and processes (driving forces for these conditions). Those working in the field point out that previously unquestioned environmental processes, concepts, ideas or entities often work or help to secure the power of an élite community (Robbins, 2004). Therefore, an important political project is raising awareness that the specific environmental entity, process, concept or idea need not have existed, or could have existed in another form (Hacking, 1999). Arguably, the reinvention or change of our understanding of issues can provide the basis for a more sustainable future.

In the case study of Mexico City in Chapter 10, the author is sceptical of statistical data as a means of identifying environmental priorities or assessing environmental agendas. Reacting to suggestions that Mexico City's air pollution is the worst in the world, and therefore demands immediate attention, she takes a political ecological approach, and de-constructs the claim. Through an analysis of the construction of burdens, agendas and policies, she uncovers the diverse group of actors and institutions involved in attempts to address the city's environmental challenges. Her historical analysis of how the environmental agendas for water and air pollution policies have been defined demonstrates how, at least in certain critical arenas, a technology-driven clean air agenda gained primacy. At the same time, within both the water supply and air-pollution agendas, certain issues and options have been displaced, not so much physically as socially. This does not make them any less urgent. Rather, the displacement of environmental impacts of Mexico's water supply system, as well as lack of attention to the impacts of vehicle growth other than on air pollution, have been crucial hidden burdens for the city. It is not coincidence that these burdens affect disempowered groups, and that the dominant agendas are framed in such a way as to obscure them.

Chapter 11's overview of the environmental burdens of more affluent cities is also critical of taking environmental statistics at their face value, and of the simple dichotomies (city/country, global/local, affluent/poor) that so often structure environmental debates. Rather than strengthening the statistical basis for an empirical analysis of urban environmental transitions, the author concentrates on their political dimensions. In particular, the chapter focuses on the political economy of risk transference, within and beyond urban boundaries. This typically involves the externalization of harms associated with economic development, and the political minimization of their impacts. Increasingly, this involves markets: as part of their logic, cities integrated into the global economy capitalize nature, and attempt to use markets to allocate, distribute and even resolve environmental burdens. The social and economic dynamics inherent in this political economy are largely ignored by measures of environmental burden, such as EFs, which also tend to disregard the benefits that urban consumption can bring.

How does this change our understanding of urban environmental transitions?

The urban environmental transition model provides a useful framework for examining and understanding the environmental conditions and trends within and among cities. This book generally corroborates the main elements of the transition, as outlined in McGranahan et al (2001). Local environmental health problems are concentrated in the deprived neighbourhoods of the urban centres in low-income countries. City-regional burdens such as ambient air pollution are found to have a more ambiguous relation to affluence, and are often at their worst in the urban centres of middle-income countries. Global burdens are associated with affluent urban lifestyles, which are concentrated in upper-income countries. On the other hand, the chapters combine to make it abundantly clear that this transition has numerous, changing political dimensions. Moreover, the shift in the scales of urban environmental burdens is only one of a wide range of important spatial influences in the creation and distribution of environmental burdens.

Most of the empirical work on the relationship between environmental burdens and economic status has focused on the Environmental Kuznets

Curve (EKC) hypothesis – that economic growth initially increases environmental burdens, but that after a certain point continued economic growth actually decreases these environmental burdens (Stern, 2004). Even at its simplest and most economistic, the Urban Environmental Transition (UET) hypothesis suggests a more contingent relationship – localized burdens will tend to decline, global burdens will tend to increase, and city-regional burdens are the most likely to rise and then fall. More important, there is no reason to treat the relationship as one in which economic change simply drives environmental change. Indeed, it would be very surprising if the relationship between economic status and environmental burdens were not also contingent on policy choices and environmental governance. Moreover, the notion that a higher economic status brings more spatially displaced environmental burdens is almost inherently political.

Virtually every chapter in this book illustrates how government policies influence environmental burdens, their spatial scale, and their relation to economic status. Policies matter – to water and sanitation provision in deprived settlements, to air pollution and transport patterns in motorizing cities, and to the EFs of affluent urbanites. Economic growth influences and is influenced by how policies address environmental burdens, but is not an alternative to such policies. An urban environmental transition does not just result from economic growth, but from the prevailing policy agendas and regimes.

Many of the chapters also illustrate how politics affect both the environmental policy agendas and their outcomes, and hence are also central to urban environmental transitions. The politics surrounding environmental policy choices are not just about setting priorities and identifying cost-effective measures. They are also about how environmental problems are framed, how agendas are set, who wins and who loses. Moreover, the spatial dimensions of urban environmental politics do not conform to the spatial dimensions of the environmental burdens themselves. Local environmental burdens can have a global politics – as when, for example, international agencies promote multi-national water companies as a means of improving water delivery in low- and middle-income cities. Alternatively, global burdens can, at least up to a point, have a local politics. Nevertheless, the spatial scale of the environmental burdens clearly does matter, and does influence the policy possibilities and the political responses.

Considerable attention has been devoted to the scale issues inherent in emerging global problems, including most notably climate change (Wilbanks and Kates, 1999; Bulkeley and Betsill, 2003). A multitude of local actors must change their ways if greenhouse gas emissions are to be reduced, but the nature of the challenge is such that without some form of governance structure they have no economic incentive to act. The most obvious solution is to create global governance structures through which the collective need to reduce emissions is to be agreed upon, with responsibilities then delegated down to localities. Despite what might seem to be an overriding collective interest, setting up the international governance structures has proved to be extremely difficult, as the experiences with the Kyoto protocol amply demonstrate. Cities, on the other hand, have shown a surprising willingness to engage in climate change politics, and have even formed trans-national coalitions. In the US, where the national government has been particularly unwilling to make commitments, 159 local governments have signed onto the 'Cities for Climate Protection' programme under the International Council for Local Environmental Initiatives (ICLEI).[2] Whether or not these cities manage to curb their own emissions appreciably, they have clearly demonstrated the importance of what amounts to a form of multi-scaled network politics in addressing global environmental burdens.

Considerably less attention has been devoted to the scale issues inherent in the local and city-regional burdens that still plague many urban centres around the world. Here, the challenge is different. If there is a need for global governance structures, it is not just to ensure that local action conforms to collective global needs. Better local water, sanitation, air quality and waste disposal provide immediate benefits to people in the locality. There is every reason to believe that locally driven initiatives, provided that they represent the needs of their constituents, will be required for improving local conditions. The political and economic situation is often such, however, that multi-scaled network politics can still be important to addressing these local environmental burdens. For example, some of the most successful efforts to improve the living environments in slums have been through organizations of the urban poor, which form national federations and are represented internationally through 'Slum/Shack Dwellers International' (D'Cruz and Satterthwaite, 2005). Similarly, the LA21 initiatives described in Chapter 13 represent an internationally networked attempt to address a wide range of urban environmental issues.

Notes

1 The Fordian-Keynesian period in the developed world emerged after World War II and began to diminish during the 1970s. There are several defining features of this period that are relevant to urban development including the role of cities as the centres of agglomeration for industrial production (durable goods and heavy industry), the rise of industrial labour markets; expanding suburban regions associated with a growing middle class (that increasingly depended on automobiles for mobility); declining city cores where lower class and ethnic populations resided; the increasing importance of bureaucracies, the notion that many aspects of social life could be managed in similar ways as businesses. This period set the seeds for a rise of a number of different social movements focused on social justice and the city and the sceptical viewpoint that urban planning, even when well intentioned, was largely servicing the capitalist class.
2 See ICLEI North American Participants homepage at www.iclei.org/index.php?id=1121.

References

Brenner, N. (1998) 'Global cities, global states: Global city formation and state territorial restructuring in contemporary Europe', *Review of International Political Economy*, vol 5, no 1, pp1–37

Bulkeley, H. and Betsill, M. M. (2003) *Cities and Climate Change*, Routledge, London

Champion, T. and Hugo, G. (2004) *New Forms of Urbanization: Beyond the Urban-Rural Dichotomy*, Ashgate, Aldershot

Chen, S. H. and Ravallion, M. (2004) 'How have the world's poorest fared since the early 1980s?' *World Bank Research Observer*, vol 19, no 2, pp141–169

Davies, J. C. and Mazurek, J. (1999) *Pollution Control in the United States, Evaluating the System*, Resources for the Future, Washington, DC

D'Cruz, C. and Satterthwaite, D. (2005) 'Building homes, changing official approaches: The work of the urban poor organizations and their federations and their contributions to meeting the Millennium Development Goals in urban areas', International Institute for Environment and Development (IIED), London

Folke, C., Hahn, T., Olsson, P. and Norberg, J. (2005) 'Adaptive governance of social-ecological systems', *Annual Review of Environment and Resources*, vol 30, pp441–473

Glantz, M. H. (1979) 'Science, politics and economics of the Peruvian anchoveta fishery', *Marine Policy*, vol 3, no 2, pp201–210

Hacking, I. (1999) *The Social Construction of What?* Harvard University Press, Cambridge, MA

Hassan, R., Scholes, R. and Ash, N. (eds) (2005) *Ecosystem and Human Well-being: Current State and Trends*, volume I, Island Press, Washington DC

Hines, C. (2004) *A Global Look to the Local, Replacing Economic Globalization with Democratic Localization*, Earthscan, London

Holling, C. S. (2001) 'Understanding the complexity of economic, ecological and social systems', *Ecosystems*, vol 4, pp390–405

Holling, C. S. and Meffe, G. K. (1996) 'Command and control and the pathology of natural resource management', *Conservation Biology*, vol 10, no 2, pp328–337

IMF (2000) 'The world economy in the twentieth century: Striking developments and policy lessons', in *World Economic Outlook, May 2000*, International Monetary Fund, Washington, DC, pp149–180

Jessop, B. (1994) 'Post-Fordism and the state', in A. Amin (ed), *Post-Fordism: A Reader*, Blackwell, Cambridge, MA, pp251–279

Kenworthy, J. and Laube, F. (2001) *The Millennium Cities Database for Sustainable Transport*, International Union (Association) of Public Transport, Brussels and ISTP, Perth (CD-ROM publication)

Levin, K. and Pershing, J. (2006) *Climate Science 2005: Major New Discoveries*, World Resources Institute, Washington DC

Ludwig, D. (2001) 'The era of management is over', *Ecosystems*, vol 4, pp758–764

Marcotullio, P. J. and Lee, Y.-S. (2003) 'Urban environmental transitions and urban transportation systems – A comparison of the North American and Asian experiences', *International Development Planning Review*, vol 25, no 4, pp325–354

McGranahan, G. and Budds, J. (2003) *Privatization and the Provision of Urban Water and Sanitation in Africa, Asia and Latin America*, IIED, London

McGranahan, G., Jacobi, P., Songsore, J., Surjadi, C. and Kjellen, M. (2001) *Citizens at Risk: From Urban Sanitation to Sustainable Cities*, Earthscan, London

Robbins, P. (2004) *Political Ecology*, Blackwell Publishing, Malden, MA

Songsore, J. and McGranahan, G. (1998) 'The political economy of household environmental management: Gender, environment and epidemiology in the Greater Accra Metropolitan Area', *World Development*, vol 26, no 3, pp395–412

Speth, J. G. (1992) 'The transition to a sustainable society', *Proceedings of the National Academy of Sciences of the USA*, vol 89, pp870–872

Stern, D. I. (2004) 'The rise and fall of the environmental Kuznets curve', *World Development*, vol 32, no 8, pp1419–1439

Swyngedouw, E. (1996) 'Reconstructing citizenship, the re-scaling of the state and the new authoritarianism: Closing the Belgian mines', *Urban Studies*, vol 33, no 8, pp1499–1521

Swyngedouw, E. (1997) 'Neither global or local: "Glocalization" and the politics of scale', in K. R. Cox (ed) *Spaces of Globalization: Reasserting the Power of the Local*, Guilford Press, New York, pp137–166

UN (United Nations) (1980) *Patterns of Urban and Rural Population Growth*, United Nations Department of International Economic and Social Affairs, New York

UN (2004) *World Urbanization Prospects: The 2003 Revision*, United Nations, New York

UN-Habitat (United Nations Human Settlement Programme) (United Nations) (2003) *Water and Sanitation in the World's Cities: Local Action for Global Goals*, Earthscan, London

Van den Berg, H. (2002) 'Does annual real gross domestic product per capita overstate or understate the growth of individual welfare over the past two centuries?' *The Independent Review*, vol VII, no 2, pp181–196

WHO and UNICEF (2000) *Global Water Supply and Sanitation Assessment 2000 Report*, World Health Organization and United Nations Children's Fund, Geneva and New York

Wilbanks, T. J. and Kates, R. W. (1999) 'Global change in local places: How scale matters', *Climatic Change*, vol 43, pp601–628

Williamson, J. (1993) 'Democracy and the "Washington Consensus"', *World Development*, vol 21, no 8, pp1329–1336

2
Urban Transitions and the Spatial Displacement of Environmental Burdens

Gordon McGranahan

Introduction

The environmental implications of economic growth, and the economic implications of taking action to reduce environmental burdens, have been hotly contested for centuries. Even now, many claim that economic growth will inevitably destroy the environment, while others claim that it is the environment's best hope. On this overtly political question, the empirical evidence is mixed, with some environmental burdens increasing with affluence, others declining, and still others sometimes increasing and sometimes decreasing. At least from an urban perspective, however, certain patterns are evident. Specifically, whether looking at the history of currently affluent cities, or comparing different urban settlements today, increasing affluence is associated with more extensive environmental burdens. There are good reasons to think that this relationship was forged politically, and that its interpretation is also highly politicized. Perhaps more importantly, there is no reason to assume that it is immutable, and indeed, several good reasons to hope it can be changed.

Partly because urban economic growth has been accompanied by increasingly extensive environmental burdens, urban environmental burdens tend to amplify economic inequalities. This is not merely because the affluent are better able to buy protection from environmental hazards, and to buy access to distant environmental goods and services. Such market purchases undoubtedly exist, but for the most part they simply reflect rather than amplify economic inequalities. The spatial patterns of extra-market environmental burdens do amplify economic inequalities, however. The environmental justice movement in the US has emphasized one aspect of this: how the spatial concentration of environmental risks (e.g. in the form of hazardous waste sites) places a disproportionate burden on already disadvantaged national groups such as African Americans. This chapter emphasizes another aspect: how the displacement of environmental risks from affluent settlements, and the failure to displace them from deprived settlements, places a disproportionate burden on internationally disadvantaged groups such as urban slum dwellers.

The first section of this chapter provides a summary account of how environmental burdens shift with affluence, building on a presentation made some years ago in the context of a comparative study of environmental burdens in Accra, Jakarta and Sao Paulo (McGranahan and Songsore, 1996; McGranahan et al, 2001). The basic thesis is that increasing economic activities not only create more extensive environmental burdens, but also provide the capacity to address local environmental burdens – or at least to provide protection for the more affluent residents. In some cases addressing the local environmental burdens adds to the larger scale burdens, as when sewers are used to transfer waste from urban neighbourhoods to local waterways, smokestacks are installed to disperse smoke, waste is collected and then dumped on the outskirts of a city, or resources are sourced from greater distances to prevent local degradation. The more extensive environmental burdens are more delayed in their impacts, and are more likely to undermine life support systems than to affect human health directly.

The resulting relationship is summarized crudely in Figure 2.1 (p22) with three curves, indicating how the severity of different environmental burdens changes with increasing affluence. One curve shows the decline of local burdens, a second shows the rise and fall of city-regional burdens and a third shows the rise of global burdens. The table attached to Figure 2.1 gives examples of burdens at each of these scales, relating to water, waste and air pollution. Thus, in the case of air pollution, indoor air pollution represents a local burden (and tends to decline with increasing affluence), urban ambient air pollution represents a city-regional burden (and tends to first increase and then decline as affluence increases), and carbon emissions represent a global burden (which tends to increase with affluence).

The economic literature on the relationship between economic status and the environment has focused on the second curve mentioned in the previous paragraph. This relationship has come to be known as the 'environmental Kuznets curve', after the Nobel Prize-winning economist who suggested that economic growth first increased inequality, and then helped to bring about its decline. As described in the second section of this chapter, the large empirical literature on the environmental Kuznets curve does not support the claim that all or even most environmental burdens follow this pattern. On the other hand, this literature does provide valuable insights into the variety of different relationships that environmental burdens do have with economic status. Some of these studies have also identified the tendency for the less local, immediate and health-threatening burdens to continue to increase with average income.

The next section examines an evaluation of environmental and epidemiological transitions based on the recently completed Global Burden of Disease study (Smith and Ezzati, 2005). This evaluation provides a more explicit corroboration of the claim that household environmental burdens tend to decline with affluence, that global burdens increase, and that intermediate burdens tend to first increase and then decline. By using health as the metric, the authors of this evaluation are also able to examine the aggregate incidence, at least for the subset of environmental burdens for which they have evidence. Overall, the environmental health burden is found to decline with increasing average income.

Health is only one possible metric with which to measure the severity of environmental burdens. The following section compares two often competing perspectives (economic and ecological) and conjectures about how the aggregate burdens would look if the metrics of economic costs and ecological footprints (EFs) could be applied to the various urban environmental burdens. Urban EFs clearly increase with affluence, while it is plausible that the aggregate economic cost of ecological burdens may first increase and then decline. In effect, the three curves for local, city-regional and global burdens, illustrated in Figure 2.1, could be mimicked by curves of the aggregate urban environmental burden from the perspectives of health, economy and ecology, respectively.

The following section tries to bring the politics more explicitly back into a discussion that may often be politically motivated, but is typically expressed in empiricist terms. Firstly, the curves relating economic growth and environmental burdens often reflect political struggles, and their successes and failures. Secondly, they reflect various forms of environmental displacement, with different political and ethical implications – there is clearly a difference, for example, between cross-border pollution, driving polluting enterprises to cross borders through environmental regulations, and creating an international trade regime that shifts the competitive locations for polluting enterprises. Thirdly, they reflect a process through which global 'commons' are being appropriated, leaving little environmental space for economic laggards.

The final three sections briefly review the sanitary revolution of the 19th century, the pollution revolution of the 20th century, and the global urban environmental challenges of the 21st century. These accounts illustrate how 'turning the curves' is indeed a political process. Also, depending on one's viewpoint, the sanitary and pollution revolutions can be seen as stunning victories in the face of enormous urban environmental challenges, partial victories in wars no longer vigorously fought once the benefits to the affluent had been secured, or abject failures that have helped to create new and bigger environmental burdens. There is some truth in each of these viewpoints. Much can be learned from the sanitary and pollution revolutions, but they provide no easy answers. Indeed, this is certainly not the time for complacency, or pinning one's hopes on economic growth. Looking to the future, not only is a new global challenge fast approaching, but the conventional approach to addressing local and city-level challenges is being undermined.

A stylized account of urban environmental transitions

The urban environmental transition characteristic of the cities that became affluent over the course of the 19th and 20th centuries can be represented as a stylized set of empirical observations, as in Figure 2.1 (McGranahan and Songsore, 1994; McGranahan et al, 2001). Along the horizontal axis is economic affluence, which can be represented roughly by the average income in the city. Along the vertical axis is the severity of environmental burdens, which can be measured with a number of very different metrics, such as the health burden, the economic costs or the EF. The shapes of the curves are only meant to represent general direction, distinguishing between environmental burdens that decline with affluence, those that increase, and those that first increase and then decline. Similarly, the metrics of severity are not intended to be comparable across the curves.

In the table attached to the bottom of Figure 2.1, the left column provides examples of local burdens that decrease as income rises, the right column shows global burdens that steadily rise with per capita income and the middle column shows burdens that go through an inverted 'U'.

The empirical claim we are trying to make with Figure 2.1 is that household and neighbourhood environmental burdens tend to decline with increasing affluence, while international and global environmental burdens tend to increase, and intermediate-scale burdens are inclined to increase and then decline. We suggest that these forms apply both to the historical trajectories of currently affluent cities and to the differences among contemporary cities, though the details of the relationships differ among cities, and change over time.

An environmental burden can be defined as a threat to human well-being, resulting from human activities that damage the environment (IIED, 2001), and the scale of an environmental burden can be taken to relate to the typical distances between the damaging activities and their human consequences. Commonly cited household and neighbourhood scale environmental burdens include inadequate sanitation (allowing faecal-oral diseases to spread locally), indoor air pollution, work-related environmental hazards, and accumulations of waste in and around people's homes. Commonly cited city and regional burdens include ambient air pollution, surface and groundwater pollution and overuse, and open waste dumping. Commonly cited international and global burdens include greenhouse gas emissions, the emissions of ozone-depleting substances, the depletion of globally accessible resources, the international movement of hazardous wastes, biodiversity loss and virtual water consumption.

Defining and measuring these relationships precisely is very difficult, but for a first approximation the forms presented in Figure 2.1 conform to conventional wisdom: the 19th and 20th centuries saw a steady improvement in household and neighbourhood environments, a steady increase in activities contributing to international and global environmental burdens, and more

22 Scaling Urban Environmental Challenges: From Local to Global and Back

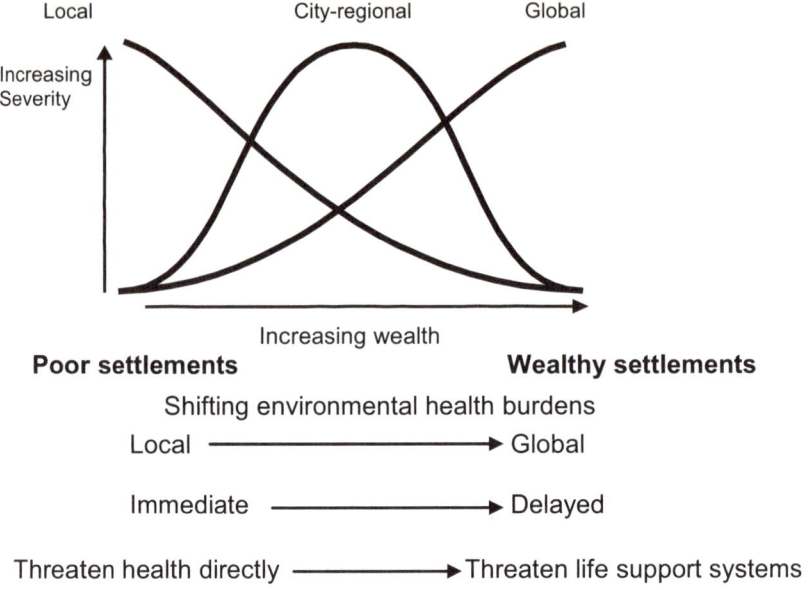

Note: *The global burdens are defined here as the water consumed, carbon emitted and waste generated in producing and supplying all the goods and services consumed in an urban centre. Virtual water consumption has been defined to include such consumption, and to follow the same terminology, one could refer to virtual carbon emissions and virtual waste generation.

Source: Adapted from McGranahan et al (2001)

Figure 2.1 *A stylized urban environmental transition*

ambiguous (and some would say first rising and then falling) changes in urban and regional environmental burdens (for evidence from the US, see Tarr, 1996; Melosi, 2000). They also conform to conventional wisdom concerning international differences in urban environmental burdens. Indeed, the initial curves upon which Figure 2.1 was originally modelled (which displayed the share of urban population without sanitation, urban concentrations of sulphur dioxide and carbon dioxide emissions) were based on cross-country data (World Bank, 1992).

We also argue, as indicated at the bottom of Figure 2.1, that the spatial shifts described by the curves also involve shifts through time, and shifts from burdens that harm human health directly to burdens that undermine the often more distant life-support systems upon which humans depend. Both of these claims also conform to conventional wisdom. Ecologists working with scale issues have observed that 'big' processes also tend to be 'slow' (see Millennium Ecosystem Assessment, 2003, chapter 8, 'Dealing with Scale'), so it is to be expected that global processes such as global climate change should involve delayed impacts. What is perhaps more surprising, and may reflect the social side to these shifts, is that even environmental health impacts that have not involved such obvious shifts in scale tend to be more delayed in more affluent settings, with long-term risks of exposure to carcinogens taking over from short-term risks from infectious diseases. More generally, where there have been scale shifts, it also makes sense that there should be a shift from predominantly environmental health risks towards risks that are more threatening to life-support systems: more dispersed urban pollution or resource demands are less concentrated in and around humans, so a larger share of their impacts does not affect people directly.

Unfortunately, as they are formulated, it is difficult to evaluate these claims in a rigorous manner. There are insufficient data on the economic and environmental history of cities to undertake a detailed statistical analysis of the environmental transitions of economically growing cities in the 19th and 20th centuries. Moreover, several of the concepts are not well defined; identifying what constitutes an urban environmental burden, and how its severity ought to be measured, is open to debate. However, there has been an enormous amount of empirical work undertaken on what has come to be termed the environmental Kuznets curve – the inverted U-shaped curve associated with city-regional burdens in Figure 2.1. In addition, there has been important empirical work on environmental health transitions, which relates closely to several of the claims of Figure 2.1. The three curves displayed in Figure 2.1 suggest an alternative to the single environmental Kuznets curve, and to the conventional environmental risk transition with its two (traditional and modern) curves.

From an environmental Kuznets curve to multiple curves of urban environmental transition

The single-minded pursuit of economic growth is often criticized for causing inequality and environmental damage, though proponents claim that this is a temporary aberration. In the middle of the 20th century, Nobel prize-winning economist Simon Kuznets observed, on the basis of what he described as 'perhaps 5 per cent empirical information and 95 per cent speculation', that in the course of economic development income inequalities first increased and then decreased (Kuznets, 1955, p26). He suggested that Marxists' concerns about the immiseration of the working classes in the 19th century were based on a misguided extrapolation of rising inequality: they failed to foresee that with continued economic development, inequality would eventually decline. The term 'environmental Kuznets curve' was coined in the 1980s in response to the finding that (some) environmental problems display a similar pattern, initially increasing with economic growth and then declining. It was controversial because some proponents of the environmental Kuznets curve, and even more opponents, saw it as an attack on environmentalists – who, like Marxists, could be accused of attacking capitalism and economic growth on the basis of a misleading extrapolation of the rising part of an inverted U-shaped curve.

The environmental Kuznets curve has been the topic of over 100 peer-reviewed publications (Yandle et al, 2004). It has been applied to a wide variety of environmental burdens, and in the course of these studies it has been confirmed, rejected, interpreted, reinterpreted, explained, and explained away in innumerable different ways (for recent reviews, see: Stern, 2004; Yandle et al, 2004; Nahman and Antrobus, 2005).

From the time when the environmental Kuznets curve was first popularized in the World Development Report of 1992 (World Bank, 1992), it should have been evident that not all environmental burdens exhibit the same type of relationship with per capita income, either cross-nationally or over time. Indeed, as noted above, the contrasting curves for sanitation, urban air pollution and carbon emissions first employed in developing Figure 2.1 were taken from that report (McGranahan and Songsore, 1994; McGranahan et al, 2001). Similarly, it should have been clear that there is a great deal of variation in environmental burdens that is not related, either causally or statistically, to economic status. It would be foolish to estimate the environmental burdens of a city or town on the basis of its economic status alone, or to expect the relationship between environmental burdens and economic status to be stable over time.

More specifically, as the basis for claiming that economic growth is inherently good for the environment, or as a representation of how economic growth affects the environment generally, the environmental Kuznets curve has been largely discredited (Ekins, 1997, 2000; Stern, 2004). On the other hand, there is now a rich empirical literature, much of which is of interest even if the notion that there is a single curve, or a single shape of curve, is rejected. Indeed, the analysis generated in the course of these studies provides numerous insights directly relevant to the claims of Figure 2.1. In particular, it has been observed that the burdens that increase with income, and show less sign of declining even at comparatively high income levels, tend to be of larger scale (Cole et al, 1997), have more delayed impacts (Lieb, 2004), and do not affect health as directly (Gergel et al, 2004).

Perhaps the only empirical analysis to compare explicitly environmental indicators for burdens at different scales found that 'meaningful EKCs [environmental Kuznets curves] exist only for local [city-level] air pollutants whilst indicators with a more global, or indirect, impact either increase monotonically with income, or else have predicted turning points at high per capita income levels with large standard errors' (Cole et al, 1997, p401). Examining other empirical studies reinforces this conclusion (though it should be noted that the same or overlapping data sets are often used in different studies). For example, in a table summarizing 34 different empirical studies (Lieb, 2004), environmental Kuznets curves were found in most of those, examining urban-regional pollutants such as sulphur dioxide (15 out of 23), particulates (13 out of 15), oxides of nitrogen (11 out of 12), carbon monoxide (5 out of 6) or river pollution (6 out of 7). In contrast, most of those examining carbon dioxide (11 out of 17) found monotonically rising emissions. Similarly, all three studies that examined aggregate waste generation found a monotonically rising burden.

These same results have also been interpreted in terms of the timing of their impacts. Indeed, the table of 34 empirical studies referenced in the previous paragraph was actually constructed to provide evidence for the claim that the more immediate flow pollutants conformed to the environmental Kuznets curve, while the more delayed stock pollutants did not (Lieb, 2004). They could also be distinguished in terms of their direct health impacts, with the more local and immediate burdens having more immediate health impacts than the more global and delayed burdens.

What are missing from these examples are the very local, immediate and health threatening environmental burdens, such as bad sanitation and indoor air pollution. Although downward sloping curves were estimated for both household sanitation and household water supplies in the early analysis presented in the 1992 World Development Report (World Bank, 1992), there has been little subsequent analysis of the household environmental burdens (an exception being Kumar and Viswanathan, 2004). This may be because few would dispute that they generally decline with economic growth. However, as described in the following section in relation to the sanitary revolution, the evidence that economic growth automatically brings environmental improvement is not compelling even for these household burdens. There is evidence, for example, that in many of the early industrializing cities, economic growth was accompanied by increasingly unhealthy living environments until local groups and governments took vigorous action (Szreter, 2005). Moreover, as with other environmental burdens, the smooth curves hide a great deal of variation, much of which almost certainly reflects policies and actions representing the interests of those worst affected, or designed explicitly to curb environmental damage. However, as an empirical description rather than a causal attribution, it is safe to say that the most serious and health threatening household environmental burdens have tended to decline as average incomes have increased.

A health perspective – environmental health burdens and risk transitions

If the empirical literature on environmental Kuznets curves implicitly supports the stylized facts of Figure 2.1, a more comprehensive and explicit empirical corroboration comes from a different quarter: an analysis of the epidemiological and environmental risk transitions based on the Global Burden of Disease database and the accompanying Comparative Risk Assessment project (Smith and Ezzati, 2005). Work on the Global Burden of Disease estimates has been going on since the early 1990s. The database has been adopted by the World Health Organization (WHO), and is the first consistent global set of data on mortality and morbidity, including estimates of the burden of disease for more than 150 causes of death and illness. The comparative risk assessment brought together over 100 investigators in an attempt to attribute burdens of disease to 26 important risk factors, using the Global Burden of Disease database. A number of these risk factors were environmental, and allow environmental risks to be compared across areas of different average income.

The epidemiological transition, as popularized in the 1970s, focused on a shift from infectious (traditional) to non-infectious (modern) diseases said to accompany development, but has since been elaborated into more complex categories. The environmental risk transition was proposed more recently, and was initially based on two curves representing traditional and modern risks (Smith, 1990; Smith and Lee, 1993), corresponding at least roughly to traditional and modern diseases. In the more recent analysis being examined here (Smith and Ezzati, 2005), the environmental risk transition has been extended by employing a three-fold framework, adapted from Figure 2.1, distinguishing between household, community and global health risks. The health risks estimated and attributed disease burdens were:

Household environmental risks – poor water sanitation and hygiene; indoor air pollution from solid fuel use; exposure to malarial mosquitoes.
Community environmental risks – urban outdoor air pollution; lead pollution; occupational risks; road traffic accidents.
Global environmental risks – climate change.

The authors found that 'the simplistic conclusions commonly drawn about the epidemiologic transition, in particular the increase in chronic diseases with development, are not supported by current data; in contrast, the conceptual framework of the environmental risk transition is broadly supported in a cross-sectional analysis' (Smith and Ezzati, 2005).

What makes this presentation qualitatively different from that implied by Figure 2.1 is that there is a common measure of the severity of environmental burdens – the health burden. This allows the different risks to be directly compared and summed.

The resulting estimates indicate that:

- **household environmental health risks** account for much the largest burden, and decline with per capita income;
- **community environmental health risks** tend to be highest at middle incomes, and lower at the extremes, conforming roughly to an environmental Kuznets curve;
- **global environmental health risks** account for a very small share of the overall risk, and their **contribution** increases with income, while their **impact** (where the health risks are encountered rather than where the emissions occurred) declines;
- the **combined contribution** declines with income, and the **combined impact** declines even more steeply.

Source: Based on Smith and Ezzati (2005)

While even the health analysis presented above is far from comprehensive (as it omits a great many risks) and far from accurate (as it is based on some very rough estimates), it is worth conjecturing how the results would be likely to change if a different metric were used to measure the severity of environmental burdens.

Economic and ecological perspectives and alternative metrics

Two of the more common perspectives on environmental burdens are the economic and the ecological. From an economic perspective, the obvious metric with which to measure environmental burdens is economic cost (where the burdens are valued in monetary units, using market prices or their closest surrogate). From an ecological perspective the metric of choice is more likely to be physical, as with, for example, the EFs (where burdens are valued in hectares of productive land, using physical estimates of appropriated or compensatory land requirements). Before examining how applying these metrics might alter the result suggested by the health metric, that the aggregate burden actually falls with increasing affluence, it is worth considering some of the differences between economic and ecological perspectives. Despite the efforts of many researchers, and journals such as *Ecological Economics*, communication between these disciplines tends to be poor, with environmental burdens identified and measured very differently.

Humans and their markets are at the centre of the economist's world, and peripheral to the ecologist's, and their views of why and where environmental problems arise vary accordingly. For most ecologists, markets are **external** forces that drive people to disrupt, disturb and degrade natural ecosystems: in effect, market-based activities are **externalities**, and urban settlements and the people in them are external to the natural ecosystems they threaten. For most economists, on the other hand, it is the absence of markets that explains environmental problems. Thus, the concept of **externalities**, which practising economists regularly use to explain environmental problems, refers to the impacts one person's activities have on another person's well-being that are 'external' to markets, and hence are not reflected in market prices or incorporated in negotiations.

Similarly, ecologists and economists have very different conceptions of the spatial dimensions of production. From an economic perspective, most production now takes place in urban areas. Urban activities account for most value added, and for most final consumption. From an ecological perspective, on the other hand, the bulk of production is taking place in rural areas. William Rees compares urban settlements to cattle feedlots and anthills, maintaining their keystone species at very high densities, but only containing a small share of the biophysical processes needed to sustain them (Rees, 2003). These settlements have what he termed a large 'ecological footprint'. Thus, it has been estimated that on average the people living on a square kilometre (km^2) of built-up land rely on about $25km^2$ of other land and water (of average biological productivity) to provide themselves with the food, fibre, timber and energy (WWF, 2004).

In short, the difference between economic and ecological perspectives is such that the different definitions of a burden are as important as the different metrics through which burdens are measured. From an ecological point of few, urban areas consume far more than they produce, and this excess consumption is the most obvious environmental burden urban areas impose. For a mainstream economist, urban centres engage in trade that benefits all parties, and environmental burdens are likely to be associated with un-traded and un-negotiated environmental impacts or externalities. Thus, from an ecological perspective, virtually all the resources an urban area consumes, as well as the wastes it produces, can be considered a burden (and the scale of these burdens depends on how far the resources come from), while from an economic perspective the clearest environmental burden is pollution (and the scale depends on how far the pollution travels).

In terms of the three curves in Figure 2.1, this helps to explain why it is ecologists who are associated with the claim that environmental burdens increase with affluence, and economists who are associated the environmental Kuznets curve.

A measure of environmental impacts closely associated with ecologists is the equation sometimes referred to as the Commoner-Ehrlich equation or IPAT: Impact = Population times Affluence times Technology (Ekins, 2000; York et al, 2003b). Unless affluence is associated with much lower populations or much more efficient technology, the equation will describe an unambiguously positive association between an affluence indicator such as gross domestic product (GDP) per capita, and the environmental impact. This is not so much because of the form of the equation, or even the causal relations implied, but because the intention is to measure the physical size of the impact, rather than the size of its consequences on human health or the economy. Local burdens such as bad sanitation and indoor air pollution involve changing environmental conditions, but only to a relatively small degree: what gives them a large health impact is that they are concentrated in the same locations as people.

As with an ecological perspective, though for different reasons, an economic perspective will tend to give less emphasis to local environmental burdens that concentrate in poor areas than does a health perspective. Local environmental burdens do often involve economic externalities (one household's bad sanitation is another household's health risk) but also directly reflect income poverty (people living on insufficient incomes cannot afford sufficient water and sanitation, partly for the same reason they cannot afford sufficient shelter and food). From an economic perspective, there is little to be gained in labelling a symptom of poverty an 'environmental' burden if it simply reflects a lack of income. Moreover, in contrast to a health metric, which always gives the same health risk the same weight regardless of the income of those affected, an economic metric implicitly gives less weight to risks faced by poor people, who have fewer economic resources with which to express their preferences. (Most people rightly object to the fact that economists tend to value the health and lives of the poor less than those of the rich, but it must be recognized that such valuation is implicit in other market prices, in the market-based decisions that result from these prices, and in public decisions based on narrow cost–benefit analysis alone.)

Unlike an ecological perspective and ecological impact measures, an economic perspective and economic costing does not give a great deal of weight to global environmental burdens. As noted above, economics is inclined to treat resource extraction for foreign consumption as benefiting the exporting as well as the importing countries, even if economists do recognize a 'natural resource curse' which often afflicts exporting countries (Sachs and Warner, 2001; Bulte et al, 2005). Thus, economics sees little or no burden, where from an ecological perspective there is a major global burden. Moreover, by discounting future costs, economics implicitly gives less weight to burdens whose impacts are long term, which includes a disproportionate share of global burdens.

In short, if aggregate burden estimates were to be constructed using disability adjusted life years (DALYs) lost, EFs and monetary units, one would expect differences in the resulting curves not only because of the different metrics, but also because of the perspectives with which they tend to be used. To a first approximation, the health metric is the most likely to fall with increasing income, the ecological metric to be the most likely to rise with increasing income, with the economic metric somewhere in between, and most likely to first rise and then fall.

These are only some of the differences that can result from applying different perspectives or metrics. Even if the same burden is measured in the same units, significant differences can result from decisions about how burdens should be

Table 2.1 Carbon dioxide emissions per capita for Beijing and Tokyo

	Beijing (1997)	Tokyo (1995)
CO_2 emitted in city (tonnes per capita)	6.4	4.9
CO_2 emitted in providing goods and services consumed in city (tonnes per capita)	8.3	12.1

Source: Data provided by Shobhakar Dhakal, based on research presented in Dhakal (2004)

attributed. For example, a city can be allocated all of the carbon emitted within its boundaries, or the carbon emitted in producing all of the goods and services consumed within its boundaries. As illustrated in Table 2.1, the choice matters is how cities are ranked. On a per capita basis, about 30 per cent more carbon is emitted in Beijing than in Tokyo. On the other hand, the carbon emitted in supporting the consumption that takes place in Tokyo is about 46 per cent more than for Beijing. The choice of which accounting procedure to use has a practical dimension (e.g. it depends on whether one is looking for measures to curb carbon-intensive consumption patterns, or measures to replace carbon-intensive technologies). But it also has a political dimension.

Bringing politics back in

There is an unfortunate tendency for curves such as those in Figure 2.1, and indeed many accounts of urban transitions, to obscure rather than illuminate human struggles, triumphs and hardships. Development trajectories and transitions have a ring of inevitability about them. They are suggestive of stages of development, and theories that map out teleological processes, which people can support or resist, but cannot ultimately change. Moreover, they tend to downplay relations between urban centres, and rural-urban relations that are not explicitly represented.

Of course the curves themselves are often manipulated for political effect. Economic growth advocates emphasize local or city-regional environmental burdens so as to ascribe environmental benefits to growth, while critics point to the ever-increasing global burdens to castigate economic growth. The choice of metric can also be political. A health metric is very egalitarian, at least among those whose risks are considered (which often excludes future generations). An ecological metric will tend to give more weight to future generations and non-human species, and will tend to ignore some of the most critical environmental problems of the urban poor. And a conventional economic metric can be misused to privilege the wealthy, by costing burdens using market prices or their equivalent, but ignoring issues of compensation and redistribution.

Equally important, such curves have political interpretations and implications. To at least some degree, the curves reflect the outcome of political struggles involving both economic growth and the environment. For many currently affluent cities, the 19th century sanitary movement and the revolution it engendered explain at least part of the declining sanitary burden. For individual settlements it is often possible to plot out the improving sanitary conditions, and their relation to local and sometimes national politics and policies. Similarly, in many of the same cities, the 20th century environmental movement helped bring about the turning of the curve for many urban and regional burdens. And if the 21st century is to be the century of sustainability, this will undoubtedly require further struggles, scientific innovations and ideological battles. Such a temporal sequencing begs the question of why, given that much of the world still lacks adequate sanitation, the sanitary revolution is typically associated with the 19th century. But it is not the curves that imply that progressive politics is the monopoly of the affluent front-runners – this interpretation is based on seeing the curves through the lenses of the dominant development narrative.

It is impossible to do justice to urban environmental politics in statistics, let alone the national and international statistics typically used in the analysis of environmental Kuznets curves. Nevertheless, it is revealing that researchers estimating or critiquing environmental Kuznets curves have often found that factors that would be expected to influence environmental politics also influence the shape or position of the curve for a number of the local and city-regional burdens. There is less evidence of such influence when it comes to global ecological burdens.

A study focusing on urban air quality and surface water quality found that literacy, political rights and civil liberties are strongly associated with reduced environmental burdens, particularly at low-income levels (Torras and Boyce, 1998). It has also been suggested on the basis of urban air pollution data that the major force behind the declining air pollution concentrations since the 1970s has been public support for environmental protection and the resulting policy measures, and that the declines have been too large to be consistent with income-generated economic shifts alone (Deacon and Norman, 2004). A more recent study looked explicitly at environmental lobbying and the lead content of petrol, and found that environmental lobby groups did affect environmental policy stringency, as did political competition 'particularly where citizens' participation in the democratic process is widespread' (Fredriksson et al, 2005). Other studies have also found political factors to be significant in areas as diverse as deforestation (Bhattarai and Hammig, 2004) and access to water (Deacon, 1999).

On the other hand, a recent study estimating aggregate EFs econometrically found that while both income and the level of urbanization were positively associated with the size of a country's footprint, political rights, civil liberties and state environmentalism were insignificant (York et al, 2003a). This is consistent with the notion that most global environmental burdens are not the object of serious urban or national environmental politics, and that in the

32 *Scaling Urban Environmental Challenges: From Local to Global and Back*

absence of support through global governance mechanisms are not likely to become so.

The fact that the urban environmental burdens of wealthier cities and countries are more likely to cross international borders also has political implications. The transfer of environmental burdens over space and through time is highly political, even if it is not always recognized and subject to negotiation. According to economic reasoning, just as the shape of the local and city-regional curves ought to depend particularly heavily on local and city-regional policy processes, so the shape of global curves ought to depend particularly heavily on global policy processes, and whether they are effective. A revealing example of this, illustrated in Figure 2.2, is in the changing emissions of halons and chlorofluorocarbons (CFCs) in the wake of the Montreal Protocol controlling ozone-depleting substances, often cited as one of the most successful international environmental agreements. In 1986, just prior to the adoption of the Protocol, emissions of both halons and CFCs fit the profile of global environmental burdens: emissions tended to be considerably higher in wealthier countries. By 1990 the relationship conformed more closely to a gentle environmental Kuznets curve, first rising and then falling. This sort of transformation is probably unique, and reflects the ease with which affluent

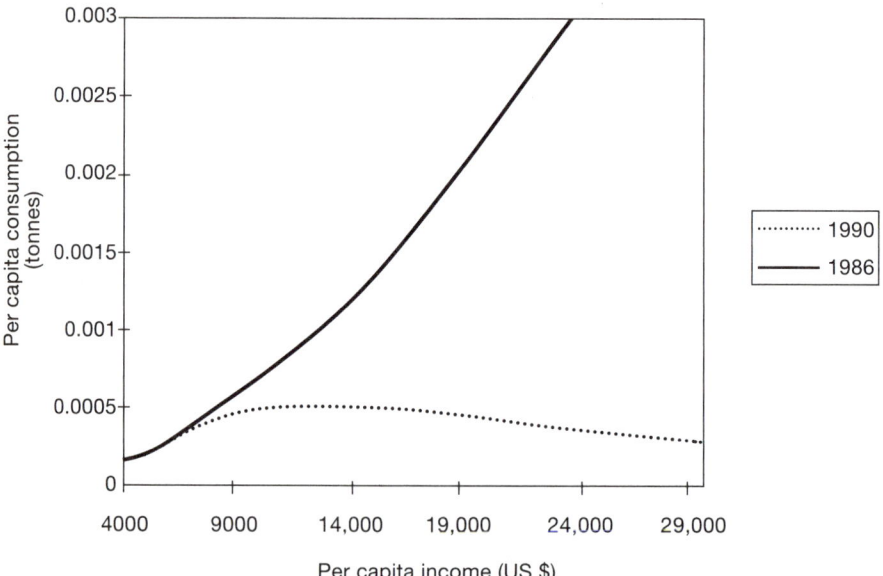

Source: Cole et al (1997)

Figure 2.2 Consumption of CFCs and halons pre- and post-Montreal Protocol

countries controlled emissions. Just as most global environmental burdens tend to be larger in wealthier countries, so they tend not to have been greatly influenced by global governance mechanisms.

More generally, there are a number of ways that pollution can be displaced internationally over space (and through time), which for the most part shift pollution from more to less affluent locations:

1   pollutants with international dispersion;
2   environmental regulations that lead local-scale polluters to move to less effectively regulated locations;
3   changing trade patterns independent of environmental regulation that favour the export of less local-pollution-intensive products (e.g. light manufacturing and service products) and the import of more pollution-intensive products (e.g. heavy industry products).

There is evidence for all of these shifts (see Cole, 2004, for a review of evidence on the second two), each of which has somewhat different political and ethical implications. The first, particularly when uncompensated, clearly goes against the polluter-pays principle as well as ethical principles, and is rarely defended openly, except through claims that the effects are uncertain or were uncertain in the past. The second is more ambiguous. On the one hand, Laurence Summers, then chief economic advisor at the World Bank, wrote a memorandum in which he correctly pointed out that by the economic logic conventionally applied by the World Bank, more polluting industries should be encouraged to move from affluent to poor countries. On the other hand, the furore that resulted when this 'humorous' memo was leaked, reflects how politically contentious such displacement is: the notion that when the wealthy clean up their environments, the poor suffer from pollution as a consequence also goes against a number of ethical principles. The third means through which pollution can be shifted is less contentious, in that it is not the result of intentional measures to reduce pollution in affluent locations, but of a changing comparative advantage. On the other hand, it is linked to trade, and as such does relate to political debate over the environmental impacts of current trade regimes. This distinguishes it from shifts in pollution outcomes that result from changing consumption patterns, including, for example, the increase in service consumption in affluent locations (such shifts were not included in the list on the grounds that they do not, strictly speaking, involve the displacement of pollution).

Somewhat similarly, there are a number of different forms through which resource pressures originating in affluent cities affect less affluent, but often resource-rich locations. Compared to transboundary pollution, there are few resource flows that involve no compensation – expropriating another country's resources without payment typically requires colonial relations of a particularly onerous sort. On the other hand, within the current international trade regime, resource-intensive industries may seek out locations with ill-defined property rights on environmental resources or ineffective state regulation (Chichilnisky, 1994). Moreover, one of the more convincing explanations for the 'natural resource curse', which is blamed for the slower economic growth in natural resource-intensive countries, is that a reliance on natural resource exploitation shapes political development in harmful ways (Bulte et al, 2005). Valuable but weakly controlled resources can amplify corruption and undermine the sort of good governance needed to achieve economic growth and human development (Leite and Wiedemann, 1999; Bulte et al, 2005). Thus, as with pollution, there are a variety of means through which natural resource pressures can be shifted internationally, some of which are far more damaging to the resource-providing countries than others.

Another important political aspect shaping the relationship between affluence and the scale of urban environmental burdens is that global resources and waste sinks are being appropriated and used to provide for affluent consumers, leaving little environmental space for the conventional economic development of low-income countries. Economic output per capita is a poor indicator of human well-being. The relationships plotted by the curves in Figure 2.1 are empirical generalizations rather than laws of development. Nevertheless, it is politically very salient that low-income countries and cities cannot realistically achieve conventional development unless the more affluent countries change their own consumption and production patterns. Indeed, even if the more affluent did reduce their global burdens considerably, it would be environmentally disastrous for low- and middle-income countries to all develop along the path that the now affluent countries once did. In effect, to the extent that the curves in Figure 2.1 do have implications for conventional urban development, far from implying that low-income cities must follow this pathway to succeed in the future, they imply that both low- and high-income cities must deviate from this pathway.

There are, of course, numerous other politically significant spatial dimensions to urban environmental burdens, and not all involve spatial dispersion. Resource supplies also have a spatial dimension, and in an interesting counterpoint to the tendency for waste and pollution problems to become more diffuse, there is evidence that spatially concentrated 'point resources' are more inclined to bring on the 'natural resource curse' (Isham et al, 2005). More importantly, while the general tendency has been for urban environmental burdens to become globalized, it is easy to forget that some of the most critical urban environmental burdens of the 19th and 20th centuries have actually become far more localized, a process not only driven by political forces, but also with important political consequences.

While the sanitary problems of 19th century cities were grounded in local conditions, they gave rise to city-wide epidemics and international pandemics (Cliff et al, 1998). The public nature of the health threat helped to drive the sanitary reforms. One of the outcomes of the sanitary revolution has been that, even in very low-income urban centres, sanitary health risks have become increasingly localized. In effect, parallel to the spatial expansion of the environmental burdens of urban affluence, there has been a localization of the environmental burdens of urban poverty. This is in itself a reflection of the political nature of the sanitary challenge – the sanitary health hazards would not have become localized if the threats had emanated from the affluent neighbourhoods.

Despite these qualifications, the spatial displacement of urban environmental burdens is a critical link between local, national and international environmental politics. Moreover, while other spatial issues will often be more important in particular urban centres, the systematic link between economic status and more dispersed urban environmental is one of the most serious challenges to conventional, or even non-conventional development. It is easy to define human development so that it is not closely coupled to economic growth. It is easy to call for sustainable development that is environmentally benign. But it is difficult to even imagine how this is going to be achieved in a world where the poorest urban centres lack the capacity to address their environmental problems, and the wealthy lack the incentive (Lee, 2006).

Urban environmental movements – revolutionary or palliative?

For those parts of the world where economic growth and urbanization is concentrated, the 20th century saw a continuous shifting of urban environmental priorities (McGranahan et al, 2001). What are now localized urban sanitation issues were still the critical concern at the start of the century. Urban pollution and regional resource issues had taken centre stage by the middle of the century, and by its end, global environmental issues were coming to the fore. At the start of the 20th century, industrial cities were still in the throes of a sanitary revolution; by its end environmentalists were calling for a sustainability revolution. As the following account attempts to demonstrate, these environmental initiatives were political, and their achievements were not simply the result of economic growth. On the other hand, there is a question as to whether this series of 'revolutions' should be seen as the progressive resolution of a series of environmental challenges that urban development has thrown up, or whether, to the contrary, they represent the progressive failure to do more than delay and shift urban environmental burdens across space and over time.

An urban sanitary revolution – the example of Britain

Early industrial cities were less healthy than the surrounding countryside. Urbanization and economic growth were extremely disruptive, and the expanding urban working class lived in crowded and unsanitary living environments. In Britain, where documentation is comparatively good, mortality levels in the industrializing cities reportedly 'deteriorated substantially during the second quarter of the 19th century and did not improve significantly thereafter until the 1870s and 1880s' (Szreter and Mooney, 1998; Szreter, 2004). Economic growth created the capacity to improve living environments for the majority, but even rising wages during the early part of the 19th century did not provide better water supplies, sanitary facilities and hygiene behaviour. Urban settlement is now considered an advantage for delivering environmental health services, but at the time it was a distinct disadvantage. On their own, urbanization and growth did not drive local environmental improvement: indeed, in the absence of concerted action to improve environmental conditions, they drove local environmental degradation.

The sanitary movement, which addressed a range of local environmental conditions, helped to turn the potential health benefits of urbanization and economic growth into a reality. This movement was built on the work of local organizers and international networkers, scientists and poets, private entrepreneurs and government officials. It captured the imagination of press and public, as well as of intellectual and the bureaucrat. It both benefited from and helped to encourage improvements in governance. The miasma theory of disease, which held that diseases were contracted from noxious vapours, was more important in motivating early sanitary reform than the bacterial theory of disease (Rosen, 1993). Sanitary reformers came in a variety of different political hues, and were not all sympathetic to, let alone supportive of, the working classes or the urban poor groups worst affected by unsanitary conditions (Hamlin, 1998). Yet even without a consensus on the detail of the science or the politics of sanitary reform, a growing number of committed activists and influential public figures fought to improve sanitary conditions.

Many sanitary reforms displaced rather than resolved environmental burdens. Sewers were used to carry faecal material away from local living environments, only to release it into streams and rivers, depriving land of nutrients and creating downstream pollution. Considerable attention went into designing better ventilation, heating and lighting in people's homes, but these same homes often spewed large quantities of smoke into the urban atmosphere. Solid waste was collected from homes, but then dumped on the outskirts of the urban areas.

This was not, however, because nobody was making a sufficiently eloquent case for more comprehensive environmental measures such as recycling waste and reducing pollution at source, rather than creating waste and dispersing pollution. One of the most celebrated poets of the 19th century, Samuel Taylor Coleridge, having noted that 'the river Rhine, it is well known, doth wash your city of Cologne', asked 'what power divine shall henceforth wash the river Rhine?' One of the century's most famous novelists, Victor Hugo, was influenced by Leroux's theory of circulus, which held that human excretions should be collected and used in agriculture to produce food for people to eat, creating a virtuous circle (Reid, 1993). He even prefigured environmental economics, writing in his novel *Les Misérables* that 'Each hiccup of our cloaca [sewer] costs us a thousand francs ... the land impoverished and the water contaminated' (Hugo, 1987 [1867]). Leading sanitary reformers had similar concerns. Edwin Chadwick, the leading proponent of sewers in Britain, intended the sewers to carry the excreta back to the land, where it could fertilize the farmers' fields. Sanitary reformers were also making similarly modern arguments for reducing air pollution, pointing out, for example, that smoky stacks should be seen as a symptom of inefficiency rather than of progress (Mosley, 2001).

In effect, the reason urban environmental burdens were displaced was not because nobody was concerned or was suggesting alternatives, but because people and groups promoting such concerns and alternatives were politically unsuccessful. Ardent sanitary reformers, like ardent environmentalists today, were viewed by many as anti-business. Getting people to take action to prevent future or distant environmental impacts was doubly difficult when this threatened industrial interests. But at a more mundane level, whatever the political systems, it is inevitably difficult to implement or sustain measures intended to address uncertain, future and distant problems.

Progress on urban pollution – the example of the United States

Urbanization and industrial growth have also created a number of city-regional environmental burdens, such as ambient air pollution, surface water pollution, and solid waste accumulation. The sanitary movement was sympathetic to these concerns, but prioritized threats to public health in and around people's homes and workplaces. As a result, it actually contributed to some of the city-regional burdens. Indeed, the 19th century sanitary movement set off what one environmental historian termed 'the search for the ultimate sink', and did nothing to curtail the scramble for natural resources (Tarr, 1996). For many contemporary environmentalists, the sanitary revolution was not the solution, but part of a new set of problems. Yet many of the environmental measures taken in the cities of the 20th century were simply a continuation of sanitary reform, often doing as much to displace as to remove environmental burdens, and, by removing the immediate and local risks of increasing economic activities, making it possible for urban environmental loads (e.g. quantities of resources consumed and wastes released) to increase.

Like the earlier sanitary movement before it, the environmental movement of the mid- to late 20th century built on the work of a wide range of groups of different political persuasions, drew heavily but selectively on science, and eventually captured the popular imagination. As with the sanitary reforms, urban groups and governments were initially at the centre of environmental reforms, but were supported by national and international networks. Unlike with sanitary reform, national government and legislation were critical to environmental reform, not just in supporting local action and setting common standards, but in resolving conflicting interests between different urban areas and between urban and rural areas. Even in the US, with its federal system, the environmental movement only really came of age with the national Environmental Protection Agency (EPA) in 1970, the Clean Air Act of 1970 and Clean Water Act of 1972.

In 1911 all major US cities had sewerage systems, with most (88 per cent in 1909) of the wastewater disposed of untreated in waterways. It was not until 1940 that more than half of the sewered urban population in the US had treated sewage (Melosi, 2000). Cities were loath to invest in systems that would primarily benefit downstream inhabitants. Sanitary engineers were inclined to support the view that water systems should be reworked to cope with sewage-polluted water from upstream cities (Tarr, 1996). With the help of federal funds, a number of sewage-treatment plants were constructed in the inter-war period, but these only made a marginal impact. Concern increased, however, as formerly unsuspected health hazards were identified and new industrial pollutants were added to the water.

By the 1960s, urban-regional environmental concerns were becoming a national issue. In 1972 the Clean Water Act was passed, calling for zero effluents by 1985. While this goal has still not been reached, significant improvements have been recorded in the pollutants targeted. Wastewater treatment is now almost universal. According to the statistics of the EPA, while only about 36 per cent of stream-miles in the US were safe for fishing and swimming in 1972, the current figure is more like 60 per cent (www.epa.gov/earthday/history.htm, accessed on 16 December, 2005).

The story for ambient air pollution is similar, with some of the differences reflecting the different spatial features of water and air pollution. In 1912, 23 of the 28 largest cities in the US had smoke control ordinances, which met with mixed success. Local groups, often largely comprising women, had lobbied for smoke abatement. Experts networked and shared experiences internationally and became increasingly influential in the early decades of the 20th century (Stradling, 1999). When the pressure was on, and local groups and experts collaborated effectively, they could reduce urban air pollution – 'smoke' – appreciably, but often only temporarily (see Pittman, 2003, for an account of St Louis' successful but temporary measures to reduce air pollution in the 1920s). In the 1940s and 1950s, natural gas and oil began replacing coal on a significant scale for domestic heating, as well as industrial, commercial and transport uses, giving respite in some locations and demonstrating the possibility for improvement. Urban air pollution continued to be a serious problem, however, and a series of well-publicized and documented episodes, such as the infamous London smog of 1952 that killed 4000 people, helped to spur a new environmental movement in the 1960s.

As with sanitation, many of the measures initially taken to address local air pollution problems quite literally displaced the burden. Electricity provided a clean fuel in the home, but power plants released air pollution at a distance from the end-users. As electricity grids developed, the distances increased. Further displacement measures were taken when, under the advice of air pollution experts in the 1950s and 1960s, taller emission stacks were increasingly used by the ore smelting and electrical utility industries, to avoid violating local air pollution ordinances. By 1963, some utility stacks had reached a height of 213m. Other measures, including efficiency improvements and cleaner fuels, led to more fundamental reductions in environmental burdens – provided they were not matched by increasing activity levels.

The 1960s also saw growing public concern with air pollution, and a series of national acts meant to provide the basis for curbing air pollution. Environmental impact statements became an important tool. Some of the more technical pollution experts have been concerned with the 'unscientific' approach of the environmental movement, and its governmental agencies; the more industry- and economy-minded have been concerned that the costs of pollution control have been excessive. But there is little doubt that the urban concentrations of smoke and most classical air pollutants have declined considerably since the 1970s. To quote a recent commentator:

> *Since 1976, the aggregate U.S. level of urban ozone, the main component of smog, has declined 31 percent. Airborne levels of sulphur dioxide, the main component of acid rain, have dropped 67 percent. Nitrogen oxide, the secondary cause of urban smog and of acid rain, has fallen 38 percent. Fine soot ('particulates'), which causes respiratory disease, has declined 26 percent. Airborne lead, considered the most dangerous air pollutant when the EPA was founded in 1970, has declined 97 percent. The EPA's 'Pollutant Standards Index,' which measures days when air quality is unhealthy, has fallen 66 percent since 1988 in major cities.* (Easterbrook, 2002)

For both water and air pollution, there has been considerable debate over whether the environmental reforms were excessive and unnecessary, insufficient and palliative, or balanced and appropriate. On its own terms, however, it seems clear that the urban pollution revolution has been a success in most affluent cities, and was very much a political process, like the sanitary revolution before it. Yet again, the real question is not whether urban pollution has been reduced, but whether this reduction hides a more fundamental failure to address a still more extensive set of environmental burdens.

Sustainable cities – one agenda, or many?

From the perspective of affluent cities, the latest generation of urban environmental burdens is even less tangible than previous ones, and less overtly urban. Sanitary problems were at their worst in crowded, low-income neighbourhoods. Urban smog was visible, and even the pollution of local waterways and other city-regional pollution and resource burdens were easy to observe, and infringed on urban lifestyles and livelihoods. It has since transpired that the true sanitary threat came, not from the stench and the stink of urban filth, as many contemporaries believed, but from invisible pathogens that could even pollute the clearest glass of drinking water. Similarly, concerns about health and air pollution have shifted towards very fine particles, largely neglected in early efforts to reduce visible smoke pollution. Nevertheless, the links between visible and often unpleasant environmental conditions and human well-being helped to drive the urban environmental agendas, successfully reducing local and city-regional burdens. The links between urban consumption patterns and climate change, ozone depletion, the loss of distant rainforests, biodiversity loss, and other global environmental issues at least initially would seem to be far less perceptible, even if they are better founded scientifically than the concerns that drove the early sanitary and environmental movements.

The globalization of environmental burdens has become a major challenge, with global climate change – despite its unique characteristics – taken as archetypal of the latest generation of environmental challenges. With the world population continuing to grow, human production and consumption growing more rapidly still, and economies becoming more tightly interlinked, local actions increasingly combine to create global stress. It is no coincidence that *think globally, act locally* has been the most popular environmental slogan since René Dubos coined the phrase in the run up to the first world environmental conference some 30 years ago (Eblen and Eblen, 2002). It not only captures the increasingly global scale of environmental burdens, but also the importance of addressing these problems in their unique physical, climatic and cultural contexts. Actions in the present are also increasingly seen to be affecting the environment for future generations. *Think of the future, act now* could also qualify as an environmental slogan, and is implicit in most definitions of sustainable development. Again, global climate change is often used as the archetypal example, since most of the burden of current greenhouse gas emissions is likely to fall on future generations – even if the initial effects are already evident.

From a global perspective, however, it is evident that for many urban settlements local environmental burdens remain a priority, while for others the city-regional burdens predominate, and many different priorities vie for attention. Given this complex combination, it is neither possible nor desirable to identify a single environmental agenda. It is quite reasonable that in some settlements the 'brown' environmental health agenda should predominate, in others the 'grey' pollution agenda, and in others the 'green' sustainability agenda (McGranahan et al, 2001; Marcotullio and Lee, 2003). This does, however, create a serious challenge to the emergence of a global movement comparable to the sanitary movement of the 19th century or the predominantly Western environmental movement of the 20th century.

In practice, *think globally, act locally* has been acted out in the economic rather than environmental arena, and from a very different political basis from the one Dubos envisaged. The sharp rise in economic globalization since 1972 has been facilitated by purposeful changes in international trade regimes, and has been driven by the local initiatives of private enterprises. It has been transnational corporations, not environmentally-minded groups or governments, that have been most successful in developing global strategies rooted in local action. It has been market signals, not public debate on the environment, that have most successfully adapted to the new communications technologies. It has been as financial centres, not as models of environmental sustainability, that cities such as New York, London and Tokyo have come to be defined as global cities (Sassen, 2000).

In retrospect, it is not surprising that the environment and development agenda has not been as successful as the free-market agenda in linking up global strategies with local action. Both agendas are inclined to claim that they will bring benefits to virtually everyone, present and future. But the most direct beneficiaries of the free market agenda are successful economic enterprises, while the most direct beneficiaries of environmental improvement ought to be future generations and people living in polluted or environmentally degraded areas – hardly the most influential groups in the global policy arena. More generally, the free-market agenda has had more powerful backers (e.g. the US), more powerful implementing agencies (e.g. the International Monetary Fund – IMF), and a more coherent strategy (e.g. structural adjustment as a condition for loans). Equally important, removing market barriers locally supports the expansion of markets globally, while improving local environments can threaten the global environment (e.g. by importing distant resources).

On the other hand, without a better multi-scaled strategy it is hard to see how the urban environmental challenge can be met. The success of past sanitary and environmental reforms provides the basis for limited optimism. As described above, sanitary reforms did manage to go against the tide of free-market liberalism in Britain in the 19th century and it was multi-scalar: the sanitary movement was international, and reforms often relied on national laws, even though the most profound changes were in urban governance. Similarly, environmental reforms went against the tide of free-market liberalism in the US in the 20th century, and were also based on multi-scaled strategies.

It is somewhat misleading to treat the full panoply of environmental burdens now generated by urban activities as an urban challenge, since their resolution does not necessarily lie in urban initiatives. On the other hand, it is a useful counterpoint to the more conventional treatment of environmental burdens, which ignores their urban dimension altogether and partly, as a result, tends to ignore their spatial dimensions. Moreover, a multi-scaled strategy must involve more than striving for global governance to address global issues, national governance to address national issues, and local governance to address local issues. While national governments and the international agencies they construct clearly remain at the centre of national and global governance respectively, urban-based groups can often play an important role at extra-urban scales as well as intra-urban scales, and networking can help bridge different scales. In relation to global burdens, networks of cities, particularly in affluent countries, have shown some promise in promoting local approaches to climate change mitigation, despite all the obstacles (Bulkeley and Betsill, 2003). At the other end of the environmental spectrum, national federations of the urban poor, also networked internationally, have shown promise in addressing the local environmental problems in low-income settlements (D'Cruz and Satterthwaite, 2005).

References

Bhattarai, M. and Hammig, M. (2004) 'Governance, economic policy, and the environmental Kuznets curve for natural tropical forests', *Environment and Development Economics*, vol 9, pp367–382

Bulkeley, H. and Betsill, M. (2003) *Cities and Climate Change: Urban Sustainability and Global Environmental Governance*, Routledge, London

Bulte, E. H., Damania, R. and Deacon, R. T. (2005) 'Resource intensity, institutions, and development', *World Development*, vol 33, no 7, pp1029–1044

Chichilnisky, G. (1994) 'North south trade and the global environment', *American Economic Review*, vol 84, no 4, pp851–874

Cliff, A., Haggett, P. and Smallman-Raynor, M. (1998) *Deciphering Global Epidemics*, Cambridge University Press, Cambridge

Cole, M. A. (2004) 'Trade, the pollution haven hypothesis and the environmental Kuznets curve: Examining the linkages, *Ecological Economics*, vol 48, no 1, pp71–81

Cole, M. A., Rayner, A. J. and Bates, J. M. (1997)'The environmental Kuznets curve: An empirical analysis', *Environment and Development Economics*, vol 2, no 4, pp401–416

D'Cruz, C. and Satterthwaite, D. (2005) 'Building homes, changing official approaches: the work of the Urban Poor Organizations and their Federations and their contributions to meeting the Millennium Development Goals in urban areas', International Institute for Environment and Development (IIED), London

Deacon, R. T. (1999) *The Political Economy of Environment-Development Relationships: A Preliminary Framework*, The University of California Santa Barbara Department of Economics, Santa Barbara, CA

Deacon, R. T. and Norman, C. (2004) *Is the Environmental Kuznets Curve an Empirical Regularity?* Departmental Working Paper 22-03, University of California, Santa Barbara, CA

Dhakal, S. (2004) *Urban Energy Use and Greenhouse Gas Emissions in Asian Mega-Cities*, Institute for Global Environmental Strategies, Kitakyushu, Japan

Easterbrook, G. (2002) 'Environmental Doomsday: Bad news good, good news bad', *The Brookings Review*, vol 20, no 2, pp2–5

Eblen, R. A. and Eblen, W. (2002) *The Encyclopedia of the Environment*, Houghton Mifflin Company, Boston, MA

Ekins, P. (1997) 'The Kuznets curve for the environment and economic growth: Examining the evidence', *Environment And Planning A*, vol 29, no 5, pp805–830

Ekins, P. (2000) *Economic Growth and Environmental Sustainability*, Routledge, London

Fredriksson, P. G., Neumayer, E., Damania, R. and Gates, S. (2005) 'Environmentalism, democracy, and pollution control', *Journal of Environmental Economics and Management*, vol 49, no 2, pp343–365

Gergel, S. E., Bennett, E. M., Greenfield, B. K., King, S.,. Overdevest, C. A. and Stumborg, B. (2004) 'A test of the environmental Kuznets curve using long term watershed inputs', *Ecological Applications*, vol 14, no 2, pp555–570

Hamlin, C. (1998) *Public Health and Social Justice in the Age of Chadwick: Britain, 1800-1854*, Cambridge University Press, Cambridge

Hugo, V. (1987 [1867]) *Les Misérables*, Signet Classic, New York

IIED (International Institute for Environment and Development) (2001), *Urban Environmental Improvement and Poverty Reduction*, IIED for Danida, London

Isham, J., Woolcock, M., Pritchett, L. and Busby, G. (2005) 'The varieties of resource experience: Natural resource export structures and the political economy of economic growth', *The World Bank Economic Review*, vol 19, no 2, pp141–174

Kumar, K. S. K. and Viswanathan, B. (2004) *Does Environmental Kuznets Curve Exist for Indoor Air Pollution? Evidence from Indian Household Level Data*, Madras School of Economics, Chennai

Kuznets, S. (1955) 'Economic growth and income inequality', *American Economic Review*, vol 45, no 1, pp1–28

Lee, K. N. (2006) 'Urban sustainability and the limits of classical environmentalism', *Environment and Urbanization*, vol 18, no 1, pp9–22

Leite, C. and Wiedemann, J. (1999) *Does Mother Nature Corrupt? Natural Resources, Corruption and Economic Growth*, IMF Working Paper, International Monetary Fund, Washington, DC

Lieb, C. M. (2004) 'The environmental Kuznets curve and flow versus stock pollution: The neglect of future damages', *Environmental & Resource Economics*, vol 29, no 4, pp483–506

Marcotullio, P. J. and Lee, Y.-S. (2003) 'Urban environmental transitions and urban transportation systems – A comparison of the North American and Asian experiences', *International Development Planning Review*, vol 25, no 4, pp325–354

McGranahan, G., Jacobi, P., Songsore, J., Surjadi, C. and Kjellén, M. (2001) *The Citizens at Risk: From Urban Sanitation to Sustainable Cities*, Earthscan, London

McGranahan, G. and Songsore, J. (1994) 'Wealth, health, and the urban household: Weighing environmental burden in Jakarta, Accra, São Paulo', *Environment*, vol 36, no 6, pp4–11, 40–45

McGranahan, G. and Songsore, J. (1996) 'Wealth, health and the urban household: Weighing environmental burdens in Accra, Jakarta and São Paulo', in S. Atkinson, J. Songsore and E. Werna (eds), *Urban Health Research in Developing Countries: Implications for Policy*, CAB International, Wallingford, pp135–159

Melosi, M. V. (2000) *The Sanitary City: Urban Infrastructure in America from Colonial Times to the Present*, The Johns Hopkins University Press, Baltimore

Millennium Ecosystem Assessment (2003) *Ecosystems and Human Well-being: A Framework for Assessment*, Island Press, Washington, DC

Mosley, S. (2001) *The Chimney of the World: A History of Smoke Pollution in Victorian and Edwardian Manchester*, White Horse, Cambridge.

Nahman, A. and Antrobus, G. (2005) 'The environmental Kuznets curve: A literature survey', *South African Journal Of Economics*, vol 73, no 1, pp105–120

Pittman, W. E. (2003) 'The one-hundred year war against air pollution', *Quarterly Journal of Ideology*, vol 26, nos 1 and 2, 23pp

Rees, W. E. (2003) 'Understanding urban ecosystems: An ecological economics perspective', in A. R. Berkowitz, C. H. Nilon and K. S. Hollweg (eds), *Understanding Urban Ecosystems: A New Frontier for Science and Education*, Springer-Verlag, New York, pp115–136

Reid, D. (1993) *Paris Sewers and Sewermen: Realities and Representations*, Harvard University Press, Cambridge, MA

Rosen, G. (1993) *A History of Public Health*, Johns Hopkins University Press, Baltimore, MA

Sachs, J. D. and Warner, A. M. (2001) 'The curse of natural resources', *European Economic Review*, vol 45, nos 4–6, pp827–838

Sassen, S. (2000) *Cities in a World Economy*, Pine Forge Press, Thousand Oaks, CA

Smith, K. R. (1990) 'The risk transition', *International Environmental Affairs*, vol 2, no 3, pp227–251

Smith, K. R. and Ezzati, M. (2005) 'How environmental health risks change with development: The epidemiologic and environmental risk transitions revisited', *Annual Review of Environment and Resources*, vol 30, pp291–333

Smith, K. R. and Lee, Y.-S. F. (1993) 'Urbanization and the environmental risk transition', in J. D. Kasarda and A. M. Parnell (eds), *Third World Cities: Problems, Policies, and Prospects*, SAGE Publications, London, pp161–179

Stern, D. I. (2004) 'The rise and fall of the environmental Kuznets curve', *World Development*, vol 32, no 8, pp1419–1439

Stradling, D. (1999) *Smokestacks and Progressives: Environmentalists, Engineers and Air Quality in America, 1881–1951*, Johns Hopkins University Press, Baltimore, MA

Szreter, S. (2004) 'Industrialization and health', *British Medical Bulletin*, vol 69, pp75–86

Szreter, S. (2005), *Health and Wealth: Studies in History and Policy*, University of Rochester Press, Rochester, NY

Szreter, S. and Mooney, G. (1998) 'Urbanization, mortality, and the standard of living debate: New estimates of the expectation of life at birth in nineteenth-century British cities', *Economic History Review*, vol 51, no 1, pp84–112

Tarr, J. A. (1996) *The Search for the Ultimate Sink: Urban Pollution in Historical Perspective*, The University of Akron Press, Akron, OH

Torras, M. and Boyce, J. K. (1998) 'Income, equality and pollution: A reassessment of the environmental Kuznets curve', *Ecological Economics*, vol 25, no 2, pp147–160

World Bank (1992) *World Development Report 1992: Development and the Environment*, Oxford University Press, New York

WWF (2004) *Living Planet Report 2004*, World Wide Fund for Nature International, Gland, Switzerland

Yandle, B., Bhattarai, M. and Vijayaraghavan, M. (2004) *Kuznets curve for the environment and economic growth*, Poverty and Environment Research Centre (PERC), Bozeman, Montana

York, R., Rosa, E. A. and Dietz, T. (2003a) 'Footprints on the earth: The environmental consequences of modernity', *American Sociological Review*, vol 68, no 2, pp279–300

York, R., Rosa, E. A. and Dietz, T. (2003b) 'STIRPAT, IPAT and ImPACT: Analytic tools for unpacking the driving forces of environmental impacts', *Ecological Economics*, vol 46, no 3, pp351–365

3

Variations of Urban Environmental Transitions: The Experiences of Rapidly Developing Asia-Pacific Cities

Peter J. Marcotullio

Introduction

Accounts of environmental challenges in cities in the developing world often include comparisons to situations already experienced by now-developed cities. For example, a 1998 article in *The Economist* on development and the environment starts out with a description of environmental problems in mid-19th century English cities, then suggests that environmental conditions in developing cities today are very similar to those of Manchester, UK, in the 1800s (Litvin, 1998).

References to the underlying causes for current environmental trends in the developing urban world also include similar factors as those experienced by developed cities. Reports on forces creating environmental problems include pollution trends from industries, rapid urbanization, poverty, socio-economic structural factors and inadequate environmental management (USAID, 1990). The aforementioned article identifies national trends – such as urbanization rates and industrialization – as the main culprits creating these problems. Other sources, such as the Brundtland Commission (World Commission on Environment and Development, 1987), state that poverty is a major cause and effect of environmental problems. Scholars studying developed world cities suggest similar circumstances in terms of urban growth (Preston, 1979) and writers have elaborated on the social income disparities, poverty and environmental impacts of industrialization processes (e.g. see Dickens, 1995 [1854]). Moreover, fragmented local and inadequate regional governmental structures, staffing and funding problems, and misinterpretations of environmental relationships are common experiences among developed world cities (Rosenbaum, 1995). Given both the comparisons of conditions within cities of different periods, and the argument that their environments are impacted by similar forces, it is no wonder that models for environment and development often consider developing world economies on the same trajectory as that of the now developed world (Grossman and Krueger, 1995; Lomborg, 2001).

This chapter argues differently. It suggests that there is evidence of significant differences between the experiences of rapidly developing world cities and those of developed world cities in terms of their urban environmental transitions. The goal of this chapter is to expand and elaborate on these points. It does so by introducing the notion of *time–space telescoping*. The time–space telescoping of development suggests that both time- and space-related effects have created a unique context for developing countries. Time-related effects are induced by historical changes in the speed and efficiency of human activities. They tend to shift the speed and emergence of environmental burdens so that issues appear at lower levels of economic wealth and change faster over time. Space-related effects are induced by increased availability and spatial concentration of diverse social, economic, political and technological phenomena.

Outcomes from the process of time–space telescoping help to distinguish the experiences of developed and rapidly developing cities in terms of sets of urban environmental conditions and the temporal and spatial scales at which they occur.[1] Specifically, in terms of shifts in temporal scales, the perspective suggests environmental challenges in developing cities are occurring *sooner* (at lower levels of income), rising *faster* (over time for similar ranges of income) and emerging *more simultaneously* (as sets of problems) than previously experienced by developed cities. In terms of spatial scales, we find greater variety of environmental impacts, concentrated at smaller scales than seen previously.

These types of generalization beg the question of how 'development' can be measured and meaningfully compared across nations and across time. To many, development is equated to the increasing capacity of the national economy to generate increases in production. Using per capita indicators demonstrates that increases in economic growth are different than that of population growth. Using purchasing power parity (PPP) per capita indicators further refines comparisons by allowing incomes standardization across goods and services between countries. That is, a PPP value for income in one country will match the ability of citizens to purchase the same amount of an exact set of goods and services in another country as well as their own. While using PPP per capita values is more appropriate than simply comparing gross domestic product (GDP) or GDP per capita figures, it still leaves out a lot. Economic growth does not speak to changes in social or political structures. Therefore, purely economic indicators are often supplemented with social indicators including emphasizing material possessions such as telephones, televisions and radios, the use of banks, schools and cinemas, and provision of housing, medical or educational services (e.g. see any *Human Development Report* of the United Nations Development Programme – UNDP).

A further emphasis has been on the reduction or elimination of poverty, inequality and unemployment in the context of economic growth. Development, in this view, should be concerned with equity and distributive justice at all scales. Recently, economists have turned toward definitions that include improving the quality of life for citizens, broadly defined and especially the poor (World Bank, 1991). These various definitions suggest that '[d]evelopment must therefore be conceived a multidimensional process involving major changes in social structure, popular attitudes and national institutions, as well as the acceleration of economic growth, the reduction of inequality and the eradication of poverty' (Todaro, 1997, p16).

The complexities of development make it difficult to measure. Rather than directly measuring aspects of development and presenting a comparable index (see, e.g. the United Nations Development Programme's Human Development Index) with which to compare cities, this chapter focuses on variations of experiences during economic growth through a space for time substitution. That is, by looking at a cross-section of cities with different income levels, we can get a picture, blurry as it may be, of the process of development in this part of the world. These variations from the western model in development patterns, although limited to environmental conditions, arguably represent differences in development pathways. The emphasis is not that income levels equal specific development levels, hence development–environment relationships, but, on the contrary, that income levels are increasingly less helpful in identifying urban environmental conditions. The findings suggest that among rapidly developing Asian-Pacific cities common patterns emerge and an important aspect of these patterns is that previously observed stages are less evident.

The implications of these differences are important: if the current context of development is creating situations that are truly different from those of the past, then responses to these challenges should also be different from those of the past. In other words, the types of responses that governments in developed countries previously implemented for their environmental problems may not be appropriate for the developing world today.

The next and second section of this chapter discusses the various perspectives used to analyse the differences between the developed urban world experience and those of the developing urban world. The third section presents an analysis of the environmental conditions in selected sets of cities in the rapidly developing Asia-Pacific region. The conclusion elaborates on the implications of this perspective, in terms of both theory and practice.

A framework for understanding the time–space telescoping of urban environmental transitions in the Asia-Pacific region

This section explains why urban environmental transitions have changed over time. The first sub-section describes urban environmental transition theory. The next sub-section explains the historical relationship between urban environmental transitions and development patterns of the developed world. The third sub-section presents a framework for analysis of both the drivers of change and the associated urban environmental impacts in the rapidly developing Asia-Pacific region.

Urban environmental transition theory

As made clear in Chapter 2, Urban Environmental Transition (UET) hypothesis suggests that there is a series of distinct environmental challenges that cities experience during development (see also McGranahan et al, 2001). This theory suggests that as cities become wealthier, their environmental impacts shift in nature, being primarily those associated with lack of access to clean water and sanitation, to those that include chemical air and water pollution, to those that threaten ecosystems and life-support systems – otherwise described as moving from brown to grey to green environmental agenda challenges (McGranahan and Satterthwaite, 2000; Marcotullio and Lee, 2003).

Beyond simply identifying the types of environmental challenge that citizens experience at different levels of development, the UET model also injects issues of temporal and geographic scale into the shifts in environmental burdens. As cities develop, for example, brown issues that have immediate impact are overcome, while environmental challenges increase in geographic scale from the household and neighbourhood levels to city-wide regions. Challenges pertaining to local water access, sanitation and indoor air pollution may be overcome, but are replaced by metro-wide chemical air and water pollution (grey issues). Other, wealthier, cities have overcome this second phase, and instead struggle with the green issues. In this case, the dominant environmental impacts of urban-based activities are geographically regional if not global: for example, greenhouse gases, acid rain and ozone-depleting emissions. Moreover, the impact of these green issues is much delayed and in some cases still not well understood.

Three aspects of the UET theory are especially noteworthy. Firstly, its model defines a relationship between development (wealth) and the urban environment by identifying shifts in *types* of environmental challenges. Secondly, it points out that cities undergo a *series* of environmental challenges that shift according to impact and timing; some of these challenges are missing in the global sustainable development agenda (McGranahan et al, 1996). Thirdly, UET theory has placed the *scale* of environmental impact at centre stage of the policy engagement, thereby facilitating the differentiation of distinct environmental conditions in different cities.

Others have used the UET model to study cities in East and Southeast Asia (Webster, 1995; Bai and Imura, 2000). These studies distinguish cities by their income levels and explore differences in both the types and scales of environmental impacts associated with urbanization and rising wealth. Neither of these studies explores in any detail the differences in geographic or temporal scale of environmental impacts associated with cities of the Asia-Pacific region and those of elsewhere. Indeed, Bai and Imura (2000) suggest that, in general, all cities undergo environmental trends in stages and they advanced an evolutionary staged model. They note that cities in this region undergo faster transitions than those elsewhere, but emphasize the simultaneous nature of transitions (one transition occurs after the other). They also mention that cities could jump between stages in their environmental evolution; but failed to provide empirical evidence of any city that had done so. This study suggests a patterned difference between environmental transitions experienced in cities undergoing rapid development. The driving forces changing transitions are the shifting contexts framing development.

Long waves of development and the sequential pattern of the developed world urban environmental experience

The theory of long waves of economic development was derived from the work of Nikolai D. Kondratieff (1979, p519), who popularized the study of 'the dynamics of economic life in the capitalistic social order' by demonstrating the existence of secular trends that were structurally linked to overall changes in the economic development of the particular society. While Kondratieff studied three types of wave of varying length (approximately 50 years' duration, 7 to 10 years' duration, and 3 to 4 years' duration), he was primarily interested in the longest waves. These were characterized by accelerating rates of price increases from deflationary depression to inflationary peaks, followed by decade-long plunges from the peaks to troughs, again followed by weak recovery and then growth. Debate continues over the waves' underlying causes, or whether another will occur again in the developed world (Maddison, 1991),[2] but general consensus suggests that capitalist national economies have experienced long-term fluctuations of some 50 to 60 years' duration.

Associated with these long-term trends in price cycles (or growth rates) are shifts in technologies (Schumpeter, 1961). Brian Berry (1997) suggests that US history, for example, is marked by the rise and fall of a succession of techno-economic systems, defined by interrelated sets of technologies, sets of raw materials, energy sources and infrastructure networks. The first techno-economic system included the use of wind and animal power (sails, wagons), the emergence of cotton textile industries and the use of iron. This was followed by the era of coal, steam and iron rails used for trains. Thereafter began the era of steel, kerosene and electricity. Petroleum, chemicals and the internal combustion engine define the fourth wave. The current fifth wave is driven by the service sector, technologies such as computers, telecommunications, biotechnologies and new materials. Scholars have associated technological and economic development with urban growth phases in now developed countries

(Borchert, 1967; Berry, 1997; Yeates, 1998). Although there is debate over the exact timing of the periods, four can be defined, with each long wave cycle resulting in a distinct urban pattern of development largely associated with the implementation of different sets of technologies.[3]

Melosi (2000) has built upon this line of thinking through his historical analysis of the emergence of significant urban environmental issues and their solutions in the US. His narrative is comparable to 'urban environmental transitions,' as it explores the shifts in environmental challenges and the increasing scale of impacts associated with new solutions. Further adding to the notion that economic and technical factors force change, Melosi (2000) also focuses on the importance of changes in scientific understanding of environmental problems. Developing this perspective, other scholars have noted other important influences, such as shifts in values, in explaining the adoption of certain infrastructure and technologies (McShane, 1979, 1994; McKay, 1988). In terms of the revolution in street paving, for example, scientific and technical advances alone do not explain the enormous effort to pave US urban streets from 1880 to 1924. Though progress in chemistry and pressure from the rapid rise in automobiles were important, the emergence of paved streets was more for social and cultural reasons, reflecting shifts in housing preferences accompanied by different perceptions of the street use and new municipal paving policies (McShane, 1979). Certainly, there are economic, technological, social, political and crisis-related influences that affect the timing of environmental transitions.[4] What is agreed upon, however, is the sequential nature of the process that the developed world experienced. This has allowed for successful response scenarios to develop, because each set of issues was dealt as they arose, with a 'first things first' approach (Warner, 1955).

Time- and space-related effects and the shifts in emergence, timing and speed of environmental transitions

The context for the emergence and timing of environmental challenges has shifted over the years. Such trends can be described in general terms as time-related and space-related effects of development. Time-related effects include changes in context that facilitate faster and more efficient human activities. Hence, at the turn of the 20th century, people were not able to travel as fast or as far as they can today. Moreover, declining real-dollar costs for owning and using a car, for example, have led to a rise in personal vehicle ownership and usage around the world.

With these changes have come shifts in the environmental impacts related to activities. In some cases, technologies have facilitated more efficient production and hence lower environmental impact. In addition shifts in speed of development have increased the intensity of environmental impact as more industrial activity, for example (in terms of both the size of a firm and the absolute size of output of the manufacturing sector within an economy), has emerged.

Space-related effects include processes that concentrate increasingly diverse phenomena in geographically uneven patterns. This concentration may be of economic processes, of people, activities, infrastructure and so on. Globalization, for example, has space-related effects by concentrating certain types of infrastructure (communication, transportation, financial and business services, headquarters of transnational companies) in specific locations around the world (e.g. world cities) (Friedmann and Wolff, 1982; Friedmann, 1986; Sassen, 1991; Lo and Yeung, 1998). This process has been called the re-territorialization of capital (Brenner, 1999).

Previous studies examining a combination of these two (time- and space-related) effects suggest changes in the constraints placed upon human activities by both and thus how human activities affect the environment and are affected by it. Janelle (1968, 1969), for example, identified the increasing speed at which people move across space and its effects on economic activity and social relations. His term, 'time–space convergence', defined the process of decreasing the friction of distance between places, resulting in decreasing average amounts of time needed to travel between them. Processes creating time–space convergence have not only made the world smaller, but also have increased our ability to impact a larger number of different environments around the world, at a more intense rate.

Harvey (1989) focused on what he called time–space compression as underpinning the emergence of the post-modern condition.[5] Time–space compression includes processes that revolutionize the objective qualities of space and time and alter, sometimes in quite radical ways, how we represent the world to ourselves. Waves of time–space compression have altered the way we understand the world around us and hence, how we react to it.

Other studies that work time- and space-related effects implicitly into analyses suggest the emergence of overlapping environmental challenges in the developing world. Smith (1990) and Smith and Lee (1993) have identified changing patterns of risk over time, as traditional risks (e.g. those associated with local indoor air pollution) have now combined with more modern risks (e.g. those associated with inhaling pesticides) generating a new genre of mixed risks in low-income countries. These notions highlight changes in the type, speed and location of activities over time, which have thus transformed human behaviours and their resultant environmental impacts.

This chapter suggests another time–space effect that shapes the contemporary relationship between development and the environment in the rapidly developing world. The processes that create these shifts have been called time–space telescoping, and their impacts are multi-fold (Marcotullio 2005; Marcotullio et al, 2005). Of note are three empirically testable outcomes including a collapsing, compression and telescoping of previously experienced development patterns, such that they now occur *sooner* (at lower levels of income) increase *faster* (over time) and emerge *more simultaneously* (as compared to the sequential patterns experienced by the now developed world).

The three results related to the time–space telescoping of development can be explained as follows. The drivers of change have created the conditions within which environmental trends follow the patterns described. In some cases, they actually mirror the sooner, faster and more simultaneous pattern. In other cases, they set the context for more diverse socio-economic and technical features within contemporary cities. For example, within almost any city or large urban area of the rapidly developing Asian world, it is not uncommon to have populations without running water, sanitation and solid waste disposal living close to modern high-rise apartments with all the latest and most modern conveniences and late model BMW cars parked next to rickshaws.

Overview of the impact of time–space telescoping on urban environmental transitions in the Asia-Pacific region

This section presents an analysis of the environmental conditions within different categories of cities of the Asia-Pacific region based upon the World Bank's country group levels. The basis of the urban typology presented here includes cities in low-income countries (Vietnam, Cambodia, Mongolia, Myanmar and Laos), cities in middle- and upper-middle income countries (China, Indonesia, Malaysia, Thailand, the Philippines and so on), and cities in high-income countries (e.g. South Korea, Japan). Also included in this typology are entrepôts (Hong Kong and Singapore) as a separate group, given their special status.

This typology matches that of the functional regional city system created through global flows of finance, trade, investments, people and information (Lo and Yeung, 1996; Lo and Marcotullio, 2001). Arguably, variables representing the level of national and urban economic wealth, the type and intensity of connections that countries and cities have to the global and regional economic system, and the degree to which these entities undergo time–space telescoping are related. It is perhaps the driving forces of globalization that move against perceptions that stages of development continue. Certainly, the idea of overlapping burdens makes it difficult to identify transitions.

At the same time that stages are blurred, a distinct pattern can be ascertained from this analysis. Within the first category of low-income cities, which are unconnected or only recently connected to the regional city system, one finds mainly either mainly brown issues or a combination of traditional brown issues and also grey issues arising. In the second category, industrializing cities have achieved their growth through connections to the regional city system, and thus have the greatest mix of environmental challenges. At the same time, they suffer largely from challenges associated with industrialization and motorization.

The green issues increasingly challenge cities at the high end of the income continuum. These cities are playing the role of the capital exporters of the region's international system, although it is interesting that every one of them still has unresolved grey and even brown issues. Given the rapid rate within which these cities developed, this overlap between the three types of environmental agenda is not surprising.

Some cities, such as Singapore, exemplify how economic geographies and an emphasis on environmental governance can overcome many urban environmental challenges. This type of city is the exception that proves the rule. Because of the government's emphasis on tackling environmental issues, and perhaps particularly because it is a city-state, Singapore has been able to overcome many of its local environmental challenges. The typology, therefore, also reflects the impact of urban environmental management capacities.

Cities in low-income countries

Many cities within the Asia-Pacific region may be considered low income – that is, a significant share of their population has an income of no more than US$1 a day. Examples of cities within this category are Phnom Penh, Hanoi, Ho Chi Minh City, Vientiane, Yangon and Ulaanbaatar.

Within this group there are at least two types of development trend. Some cities, such as Yangon, Vientiane and Ulaanbaatar, remain largely outside of the regional city system flows, and thus have significantly different development paths than the other cities in this category. Those that have recently begun to connect to the international system are undergoing rapid development. These latter cities are experiencing an overlap of brown issues with grey issues. Moreover, they are also experiencing a sooner emergence of some environmental impacts than experienced by the developed world (particularly chemical pollution trends associated with motorization and industrialization). The differences in these cities' environmental conditions, compared to previous eras, are becoming increasingly stark as their nations embrace globalization at the national level, since they do not have the urban planning systems in place nor the capacity to alleviate the negative aspects of this type of growth. Furthermore, because these cities grew slowly in the past, the urban fabric that was generated over a long period of time remains largely intact, despite rapid growth in other areas, giving rise to new environmental concerns. For example, Barter (1999) points out that within cities of East and Southeast Asia, narrow streets create greater exposure levels to automobile exhaust for inhabitants of these cities than those living in developed world cities, despite the lower numbers of vehicles on Asian streets.

The low-income and largely unconnected cities are struggling with the most basic of brown agenda concerns. For example, in Laos, the only city with a reliable piped water supply is Vientiane (where coverage is 80 per cent). Sewerage in the country is almost non-existent, with only a portion of the capital enjoying piped sewerage services. Without a centralized wastewater treatment system, households and commercial premises generally have no on-site flush latrines.

Within Vientiane, about 50 per cent of the solid waste is collected, while the other half is burned, reused or disposed of by the owner. Though waste dumping is leading to the disappearance of wetlands and open spaces, Vientiane is still considered to have a green environment, and except for issues with dust, air pollution has not yet risen to problem levels. One of Vientiane's main environmental challenges is frequent flooding (Viengsavanh and Yongnou, 2001).

In Myanmar, urbanization is not strongly associated with industrialization (Thein, 2001). The capital, Yangon, grew from 2.5 million in 1983 to 4.2 million in 2000, with a substantial additional population that visits the city daily for jobs and other economic opportunities. Environmental challenges are largely localized and related to the meagre provision of infrastructure. For example, Yangon's water supply falls short of the population's needs, causing shortages. Only East Yangon enjoys 24-hour service (Oo, 2001). Furthermore, only 7.3 per cent of the city's population is served by piped sewerage, while 18.5 per cent has septic tanks. The remaining population either has protected latrines (46 per cent), unprotected latrines (27.9 per cent) or other types of treatment (Thein, 2001). The main disposal practice for the city's solid waste is open dumping and burning. Collection service varies; while the solid waste for approximately 40–55 per cent of the entire Yangon population is collected via a municipal service, almost 90 per cent of the downtown residential and commercial area of the western district receives these services (Han, 1999).[6] Like Vientiane, the city has considerable green space, particularly in the north, and it is attempting to conserve this as a 'green belt', under the concept of 'Yangon as a Garden City' (Oo, 2001; Thein, 2001).

These conditions have a significant impact on human health. In Mandalay, for example, there are frequent outbreaks of diarrhoea and dysentery, and dengue hemorrhagic fever is not uncommon (Nyunt, 2002). Female life expectancy (which is typically higher than males) in Laos and Myanmar is 55.4 and 56.7 years, respectively.

The other cities within this category are increasingly articulated to the regional economic system and beginning to experience several types of environmental problem simultaneously. Typically, these cities have retained the structure of their colonial heritage (including basic infrastructure), have a sizable agriculture base, and have grown rapidly over the past few years. Their physical size has pushed past the adequate provision of services. These cities – Hanoi, Ho Chi Minh City and Phnom Penh, for example – are experiencing rapid growth under trade, tourism and Foreign Direct Investment (FDI)-led industrialization. Despite the economic downturn of 1997, Cambodia and Vietnam continued to grow. In 2001–2005, Vietnam averaged over 7.5 per cent growth, reaching a peak of 8.4 per cent in 2004. Economic growth is evident in Hanoi and Ho Chi Minh City, as automobiles, for example, are an increasingly common site (*The Economist*, 2005). Ho Chi Minh City is the industrial centre of Vietnam, accounting for up to 25 per cent of output. At Song Than industrial zone, outside the city, foreign investments in factories reached US$5.8 billion in 2005 (*The Economist*, 2006). Meanwhile, Phnom Penh municipality, the commercial and industrial base for Cambodia, grew by 10.3 per cent annually from 1992 to 1994 (Sarin, 1998). Cambodia's economy grew by 7 per cent in 2004. These locations boast cheap labour and have thus attracted low-skilled labour industries.

The most critical environmental problems for these cities are increasingly mixed, including those related to water supply quantity and quality, wastewater removal, solid waste services, inadequate electricity, vehicle traffic and related chemical pollution, and hazardous waste from industry and hospitals. An example of the mix of brown and grey environmental issues can be seen in Ho Chi Minh City and Phnom Penh. In Ho Chi Minh City, the government suggests that 80 per cent of waste is collected, but treatment consists largely of open dump landfills. At the same time, in both these cities vehicle transport and related air pollution issues are a growing concern. Ho Chi Minh City and Phnom Penh are known for their high numbers of motorcycles and scooters (and recently automobiles). Given the small and narrow streets, air pollution from these vehicles is an increasing health problem (Barter, 1999). It is also interesting to note that the income levels of these nations are far below those of the US and Western European nations during the rise of the automobile at the turn of the century (Table 3.1).

Cities in middle- and upper-middle income countries

Industrial manufacturing processes are vitally important to the region's growth and development, and hence manufacturing cities play an important role in the regional economic system. Global integration has affected the pattern of development by encouraging the concentration of manufacturing plants in a doughnut-like ring around the city (Lo and Marcotullio, 2001; Webster, 2002). At the same time, modern commercial centres appear in the city core, often next to slums and highly degraded areas. These cities' rapid industrialization and uncontrolled population growth have contributed to a pattern of unsustainable development processes affecting the environment

Table 3.1 *Comparative dates of similar GDP per capita levels, selected Asia-Pacific economies the US, UK and France, 2000*

Asian country	Asian GDP per capita 2000	Year similar Asian 2000 GDP/capita reached by US	Year similar Asian 2000 GDP/capita reached by UK	Year similar Asian 2000 GDP/capita reached by France
Cambodia	1087	Before 1820	Before 1820	Before 1820
Laos PDR	1173	1820	Before 1820	1820
Mongolia	1085	Before 1820	Before 1820	Before 1820
Myanmar	1353	1820	Before 1820	1840
Vietnam	1790	1850	1820	1861

Note: GDP per capita in Geary-Khamis International dollars.

Source: Maddison (2001)

(Douglass, 1991). As manufacturing production has become an increasingly important part of these urban economies, levels of air and water pollution and concentrations of hazardous wastes have also risen. At the same time, brown issues remain unresolved.

Air pollution is a major problem in industrial cities, and motor vehicles (particularly two-stroke motorcycles, three-wheel taxis, and diesel buses and trucks) are significant contributors (see also Chapters 8 and 9).[7] Cities such as Jakarta, Bangkok and Shanghai are currently struggling to overcome these types of problem, but increases in vehicle ownership outpace even economic growth (Marcotullio and Lee, 2003). Air quality continues to deteriorate in cities such as Bangkok and Jakarta as total suspended particles (TSP) and carbon monoxide increase, primarily from traffic congestion (Webster, 1995). The resultant health and productivity effects of pollution cost these cities billions of dollars a year. For example, the annual cost of air pollution is estimated at US$1.3–3.1 billion in Bangkok, US$1.0–1.6 billion in Kuala Lumpur and the Klang Valley, and US$400–800 million in Jakarta (World Bank, 1992; Brandon, 1994).

In industrializing cities in China, coal-driven electrical generators – used for about three-quarters of the country's energy consumption – have had tremendous impacts on the urban air quality. For example, sulphur dioxide concentrations in cities such as Chongqing, Taiyuan, Qingdoa and Guiyang substantially exceed the high end of the World Health Organization (WHO) guidelines for 24-hour exposure safety scenarios (Lo and Xing, 1998). Shanghai's level of sulphur dioxide is well over the WHO's annual standard for maximum exposure levels for safe health.

Providing adequate, clean water has been problematic for many of these cities. The Municipal Water Works of Jakarta provides raw water to approximately 30–40 per cent of the population, but much of this water is contaminated with E. coli (Bianpoen, 2001). For the Bangkok Metropolitan Region (BMR), the Metropolitan Waterworks Authority's (MWA) main water system supplies water to the city core from the Chao Phraya River and therefore requires treatment. However, during the late 1990s, only 1.6 million (18 per cent) of the 8.8 million potential customers were being served by the MWA (JICA, 1997). A number of water supply improvement projects begun during the 1990s and planned for completion early in the 21st century will hopefully achieve the intended goal of 93 per cent coverage. Unfortunately, the 1997–1998 financial crisis has slowed down infrastructure development.

For rapidly developing cities, water pollution is primarily caused by domestic wastewater flowing into open pits and canals without treatment, but increasingly, industries are contributing to the problem. In both Jakarta Bay (Jellinek, 2000) and around Bangkok (Phantumvanit and Liengcharensit, 1989), harmful industrial wastes are being dumped into waterways. The Chao Phraya River running through Bangkok no longer supports life; the dissolved oxygen concentrations in parts of the river approach zero at certain times of the year (Setchell, 1995). The city's *klongs*, a storm drainage system of canals, are extremely dirty as much of the city's wastewater makes its way directly into these waterbodies without treatment of any kind (Daniere, 1996). The Surabaya River in East Java is considered the most polluted on the island, and up to 60 per cent of its pollutant load is from industries (BAPEDAL, 2001).

In Bangkok there is a serious lack of sewer service. In the mid-1990s sewers served only 2 per cent of the city's population (Webster, 1995), although plans to construct wastewater treatment facilities handling up to 30 per cent of the population's wastewater were underway (Asian Development Bank, 1994). In Shanghai, river pollution has cost the city at least US$300 million, since municipal water intakes had to be moved 40km upstream. In the currently developing Pudong New Area of Shanghai, surface water pollution levels are already very serious (Wu, 1998).

For rapidly developing cities, as also experienced by their predecessors, solid waste has become a problem associated with inadequate garbage collection. In Jakarta, for example, between 20 and 25 per cent of the garbage is collected by the public sector or private conveyors (Pernia, 1992; Webster, 1995). In the BMR, about 20 per cent of solid waste goes uncollected (Daniere, 1996).[8] Sanitary landfills are rare in Kuala Lumpur, Manila, Jakarta and Bangkok; much of the municipal garbage is eliminated by less sanitary means, such as open burning, dumping into rivers and canals, or into abandoned mine sites and swamp areas (Lee, 1994). BAPEDAL (2001) suggests that all waste is openly dumped in Surabaya. Also, like cities in the past, local communities and scavengers collect an additional 60 per cent of the waste stream (UNESCO, 2000). What is different about today's solid waste is the type of material included in the garbage stream. Generally, hazardous wastes in these cities – particularly hospital wastes and toxic substances – are not adequately handled.

Many of these major cities are found on the coast, where large rivers run into the ocean. Urban and industrial water contamination is seriously impacting coastal zone ecology (Lebel, 2002). Degradation can be seen in increased marine pollution, loss of mangrove forest and degraded condition of coral reefs. Within the Asia-Pacific region, coastal and marine pollution has increased mainly due to domestic and industrial effluent discharges, atmospheric deposition, oil spills and other wastes and contaminants from shipping as well as land development, dredging and up-stream river modifications (UNESCAP, 2000; UNEP, 2002). Sewerage effluent from urban and tourist areas in Thailand makes substantial and increasing contributions to pollutant loads in the upper Gulf of Thailand (Lebel, 2002). In Bangkok, approximately 1.5 million cubic metres of untreated domestic and industrial pollutants are discharged directly into the waterways on a regular basis, with significant adverse impacts on water quality (Kaothien, 1995; Setchell, 1995).

Among those in the regional city system, industrial cities are characterized by some of the most mixed sets of environmental conditions. Air, water and ecosystem damage are widespread and severe. Globalization flows have fuelled the deteriorating state of the environment and have helped to create new sets of risks, by-products associated with both traditional (agricultural) and modern (urban) lifestyles (Smith, 1995). These cities are the locations of the most intense time–space telescoping of the urban environmental transition.

Cities in high-income countries

High-income cities feature concentrations of trans-national corporation headquarters, multi-national banks, production and business services. They have multi-nodal structures in which several commercial and industrial districts arise in different locations within the city. At the same time, these cities are expanding outwards, leaving workers with longer commutes, since many of the jobs remain in the inner city area. Manufacturing industries have decentralized from the centre while advanced services are concentrated in the core regions of the city.

Contemporary development pressures have positioned consumption-related pollution, open-space/quality of life issues and those pertaining to urban sprawl high on the list of policy priorities. These cities are reducing industrial pollution and are also struggling with consumption-related pollution and quality of life issues.

Air pollution challenges have been a constant struggle. Cities such as Tokyo and Seoul have been able to control, to varying degrees, air pollution from point sources and have also seen reductions in sulphur dioxide emissions per capita and TSP levels, for example. But they are still attempting to control the increase in air pollution that has accompanied lifestyle changes (i.e. increased automobile usage) (Kim, 1996; Sawa, 1997; Republic of Korea, Ministry of the Environment, 1999; Tokyo Metropolitan Government 1999c).[9] In Seoul, for example, automobile ownership increased from 60,000 vehicles in 1970 to over 2.2 million in 1999, bringing with it congestion, noise and air pollution (Seoul Metropolitan Government, 2000). Most of this increase started in the mid-1980s (Barter, 1999) and has been accompanied by increases in various air pollutants (Kim, 1996).

Important regional and global air pollution emissions problems stemming from increased energy demands and waste treatment demands consist of rising amounts of carbon dioxide, a major contributor to global greenhouse warming, and increased levels of toxic and hazardous wastes in the environment. Tokyo's emissions have more than doubled since 1970, rising 2.5 per cent per year (Dhakal and Kaneko, 2002). In Japan, citizen concern is on the rise regarding dioxin levels related to waste incineration, and a recent accident at a nearby nuclear power plant.[10]

Water pollution is also a significant issue in these cities. The Han River's concentration of total nitrogen increased from 2839 to 5424 micrograms (μg) per litre, and total phosphorus increased from 116 to 261μg per litre between 1989 and 1995. Hazardous substances were also detected in 13 of 136 sites on the four main rivers in South Korea (ambient standards for these contaminants were not exceeded, however). Point discharges of domestic and industrial wastewater place a heavy burden on the country's surface waters. Between 1986 and 1994, rapid economic development caused a doubling of industrial effluent discharges in South Korean cities. About 15 per cent of the total industrial effluent volume is discharged into Seoul's Han River and its tributaries alone (OECD, 1997). Subsequently, the government invested large sums of money (27 trillion won by 2005) into controlling these sources with positive effect (Republic of Korea, Ministry of Environment, 2005b). For example, a survey of the Han River in Seoul demonstrated that the number of species of fish increased from 21 in 1990 to 50 in 1998 (Lee, 2001). However, the government has yet to meet popular demand for clean rivers (Republic of Korea, Ministry of Environment, 2005).

Also high on the policy agenda of these cities are the local issues for maintaining a high quality of life for its wealthy citizens. These policies include increasing open space, waterfront access, urban entertainment and cultural activities (Kato, 1998). Among large cities in the developed world, Tokyo's comparatively low level of land designated as public open space has prompted city planners to promote laws to increase greenery (Tokyo Metropolitan Government, 1999c). Seoul residents are also expressing dissatisfaction with the lack of parks and open space (Kwon, 2001). One current topic within the city is now that the National Green Belt Policy has been reformed, how can open space be preserved around the city?

In general, Japan and South Korea are using legal instruments to respond to individual environmental problems that were neglected in the early decades of their respective economic development. A significant portion of new infrastructure projects is dedicated to tackling the remaining brown agenda issues within these two cities. In Tokyo, several new laws – including the Tokyo Metropolitan Basic Environment Plan (1997) and the new Environmental Impact Assessment Law (1998) – are attempts to improve the city's environmental quality.

Furthermore, aware of their impact on the metropolitan area as well as the Southeast Asian region, Tokyo, Seoul and Taipei are attempting to control emissions through demand management and recycling strategies. In Tokyo, for example, the city has been combating pollution from incinerator plants through ISO 14001 certification for environmental quality performance (Tokyo Metropolitan Government, 1999a). Tokyo is also promoting the use of rainwater and the recycling of wastewater to help clean up Tokyo Bay (Tokyo Metropolitan Government, 1999b, 1999d). Seoul has recently set up a pricing system for the *Namsan* Tunnels (which feed into the central business district – CBD) in an attempt to reduce traffic. The city has also implemented a new waste management plan that focused on reducing the source of waste and increasing recycling (Seoul Metropolitan Government, 2000). In 2003, the recycling rates for the entire nation approached 45 per cent for domestic waste and 80 per cent for industrial waste (Republic of Korea, Ministry of Environment, 2005a).

Capital exporting cities are reaching a level of maturity associated with a de-concentration of manufacturing industries and population. In the case of Tokyo, it has already been through a set of environmental transitions, although there are unfinished agenda issues. The dominant concerns are to control consumption-related pollution and strike a balance among the many different urban functions and the quality of life issues demanded by the higher-income segment of society.

Contrasted with other cities in the region, the environmental challenges in Tokyo, Seoul and Taipei are relatively similar. In each city, vast amounts of public infrastructure have been developed in and around the urban regions, which has helped to put them on top of both their national and the regional urban hierarchies. At the same time, each city still retains unresolved grey issues. Though top-down centralized management of cities and city systems enabled these nations to handle rapid growth and, for example, control automobile ownership (until recently), their flexible planning systems have not been able to overcome all environmental problems associated with the different agendas.

The well-managed entrepôts

The growth of Singapore and Hong Kong is directly related to their functions as the region's entrepôts. The flows of people and goods from the city to the outlying areas accompanied an increasing level of cross-border capital flows and economic growth.

For these cities, achieving a high quality of life has become an important concern (UNESCAP, 1995). While their economies changed from labour-intensive industries in the 1960s and 1970s to high-tech, service and finance industries in the 1980s and 1990s, they have seen the emigration of manufacturing and other related activities from their borders.

One of the most successful policy areas for Singapore has been environmental management (UNESCAP, 1995; Ooi, 1995). It has certainly provided a model for traffic control (Webster, 1995). Because of these types of policy, both Hong Kong and Singapore appear on the list of 10 most liveable cities in the region (Choong, 1997). Singapore has not only overtaken the US in terms of per capita GDP,[11] but is considered Asia's cleanest city (UNESCAP, 1995).

Like other wealthy cities, Singapore's success in regulating air pollutants has been mixed, although air quality is within international standards. Some qualities – such as acidity and urban smoke – have declined, while nitrogen oxide and dust fallout levels have varied (Perry et al, 1997). This, no doubt, is due to vehicular usage. In 1995, Singapore's vehicle population stood at 584,322 (Hui, 1995), approximately a 70 per cent increase from 1980. In 2004, the total vehicle population in the city was 727,395 (Singapore and Transport Authority, 2005). Despite increases in the numbers of automobiles from 1995 to 2000, the city has also experienced reductions in emissions of ozone and particulate matter. Any reductions in emissions since 1995 have been due to an excellent public transit system, traffic management measures and policy used to control vehicle fuel consumption (Dhakal, 2002).

Other services, such as water supply, are delivered at the highest of standards. Singapore's water loss stands at 8 per cent (Briscoe, 1993), which is comparable with levels experienced by cities in the developed world. All homes receive piped potable water, and wastewater is collected and treated. A substantial portion of the island nation was set aside as water catchments for the collection, storage and protection of water resources. Wastewater from domestic, commercial and industrial premises throughout the island is channelled to six centralized sewage treatment works, where it is treated before being discharged into the sea.

Bringing the water system up to high levels was not without struggles. In the mid-1970s, the Singapore River and its tributaries degenerated into open sewers. The sources of the pollution were pig and duck farms, unsewered premises and street hawkers, river activities, vegetable wholesale activities and indiscriminate discharge of rubbish and wastewater. Within 10 years, active clean-up programmes resulted in a 90 per cent reduction of the pollution load (Hui, 1995; Perry et al, 1997).[12]

A comprehensive refuse collection system covers the entire island. Seven Environmental Health Offices provide daily street cleansing and solid waste collection services. Some commercial and industrial establishments are responsible for collection and disposal of their own solid wastes. Singapore has both incinerators and sanitary landfills for waste treatment.

Singapore has also developed an extensive park system, including popular recreation facilities and wildlife reserves. In 1993 the parks' budget topped Sing$53 million, and over the previous 20 years, the city-state invested more than Sing$700 million in these facilities (Lee, 1995).

There are two important issues to consider when summarizing Singapore's experience. Firstly, it was able to attain a clean urban environment because of the resources put into environmental policies, laws, regulations and their implementation. Singapore is the 'garden city' in the region because of a series of public campaigns for a cleaner environment that began as early as 1959. By 1995, 21 different campaigns had been initiated (UNESCAP, 1995; Ooi, 1995). Together these policies have enabled the city to increase the quality of its environment.

Secondly, the city is able to avoid some of the impacts of its activities because the city borders are the same as that of the nation. Rural-urban migration (meaning immigration to the city-state) has been tightly controlled. At the same time, the city-state has been able to move industrial firms to outside the city boundaries, a move that played a significant role in keeping the urban area clean. Furthermore, to a certain degree it has been able to avoid the larger environmental impact accompanying changes in its activities because of its small size and the marriage of national and local interests. Thus, while Singapore remains an example to the rest of the region, its unique history and geography may make it a one-of-a-kind phenomenon.

Conclusions

This chapter briefly described the underlying reasons why we should expect different forms of urban environmental transitions in rapidly developing Asia-Pacific economies, when compared to those of the developed world. Evidence for these claims, in terms of changes in the environmental drivers and their impacts on cities within the region is difficult to collect. Overviews of conditions within cities demonstrate the outlines of the so-called time–space telescoping of environmental transitions.

Differences in development patterns have significant practical implications. For example, if environmental impacts are occurring at lower levels of income, then revenues to government are also lower, translating to less money to address these issues when compared to the developed world's experience. Moreover, management of these cities is more complex now than in the past, as environmental impacts are emerging more simultaneously and at different scales, as opposed to the more sequential pattern experienced by the developed world, in which problems were at similar scales. The new context must be accounted for in developing strategic plans and in particular when considering adapting developed world planning and management programmes.

It would be a mistake, however, to interpret these results as a demonstration of worse environmental conditions within contemporary Asia when compared to the developed world at similar levels of income. Rather, cities of the 19th and early 20th centuries had much greater health risks. This can be seen by high urban mortality penalty and much lower levels of longevity. Interestingly, while management seems more complex, the quality of life for citizens now, compared to the past, has vastly improved.

Given increasing interest in the sustainable development agenda (the green issues), and the emergence of a greater diversity of technologies available to address problems, there is hope that cities in the rapidly developing world will be able to resolve their environmental challenges in a more efficient and effective way than those of the developed world. This will require, however, radically different policies than those implemented by developed countries when they faced their challenges. Importantly, cities within the region must develop policies that address multi-scalar impacts and cross-sectoral issues.

The timing of implementation of these policies is also important given the narrowing of the window of opportunity for the resolution of environmental challenges. For example, securing freshwater supplies, given rapid expansion of population into areas of groundwater and watershed storage, remains a critical issue. Given the ability of development to rapidly expand from the city centre, there is less time to consider alternatives than previously experienced. Certainly, eyes will be on cities in rapidly developing Asia to see how they address their environmental concerns. If necessity is the mother of invention, we can expect to see new and effective environmental policies emerging within cities of the region.

Acknowledgements

While Gordon McGranahan provided valuable comments on a previous draft, all mistakes remain the sole responsibility of the author.

Notes

1 Transitions are defined as periods marked by breaks or inflections in long-term trends (both in terms of quantities and rates of change) (e.g. see National Research Council, 1999).
2 Angus Maddison (1991) provides an in-depth analysis of long-term economic trends and finds no convincing evidence to support the notion of long waves, but does suggest, nevertheless, that there have been 'significant changes in the momentum of capitalist development. In the 170 years since 1820 one can identify separate phases which have meaningful internal coherence in spite of wide variations in individual country performance within each of them'. He suggested that the move from one phase to another was governed by exogenous or accidental events that are not predictable.
3 This schema includes frontier mercantile (to 1845), early industrial capitalist (1845–1895), national industrial capitalist (1895–1945), and mature capitalist (1945–present) city types. On the other hand, Borchert (1967) uses the following categories: sail–wagon (1790–1830), iron horse (1830–1879), steel rail (1870–1920) and auto-air-amenity (1920–present).
4 See, for example, Chapter 4 where Satterthwaite suggests that some cities cleaned up after major crises. He points to improved basic infrastructure in Surat, India, after a 1994 plague epidemic. He also mentions transitions in cities within Peru as they addressed the spread of cholera in 1993. The theme of crisis driving urban environmental change has also been documented for developed world cities (Burns et al, 1999)
5 According to Nigel Thrift (1995, p21), Harvey uses this idea in two main ways. He uses it to express the increasing pace of life brought about by innovations such as modern telecommunications. Secondly, it signals the upheaval in our daily experiences of life, as we are increasingly unable to map the representation of space and time.
6 In Myanmar the 'bell ringing' method remains. A collection vehicle with a crew of three or four workers moves along the pre-determined route ringing a bell, which signals for residents to bring their waste containers to the vehicles.
7 In Chapter 7, Kenworthy and Townsend note that car and motorcycle ownership, as well as other transportation indicators, are higher in low- and middle-income cities, such as Bangkok, when compared to Tokyo and Nagoya at either similar levels of income, or even at different levels. This is precisely one of the results of what this chapter calls the time–space telescoping of development patterns. These cities are experiencing much higher levels of auto ownership, for example, at lower levels of income than were experienced previously.
8 In the slum settlements, approximately 68 per cent of all households benefit from regular solid waste collection (Daniere, 1996).
9 South Korea's sulphur dioxide emission decreased by only 5 per cent to 1.5 million tonnes per year from 1990–1995, and emissions of particulates decreased by 3 per cent during that period, while Seoul's sulphur dioxide emissions have been declining from the 1980s due to low sulphur oil and liquefied natural gas. Seoul's TSP levels have roughly halved since 1988 (OECD, 1997).
10 South Korea's carbon dioxide emissions increased by 53 per cent from 1990 to 1995 (OECD, 1997).
11 In 1998, Singapore's GDP per capita was US$30,060, compared to US$29,340 in the US, giving the former the 9th highest GDP per capita in the world (World Bank, 1999).

12 Cleaning the river provided part of the rationale for the clearance of slums and squatter settlements from the central area of the city, and the relocation of people and businesses into the new, controlled environments of public housing satellite towns (Perry et al, 1997).

References

Asian Development Bank (1994) *Managing Water Resources To Meet Megacity Needs*, Asian Development Bank, Manila
Bai, X. and Imura, H. (2000) 'A comparative study of urban environment in East Asia: Stage model of urban environmental evolution', *International Review for Environmental Strategies*, vol 1, pp135–158
BAPEDAL (Badan Pengendalian Dampak Lingkungan) (2001) 'East Java', presented at the ASEAN Regional Workshop on Sustainable Urban Planning and Environmental Management, 7–9 November, Jakarta
Barter, P. (1999) 'An international comparative perspective on urban transport and urban form in Pacific Asia: The challenge of rapid motorization in dense cities', Dissertation thesis, Institute for Sustainability and Technology Policy, Division of Social Sciences, Humanities and Education, Murdoch University, Perth
Berry, B. J. L. (1997) 'Long waves and geography in the 21st century', *Futures*, vol 29, pp301–310
Bianpoen, I. (2001) 'Lessons (not yet) learned', presented at the ASEAN Regional Workshop on Sustainable Urban Planning and Environmental Management, 7–9 November, Jakarta
Borchert, J. A. (1967) 'American metropolitan evolution', *Geographical Review*, vol 57, pp301–332
Brandon, C. (1994) 'Reversing pollution trends in Asia', *Finance and Development*, vol 31, pp21–23
Brenner, N. (1999) 'Globalization as reterritorialization: The re-scaling of urban governance in the European Union', *Urban Studies*, vol 36, pp431–451
Briscoe, J. (1993) 'When the cup is half full: Improving water and sanitation services in the developing world', *Environment*, vol 35, pp7–15 and 28–37
Burns, R., Sanders, J. and Ades, L. (1999) *New York: An Illustrated History*, Alfred A. Knopf, New York
Choong, T. S. (1997) 'Special report cities: The best cities in Asia', *Asiaweek*, 5 December, pp44–48
Daniere, A. (1996) 'Growth, inequality and poverty in South-east Asia, the case of Thailand', *Third World Planning Review*, vol 18, pp373–395
Dhakal S. (2002) 'De-coupling of urban mobility need from environmental degradation: Successful experience of Singapore', Working Paper, Institute for Global Environmental Strategies (IGES), Kitakyushu
Dhakal, S. and Kaneko, S. (2002) 'Urban energy use in Asian mega-cities: Is Tokyo a desirable model?' Proceedings of the Institute for Global Environmental Strategies/Asia Pacific Network for Global Change Research (IGES/APN) Mega-City Workshop on Policy Integration of Energy Related Issues in Asian Cities, 23–23 January, Kitakyushu, Japan, pp173–181
Dickens, C. (1995 [1854]) *Hard Times*, Penguin Books, London
Douglass, M. (1991) 'Planning for environmental sustainability in the extended Jakarta metropolitan region', in N. Ginsburg, B. Koppel and T. G. McGee, (eds), *The Extended Metropolis, Settlement Transition in Asia*, University of Hawaii Press, Honolulu

Friedmann, J. (1986) 'The world city hypothesis', *Development and Change*, vol 17, pp69–83

Friedmann, J. and Wolff, G. (1982) 'World city formation: An agenda for research and action', *International Journal of Urban and Regional Research*, vol 6, pp309–343

Grossman, G. M. and Krueger, A. B. (1995) 'Economic growth and the environment', *Quarterly Journal of Economics*, vol 110, pp353–377

Han, M. T. Z. (1999) 'A system dynamics approach to environmental planning and management of solid waste: A case study of Yangon, Myanmar', Dissertation thesis, Urban Environmental Management Programme, School of Environment, Resources and Development, Asian Institute of Technology, Bangkok

Harvey, D. (1989) *The Condition of Postmodernity: An Enquiry into the Origins of Cultural Change*, Blackwell, Oxford

Hui, J. (1995) 'Environmental policy and green planning', in G.-L. Ooi (ed), *Environment and the City, Sharing Singapore's Experience and Future Challenges*, Institute of Policy Studies and Times Academic Press, Singapore

Janelle, D. G. (1968) 'Central place development in a time-space framework', *Professional Geographer*, vol 20, pp5–10

Janelle, D. G. (1969) 'Spatial reorganization: A model and concept', *Annals of the Association of American Geographers*, vol 59, pp348–364

Jellinek, L. (2000) 'Jakarta, Indonesia: Kampung culture or consumer culture?' in N. Low, B. Gleeson, I. Elander and R. Lidskog (eds), *Consuming Cities*, Routledge Press, London

JICA (1997) *The Study on Urban Environmental Improvement Program in Bangkok Metropolitan Area, Volume 3: Sector Plans and Technical Studies*, Japan International Cooperation Agency, Tokyo

Kaothien, U. (1995) 'The Bangkok metropolitan region: Polices and issues in the seventh plan', in T. G. McGee and I. M. Robinson (eds), *The Mega-Urban Regions of Southeast Asia*, University of British Columbia Press, Vancouver

Kato, K. (1998) 'Grow now, clean up later? The Japanese experience', in W. Cruz, K. Takemoto and J. Warford (eds), *Urban and Industrial Management in Developing Countries: Lessons from Japanese Experience*, The World Bank, Washington, DC

Kim, K.-G. (1996) *Urban Ecology Applied to the City of Seoul, Implementing Local Agenda 21*, UNESCO/MAB – Seoul/Korea National MAB Committee, Seoul

Kondratieff, N. D. (1979) 'The long waves of economic life', *Review*, vol II, pp519–562

Kwon, W.-Y. (2001) 'Globalization and the sustainability of cities in the Asia Pacific region: The case of Seoul', in F. C. Lo and P. J. Marcotullio (eds), *Globalization and the Sustainability of Cities in the Asia Pacific Region*, UNU Press, Tokyo

Lebel, L. (2002) 'Global change and development: A synthesis for Southeast Asia', in P. Tyson, R. Fuchs, C. Fu, L. Level, A. P. Mitra, E. Odada, J. Perry, W. Steffen and H. Virji (eds), *Global-Regional Linkages in the Earth System*, Springer, Berlin

Lee, C.-W. (2001) 'Sustainable urban management by partnership: The case of Seoul Agenda 21', in W.-Y. Kwon and K.-J. Kim (eds) *Urban Management in Seoul, Policy Issues and Responses*, Seoul Development Institute, Seoul

Lee, S. K. (1995) 'Concept of the Garden City' in G.-L. Ooi (ed) *Environment and the City, Sharing Singapore's Experience and Future Challenges*, Institute of Policy Studies and Times Academic Press, Singapore

Lee, Y.-S. F. (1994) 'Urban water supply and sanitation in developing countries', in J. E. Nickum and K. W. Easter (eds), *Metropolitan Water Use Conflicts in Asia and the Pacific*, Westview Press, Boulder, CO

Litvin, D. (1998) 'Dirt poor: Poor countries have the world's worst environmental problems. They cannot afford to put up with them', *The Economist*, vol 346, no 8060, pp Survey 3–15

Lo, F.-C. and Marcotullio P. J. (eds) (2001) *Globalization and the Sustainability of Cities in the Asia Pacific Region*, United Nations University Press, Tokyo,

Lo, F.-C. and Xing Y.-Q. (1998) *China's Sustainable Development Framework: Summary Report*, United Nations University Institute of Advanced Studies, Tokyo

Lo, F.-C. and Yeung Y.-M. (eds) (1996) *Emerging World Cities in Pacific Asia*, United Nations University Press, Tokyo

Lo, F.-C. and Yeung Y.-M. (eds) (1998) *Globalization and the World of Large Cities*, United Nations University Press, Tokyo

Lomborg, B. (2001) 'The truth about the environment', *The Economist*, vol 360, pp63–65

Maddison, A. (1991) *Dynamic Forces in Capitalist Development: A Long-Run Comparative View*, Oxford University Press, Oxford

Maddison, A. (2001) *The World Economy, A Millennial Perspective*, Organisation for Economic Co-operation and Development, Paris

Marcotullio, P. J. (2005) *Time–Space Telescoping and Urban Environmental Transitions in the Asia Pacific*, United Nations University Institute of Advanced Studies, Yokohama

Marcotullio, P. J. and Lee, Y.-S. F. (2003) 'Urban environmental transitions and urban transportation systems: A comparison of North American and Asian experiences', *International Development Planning Review*, vol 25, pp325–354

Marcotullio, P. J., Williams, E. W. and Marshall, J. D. (2005) 'Faster, sooner, and more simultaneously: How recent transportation CO_2 emission trends in developing countries differ from historic trends in the United States of America', *Journal of Environment and Development*, vol 14, pp125–148

McGranahan, G., Jacobi, P., Songsore, J., Surjadi, C. and Kjellen, M. (2001) *The Citizens at Risk, From Urban Sanitation to Sustainable Cities*, Earthscan, London

McGranahan, G. and Satterthwaite, D. (2000) 'Environmental health or ecological sustainability? Reconciling the brown and green agendas in urban development', in C. Pugh (ed), *Sustainable Cities in Developing Countries*, Earthscan, London

McGranahan, G., Songsore, J. and Kjellen, M. (1996) 'Sustainability, poverty and urban environmental transitions', in C. Pugh (ed), *Sustainability, the Environment and Urbanization*, Earthscan, London

McKay, J. P. (1988) 'Comparative perspectives on transit in Europe and the United States, 1850–1914', in J. A. Tarr and G. Dupuy (eds), *Technology and the Rise of the Networked City in Europe and America*, Temple University Press, Philadelphia

McShane, C. (1979) 'Transforming the use of urban space: A look at the revolution in street pavements, 1880–1924', *Journal of Urban History*, vol 5, pp279–307

McShane, C. (1994) *Down the Asphalt Path, The Automobile and the American City*, Columbia University Press, New York

Melosi, M. V. (2000) *The Sanitary City: Urban Infrastructure in America from Colonial Times to the Present*, Johns Hopkins Press, Baltimore, MA

National Research Council (1999) *Our Common Journey: A Transition toward Sustainability*, National Research Council, Washington, DC

Nyunt, M. T. (2002) 'Country report and city report, Union of Myanmar', in *1st ASEAN Healthy Cities Conference*, Kuching, Malaysia

OECD (1997) *Environmental Performance Reviews, Korea*, Organisation for Economic Co-operation and Development, Paris

Oo, N. N. (2001) 'Sustainable urban planning and environmental management in Myanmar', Paper presented at the ASEAN Regional Workshop on Sustainable Urban Planning and Environmental Management, 7–9 November, Jakarta

Ooi, G. L. (ed) (1995) *Environment and the City, Sharing Singapore's Experience and Future Challenges*, Times Academic Press and The Institute of Policy Studies, Singapore

Pernia, E. M. (1992) 'Southeast Asia', in R. Stren, R. White and J. Whitney (eds), *Sustainable Cities: Urbanization and the Environment in International Perspective*, Westview Press, Boulder, CO

Perry, M., Kong, L. and Yeoh, B. (1997) *Singapore, A Developmental City State*, John Wiley & Sons, Chichester

Phantumvanit, D. and Liengcharensit, W. (1989) 'Coming to terms with Bangkok's environmental problems', *Environment and Urbanization*, vol 1, pp31–39

Preston, S. H. (1979) 'Urban growth in developing countries: A demographic reappraisal', *Population and Development Review*, vol 5, pp195–215

Republic of Korea, Ministry of the Environment (1999) *Green Korea 1999: Environmental Vision for a Sustainable Society*, Korean Ministry of Environment, Kwacheon

Republic of Korea, Ministry of the Environment (2005a) *2005 Environmental Statistics Yearbook*, Seoul

Republic of Korea, Ministry of the Environment (2005b) *Green Korea 2005, Towards the Harmonization of Humans and Nature*, Korean Ministry of the Environment, Kwacheon

Rosenbaum, W. A. (1995) *Environmental Politics and Policy*, Congressional Quarterly Press, Washington, DC

Sarin, C. (1998) 'Urban solid waste disposal: A case study of the city of Phnom Penh, Cambodia', Dissertation thesis, Urban Environmental Management Programme, School of Environment, Resources and Development, Asian Institute of Technology, Bangkok

Sassen, S. (1991) *The Global City: New York, London, Tokyo*, Princeton University Press, Princeton, NJ

Sawa, T. (1997) *Japan's Experience in the Battle against Air Pollution*, Committee on Japan's Experience in the Battle Against Air Pollution, Tokyo

Schumpeter, J. A. (1961) *The Theory of Economic Development*, Oxford University Press, Oxford

Seoul Metropolitan Government (2000) *Waste Management in Seoul*, Seoul Metropolitan Government, Seoul

Setchell, C. A. (1995) 'The growing environmental crisis in the world's mega-cities: The case of Bangkok', *Third World Planning Review*, vol 17, pp1–17

Singapore Land Transport Authority (2005) *Singapore Land Transport Statistics in Brief, 2005*, Singapore

Smith, K. (1990) 'The risk transition', *International Environmental Affairs*, vol 2, pp227–251

Smith, K. R. (1995) 'Environmental hazards during economic development: The risk transition and overlap', in E. G. Reichard and G. A. Zapponi (eds), *Assessing and Managing Health Risks from Drinking Water Contamination: Approaches and Applications*, The National Water Research Institute, Fountain Valley, CA

Smith, K. R. and Lee, Y.-S. F. (1993) 'Urbanization and the environmental risk transition', in J. D. Kasarda and A. M. Parnell (eds), *Third World Cities: Problems, Policies, and Prospects*, Sage Publications, Newbury Park, CA

The Economist (2005) 'Asia: Good morning at last; Vietnam', *The Economist*, vol 380, no 8489, 5 August, 50pp

The Economist (2006) 'Asia: Trouble at the mill: Vietnam', *The Economist*, vol 378, no 8462, 28 January, 66pp

Thein, D. L. L. (2001) 'Country paper: Myanmar', Paper presented at the ASEAN Regional Workshop on Sustainable Urban Planning and Environmental Management, 7–9 November, Jakarta

Thrift, N. (1995) 'A hyperactive world', in R. J. Johnston, P. J. Taylor and M. J. Watts (eds) *Geographies of Global Change*, Blackwell, Oxford, pp18–35

Todaro, M. P. (1997) *Economic Development*, Longman, London and New York

Tokyo Metropolitan Government (1999a) 'Suginami incineration plant wins ISO environmental certification', in *Tokyo Metropolitan News, A Quarterly Journal of the Tokyo Metropolitan Government*, vol 49, p5

Tokyo Metropolitan Government (1999b) 'Tokyo announces water environment master plan', in *Tokyo Metropolitan News, A Quarterly Journal of the Tokyo Metropolitan Government*, vol 49, p5

Tokyo Metropolitan Government (1999c) *Tokyo Current Issues*, Tokyo Metropolitan Government, Tokyo

Tokyo Metropolitan Government (1999d) *Water Supply in Tokyo*, Tokyo Metropolitan Government Tokyo

UNEP (United Nations Environment Programme) (2002) *Global Environmental Outlook 3, Past, Present and Future Perspectives*, Earthscan, London

UNESCAP (1995) *State of the Environment in Asia and the Pacific*, United Nations Economic and Social Commission for Asia and the Pacific, Bangkok

UNESCAP (2000) *State of the Environment in Asia and the Pacific*, United Nations Economic and Social Commission for Asia and the Pacific, Bangkok

UNESCO (2000) *Reducing Mega-City Impacts on the Coastal Environment: Alternative Livelihoods and Waste Management in Jakarta and the Seribu Islands*, United Nations Educational Scientific and Cultural Organization, Paris

USAID (1990) *Urbanization and the Environment in Developing Countries*, United States Agency for International Development, Washington, DC

Viengsavanh, D. and Yongnou, X. (2001) 'Urban environmental management in LAO PDR', Paper presented at the ASEAN Regional Workshop on Sustainable Urban Planning and Environmental Management, 7–9 November, Jakarta

Warner, S. B. (1955) 'Public health reform and the depression of 1873–1878', *Bulletin of the History of Medicine*, vol 29, pp503–516

Webster, D. (1995) 'The urban environment in Southeast Asia: Challenges and opportunities', *Southeast Asian Affairs*, vol 22, pp89–107

Webster, D. (2002) 'Peri-urbanization: Zones of rural-urban transition', in S. Sassen (ed) 1.18. Human Resource System Challenge VII: Human Settlement Development, *Encyclopedia of Life Support Systems (EOLSS)*, Developed under the auspices of the UNESCO, EOLSS Publishers, Oxford, www.eolss.net

World Bank (1991) *World Development Report*, World Bank and Oxford University Press, New York

World Bank (1992) *World Development Report*, World Bank and Oxford University Press, New York

World Bank (1999) *World Development Report 1999/2000*, World Bank and Oxford University Press, New York

World Commission on Environment and Development (1987) *Our Common Future*, Oxford University Press, Oxford

Wu, W. (1998) 'Shanghai', *Cities*, vol 16, pp207–216

Yeates, M. (1998) *The North American City*, Longman, New York

4

In Pursuit of a Healthy Urban Environment in Low- and Middle-income Nations

David Satterthwaite

Introduction

This chapter seeks to provide an overview of the most serious environmental problems in 'unaffluent urban areas' to complement Graham Haughton's overview of environmental problems and affluent cities (Chapter 11). It also considers what factors contribute to or inhibit the reduction of the environmental health risks generally associated with unaffluent urban areas and the extent to which this is achieved by cost and risk transfers – in effect, a consideration of the transitions theme from an environmental health perspective. It highlights the large environmental health burden suffered by a large section of the urban population in low- and middle-income nations, but also considers how middle- and upper-income groups in these urban areas generally avoid these burdens and how they and the more powerful urban-based commercial and industrial enterprises have avoided contributing funding towards their resolution. It also discusses why aid agencies and international development banks have contributed little towards resolving these problems.

This chapter has difficulties comparable to those of Chapter 11 in defining which urban areas are its focus; but in addition, the difficulty is in defining unaffluent urban areas. All urban areas in what the World Bank classifies as low- and middle-income nations[1] could be taken as 'unaffluent'. But that would imply including cities such as São Paulo, Bangkok and Mexico City, and although these have large numbers of people with incomes below the poverty line and 'living in poverty' they also have very large middle classes and a proportion of the world's wealthiest families. Porto Alegre in Brazil is in a middle-income nation but has standards of provision for water, sanitation, drainage and garbage collection and an average life expectancy that compare favourably with many cities in high-income nations (Menegat, 2002).

It would be easier to concentrate on urban areas in 'low-income nations' but there are many cities and smaller urban centres in Brazil, Mexico and Thailand (and most other middle-income nations) where most of the population live in poverty. An assumption that all cities in low-income nations are 'unaffluent' is also problematic, in that cities such as Nairobi, Mumbai and Lagos, all in low-income nations, also have significant concentrations of very rich households.[2] Thus, in focusing on the most serious environmental problems in urban areas in low- and middle-income nations, there is a need to recognize that these urban areas cover a very large range in terms of size, prosperity and inequality, and in the scale and range of environmental problems.

Although there is inadequate documentation of the scale and depth of environmental health burdens in urban areas in low- and middle-income nations, as discussed later in this chapter, there is plenty of evidence that large sections of their urban populations live in very poor quality and often overcrowded housing with inadequate or no basic infrastructure (piped water, sewers, drains, paved roads and footpaths, electricity) and services (health care, emergency services, schools, provision for children's play and for pre-school children) (Hardoy et al, 1992, 2001; WHO, 1992; UNCHS, 1996; UN-Habitat 2003a, 2003b). Although the proportion of an urban centre's population suffering serious deficiencies in most or all of these will vary, as will the extent of the deficiencies, very few urban centres get close to the coverage and the quality of provision expected by urban populations in high-income nations.[3] No family in urban areas in high-income nations, however poor, expects to have to walk several hundred yards to collect water from a communal standpipe shared with hundreds of others or to have no toilet in their home.

Another striking difference between urban areas in high-income nations and low- and middle-income nations is the scale of the health burden generated by infectious and parasitic diseases. Again, the documentation is incomplete, but it is clear that infectious and parasitic diseases have a very large impact in terms of serious illness and premature death among large sections of the young populations in most urban centres in low- and middle-income nations, and very little impact among young populations in high-income nations (WHO, 1992; Satterthwaite et al, 1996; Bartlett et al, 1999). There is also some evidence of the much larger health burdens for adults (see, for instance, Bradley et al, 1991; WHO, 1992, Pryer, 2003). Thus, there is a particular interest in what causes or supports the policies and investments that underpin this shift from urban centres with high environmental health burdens from infectious and parasitic diseases to one with low environmental health burdens – or to put it another way, where diseases such as diarrhoeal diseases, acute respiratory infections (ARIs), malaria and tuberculosis do not figure among the main causes of death, and where parasitic infections and water-washed or waterborne diseases have much less influence on health status.

Physical hazards evident in the home and its surroundings feature as among the most common causes of serious injury and premature death in most urban areas in low- and middle-income nations (Goldstein, 1990; WHO, 1999) – for example, burns, cuts and scalds and injuries from falls. The health burdens these cause are particularly large in districts where housing quality is poor (especially where flammable materials are used for housing), infrastructure deficient and there are high levels of overcrowding. Large health burdens and high levels of accidental death from physical hazards are also related to the lack of provision for rapid and appropriate treatment, both from health care and emergency services (Goldstein, 1990).

This focus on what can be termed the first of the environmental transitions is distinct from the second (the control of air and water pollution arising from industrial development, power generation and motor vehicles) and the third (controls on the generation of wastes whose main impacts are regional and global, especially greenhouse gases) because of at least two aspects. The first is that it is not addressed through regulation (e.g. pollution control legislation) but through better housing conditions with provision of infrastructure and services. This immediately raises the issues of who pays for such provision and who is responsible for it. The second is that it does not fall so neatly into a spatial or temporal displacement of externalities by those generating the problem. Although infectious diseases are spread through person-to-person contact, air, water or food or through disease vectors, and the disposal of human excreta and household wastes may create environmental health risks for others, in general, middle- and upper-income groups can avoid these risks. And, as will be discussed in more detail later, many of the most serious environmental health risks faced by large sections of the population in urban areas in low- and most middle-income nations are not a threat to richer groups and their solution implies no risk transfer. However, it usually implies a cost to middle- and upper-income groups because part of the capital investment needed to address these problems has to be raised from richer groups or from urban populations in general. One of the key reasons why lower-income groups face such high environmental health burdens in urban areas is the extent to which higher-income groups and commercial and industrial enterprises avoid contributing taxes or charges that could provide the capital base for improving housing conditions and infrastructure and service provision to lower-income populations. In many cities, this reaches high levels of perversity as middle- and upper-income groups, formal enterprises and government institutions receive provision for water and sanitation at below cost, while most lower-income groups receive no public provision and many pay high prices for informal provision (UN-Habitat, 2003b).

Thus, an analysis of what does or does not drive down environmental health burdens in urban areas in low- and middle-income nations is less about how environmental costs generated by production or by middle- and upper-income groups' consumption are displaced, and more an examination of why public actions do not address the environmental health risks associated with poor quality housing and lack of infrastructure and services. To put it crudely, everyone has to defecate, every household generates wastewater and every person is potentially a source of infection for other people. Minimizing the environmental risks is only possible through a combination of prevention (e.g. toilets linked to a sewage system that ensures safe disposal and treatment of human excreta) and of rapid and effective treatment when illness or injury does occur. Managing a household's liquid and solid waste is a critical part of disease prevention – and this is also a problem generated by everyone (even if richer households generate larger volumes of liquid and solid waste) and dispersed throughout the urban areas where people live. Liquid and solid waste are also potential causes of non-point-source pollution, except where there is a sewer system to collect liquid waste from each building. Good quality storm and surface-water drainage performs this same transfer of non-point-source pollutants to a single point at which a treatment plant can be built. One large advantage of sewers and drains from an environmental perspective is the much-reduced unit cost of treating both kinds of wastewater, as they convert dispersed problems into one that can be managed centrally and treated 'end of pipe'. Of course, one large disadvantage is when inadequate or no provision is made to treat their outflows.

After reviewing the scale and range of environmental problems in urban areas in low- and middle-income nations, this chapter will examine the transitions theme by focusing on actions to reduce one important subset of environmental health problems – those linked to inadequate provision for water, sanitation and drainage. This focus is chosen both for its importance to environmental health and for the extent to which it highlights the link between the scale of environmental health burdens and the capacity and competence of urban governments.

The emerging interest in slum and squatter settlements

The inadequacies in what governments and international agencies have done to address the environmental health burdens associated with very poor housing conditions and the lack of provision for water and sanitation cannot be blamed on a lack of documentation. The fact that large sections of the urban population in low- and middle-income nations live in very poor quality housing lacking basic infrastructure and services is well known and has been a central part of the 'urban' literature on development for at least 40 years (Abrams, 1964; Turner, 1968; Mangin, 1967; Turner and Fichter, 1971; Leeds, 1974; Dwyer, 1975; Turner, 1976; Ward, 1976). This point was also recognized within the United Nations (UN) system, as can be seen by the decision to highlight the problem of 'human settlements' in the 1976 UN Conference on Human Settlements in Vancouver. This represented an international recognition of the need to elevate human settlement issues to be among the main global issues that the UN system should focus on, along with 'the human environment' (the 1972 UN Conference on the Human Environment, at Stockholm), 'population' (the 1974 UN Conference on Population in Bucharest), 'women' (the 1975

UN Conference on Women, held in Mexico City) and 'food' (the 1974 UN Conference on Food, held in Rome).

The growing volume of research and publications on urban issues in low- and middle-income nations during the 1960s and 1970s focused on the poor quality of the housing in which much of the rapidly expanding populations lived and the lack of provision for water and sanitation. The need for a much greater commitment by national governments and international agencies to improving provision for water and sanitation was one of the central themes for the 1976 UN Conference on Human Settlements, and among the recommendations formally endorsed by the 132 governments attending this conference (UN, 1976), and subsequently further endorsed at the UN Conference on Water, held in Mar del Plata in 1977. Much attention was also paid to the growth of informal or illegal settlements which had come to house an increasing proportion of the population in most cities and many smaller urban centres (UN, 1976). Among the most heavily debated issues at that time were the role of the 'informal sector' and the nature of its connections with 'the formal sector' (was it a key part of a city economy or a peripheral 'subsistence survival' mode for poorer groups unable to find real jobs?); the most appropriate form of government intervention in land markets (should governments develop land banks, how could they use property taxes to strengthen local authorities and how should they recapture the 'unearned increment' in land values that accrued to landowners?); and what attitude governments should take towards illegal settlements (bulldoze them, rehouse their population, support their upgrading). Some governments and a few international agencies took measures to support the 'squatters' who developed and built their own homes and neighbourhoods in illegal settlements, as long as they did not invade the best quality land. At this time, there was a shift away from using the term 'slum' for neighbourhoods or districts with poor quality housing and inadequate provision for infrastructure for two reasons. The first was the inaccuracy in labelling such a diverse range of settlements as 'slums' – for instance, the term came to be applied to tenement areas, squatter settlements, settlements built on informal subdivisions and areas with legal but poor quality housing. The second was the pejorative connotations; a 'slum' implied a neighbourhood where the housing should be replaced yet most 'slums' had housing and infrastructure that could be upgraded which was much cheaper than replacing them and much preferred by their inhabitants.

Many local non-governmental organizations (NGOs) were active in supporting upgrading and by the late 1970s were influencing the policies of some governments (for discussions of this, see Hardoy and Satterthwaite, 1989; also Connolly, 2004, for Mexico and Boonyabancha, 2003, for Thailand). The World Bank began its support for 'serviced sites' (which sought to provide land plots with infrastructure and services that lower-income households could afford as alternatives to illegal land occupation and development) and 'slum and squatter upgrading' in the early 1970s, and these continued to receive support up to the present (although with fluctuating support and never with a high priority). UNICEF was also pioneering more community-based upgrading during the 1970s and early 1980s (UNICEF, 1988).

A review of the housing, land and settlement policies in 17 low- and middle-income nations in the late 1970s showed considerable innovation among various governments (Hardoy and Satterthwaite, 1981). There were significant shifts in the policies of various governments away from large-scale slum clearance (although with important exceptions) and towards slum and squatter upgrading. Many governments had set up new agencies to help lower-income households fund the acquisition, construction or improvement of their homes.

The 1980s was designated by the UN General Assembly as the International Drinking Water Supply and Sanitation Decade, with ambitious goals set for reaching most of the urban and rural population with improved provision by 1990. Certainly, by the late 1970s, there were grounds for optimism that these issues were being understood and addressed – and there were also signs in many nations (especially in Latin America) of shifts, or returns, to more democratic forms of governments, including decentralization and increased local democracy – shifts that were to continue and develop in many nations during the 1980s.

But there were fewer grounds for optimism by the mid-1980s. By this time, many low- and middle-income nations were facing economic stagnation and debt repayment crises, in part linked to the changes in the world's financial systems brought about by the strong anti-inflation policies of many high-income nations that had been initiated by the Thatcher and Reagan governments in the UK and the US respectively. There was far less support for government intervention to address urban poverty (e.g. through government-funded housing programmes) among most high-income nations, and this also influenced the policies and funding priorities of the bilateral aid agencies and multilateral agencies such as the World Bank. The World Bank, in particular, began an important shift away from loans for projects to loans supporting economic and financial reforms that included rolling back the state.

Meanwhile, the recognition that too little attention had been given to rural development that arose in the early 1970s, and the idea that this was in part caused by 'urban bias' (Lipton, 1977), was also working its way through many international agencies, and many subsequently refused to work in urban areas. Many international agencies accepted Michael Lipton's critique of development as being too urban biased, and they were also recognizing the extent to which the large infrastructure projects they had supported were deteriorating.

The strong shift by international donors to policies that sought to roll back the state also meant little or no support to the kinds of government interventions in urban land markets that only a few years previously had been seen as essential to guaranteeing poorer groups access to land for housing. Meanwhile, although the international agencies may have applauded the shift to democracy in Latin America, very few actually provided tangible support for it in terms of increasing aid flows (Satterthwaite, 1990). The pioneering community-based urban projects that many United Nations Children's Fund

(UNICEF) country offices had initiated also lost favour with UNICEF's headquarters (Satterthwaite, 1998). In effect, the critical role the international funding could have fulfilled, providing the capital needed for reducing the environmental health burdens faced by lower-income groups in urban areas and supporting the local governance systems to invest and manage this, was not taken up.

Missing the environmental and health implications; the lack of an evidence base on environmental health and on urban poverty

Much of the documentation from the 1970s justified the need for interventions to improve housing and provision for infrastructure and services. Slum and squatter upgrading, serviced sites, housing finance institutions to help low-income groups acquire housing and improved provision for water and sanitation were recommended because of the very poor conditions evident in tenements, cheap boarding housing and informal or illegal settlements. This was not framed as an environmental problem or as a health problem. At this time, there was relatively little documentation of the health problems faced by those living in poor quality housing. In part, this was because most of the documentation was undertaken by those with no specialist knowledge or training in health. Perhaps it was considered self-evident – but little of the general or specialist literature on urban issues in low- and middle-income nations had any detailed consideration of health conditions (Basta, 1977, being one important exception).

UNICEF was among the first of the international agencies to try to gauge the health implications of very poor housing conditions and lack of basic services in urban areas. It collected data on infant and child mortality rates among those living in illegal settlements and compared these to the rates for those living in better quality housing or against city averages (Rodhe, 1983; UNICEF and WHO, 1984) – but as already noted , the support that UNICEF had previously provided to community-based upgrading in urban areas was much diminished, in part because of diversion of funds to large emergencies, and partly because of the new concentration on achieving scale through the 'child health revolution', which concentrated on cheap, easily implemented interventions on a mass scale. This meant far less emphasis on the kinds of community-driven upgrading programme that show how larger programmes should operate – what UNICEF termed area-based programmes (Black, 1986 1996; Satterthwaite, 1998).

There were members of staff at the World Health Organization (WHO) who recognized the need for more attention to the environmental health problems evident in cities in low- and middle-income nations. This can be seen in the setting up of the Rural and Urban (RUD) Programme within WHO and its series of technical and expert meetings and publications through the

1980s – but this never received much support from within the organization. Here, as in many other international agencies, there was still an assumption that relatively few poor people lived in urban areas; it is worth noting that this WHO programme had to call itself the 'Rural and Urban' programme to avoid this criticism, even though its work focused almost exclusively on urban issues. Meanwhile, the UN Centre for Human Settlements (Habitat)[4] considered that health issues were not within its remit and so supported no work in this area (although it was later to undertake some work on health issues – see, for instance, Clauson-Kaas et al, 1999).

It was only in the late 1980s and early 1990s that a general literature developed on the health problems associated with urban development in low- and middle-income nations (Hardoy and Satterthwaite, 1987; Harpham at al, 1988; WHO, 1989; Hardoy et al, 1990; McGranahan, 1991). Urban issues in low- and middle-income nations received increased attention within WHO during the early 1990s (WHO, 1992, 1999) but again, this was not sustained. The reasons for this disinterest in urban health issues within WHO and other international agencies are not clear – although a belief among many international agency staff that urban populations were much better off than rural populations and already privileged by 'urban biases' (or this belief among the bilateral agencies that might fund international agency programmes on this) was one of the key underpinnings of this.

Meanwhile, within the large and growing literature on environmental problems in low- and middle-income countries from the mid-1970s onwards, very little attention was given to urban environmental problems. In part, this was because the development debates at that time gave very little attention to urban issues, even though there was little evidence that the majority of the urban population was benefiting from development (or from urban biases). This was partly because of the prominence given by researchers and research institutions from Europe and North America to 'resource' scares that, with hindsight, have been shown to be based on hasty and often inaccurate diagnoses about resource use in low- and middle-income nations – for example, in regard to advancing desertification or soil erosion, or deforestation from woodfuel and charcoal use, or the assumption that rising population densities in rural areas would be associated with resource degradation.

Although urban areas were peripheral to most such discussions, when they did get considered, they were usually seen as causes or major contributors to resource problems – for instance, the assumption that cities in low-income nations would be at the centre of widening circles of deforestation, and that urban expansion was causing serious losses in agricultural land. This is not to deny that there was some evidence to back up these assertions: many cities did expand over high-quality agricultural land, and there were some cities or rural areas where deforestation could be linked to woodfuel and charcoal use. But it was the extrapolation from a few (often unusual) examples to a general diagnosis that was at fault. For example, statistics were produced for the extent to which urban expansion in Egypt significantly decreased the amount of agricultural land: but this is a highly unusual nation with 96 per cent of the national territory as desert and virtually all the population concentrated in the remaining 4 per cent living along the River Nile. In addition, case studies have shown that loss of agricultural land to expanding urban areas does not necessarily lead to decreased food production. Land-use changes around expanding cities may include not only expanded built-up areas, but also much increased food production from horticulture, pisciculture or livestock, or shifts to more valuable crops (Bentinck, 2000; van den Berg et al, 2003).

There were some exceptions to the lack of consideration given to urban issues in most discussions of environment and development – for instance, the inclusion of a chapter on urban issues within the Brundtland Commission's report *Our Common Future* (World Commission on Environment and Development, 1987) which helped set the UN on its new focus on sustainable development. But even here, the idea of a chapter in the report focusing on urban issues was strongly contested within the commission and opposed by its chair.

In addition, what little attention was given to urban environmental problems was very much seen through the perspective of environmental problems in high-income nations. The health burdens arising from infectious and parasitic diseases (fundamentally environmental, as they are transmitted through air, contact, water or disease vectors) were usually ignored. More attention was given to air pollution, toxic wastes and the loss of agricultural land to urban expansion. During the 1980s and early 1990s, much attention was also given in the literature to 'exploding cities', perhaps especially to 'ever-expanding mega-cities' (cities with populations exceeding 10 million inhabitants). However, most mega-cities and many other large cities at that time had much reduced population growth rates, partly linked to less in-migration, due to economic stagnation, and partly (for many nations) because of reduced rates of natural increase (UNCHS, 1996; Satterthwaite, 2005).

Over the last 30 years there has also been little detailed data on the scale and nature of urban poverty, especially in regard to aspects that relate to the quality of the living and working environment (including housing and living conditions and the quality and extent of provision for basic infrastructure and services). Although it is common for documents to refer to the proportion of a city's or nation's population 'living in poverty', most definitions of poverty do not include any aspects related to living conditions. Perhaps the main reason for this is that in most low- and middle-income nations, poverty is still defined through income-based absolute poverty lines derived from data on consumption expenditures – in part because the methodologies for defining and measuring poverty were transferred from high-income nations (where environmental health burdens were less severe), and partly because it is more difficult to incorporate measures of the inadequacies in provision for infrastructure and services and in housing conditions into the surveys conventionally used for measuring poverty (Montgomery et al, 2003; Satterthwaite, 2004).

During the 1970s and 1980s, there was also surprisingly little documentation on urban poverty, even though there was a large literature on certain aspects of poverty, especially the very poor housing conditions and the lack of provision for infrastructure and services. Perhaps the main reason for this is that at this time, 'poverty' was still considered to be 'lack of income' or inadequate food intake, and the broader conceptions of (urban and rural) poverty with their recognition of multiple deprivations (including very poor quality housing and lack of infrastructure and services) was yet to come. The general literature on poverty in low- and middle-income nations focused almost exclusively on rural poverty (and much of the general literature still does so). For the few publications that sought to give figures on the scale of urban poverty, the figures were absurdly low, implying that the proportion of the urban population living in poverty in low- and middle-income nations was actually lower than the proportion in high-income nations (e.g. see the estimates in Leonard, 1989, and World Bank, 1990).

During the 1990s, there was a rapid expansion in the literature on urban poverty, especially on the need to reconceptualize it in terms of the multiple deprivations that most low-income urban dwellers face (e.g. Moser et al, 1993; Moser, 1996, 1998; Satterthwaite 1995; Wratten, 1995; Rakodi 1996; Satterthwaite, 1997b) and on detailed case studies that show the scale and nature of such deprivations (Kanji, 1995; Latapí and de la Rocha, 1995; Islam et al, 1997; Maxwell et al, 1998). However, researchers remained reluctant to admit that the scale and depth of urban poverty had been under-estimated and that it had characteristics that were distinct from rural poverty. For instance, the World Development Report 2000/2001, which focused on poverty, gave no consideration to how the characteristics and causes of rural poverty differ from urban poverty (World Bank, 2001).

This reluctance of many international agencies to engage in urban issues, the blindness of many environmental specialists to the health burden generated by infectious and parasitic diseases and physical hazards, and the lack of attention by urban researchers to health issues helps explain why, for most cities and virtually all smaller urban centres in low-income nations, there is surprisingly little detailed information available about health burdens (or the main causes) or about the scale, nature and depth of poverty. The information base on these is also deficient for most cities and smaller urban centres in most middle-income nations, too. This raises important questions about why official information systems are so inadequate in regard to two of the most important aspects of development. In part, this is because of the time-lag between key issues becoming accepted (i.e. that the scale of urban poverty has been underestimated and that the health burden associated with it has been given too little attention) and changes in national data collection systems. It may also be partly due to the broader issue of inadequate attention given to health in general and/or the difficulty in assessing environmental health conditions through the conventional information gathering institutions of governments (household surveys and censuses).

According to official government statistics, between a third and half the urban population in many low- and middle-income nations have incomes below the poverty line; in some more than half are below the poverty line.[5] However, in most nations, and in most global estimates, urban poverty is underestimated for two reasons:

1 The over-concentration on the use of income-based poverty lines drawing from consumption data with little or no attention to other aspects of deprivation. These aspects include inadequate, overcrowded and insecure housing; inadequate provision for water; sanitation; drainage and basic services such as health care, emergency services and schools; and lack of the rule of law and respect for civil and political rights (Mitlin and Satterthwaite, 2004).
2 The scale and depth of 'income' poverty in urban areas is underestimated because official poverty lines make little or no allowance for the higher costs of most non-food necessities in urban areas (or particular cities) – for instance, the cost of housing, of accessing water and sanitation, transport, keeping children at school – and paying for health care and medicines (Mitlin and Satterthwaite, 2001; Montgomery et al, 2003; Satterthwaite, 2004).

Summary of what we do know about environment and health

There are sufficient studies available in particular cities or smaller urban centres or particular settlements in cities to suggest the following:[6]

1 It is common for between a third and two-thirds of an urban centre's population to live in housing of poor quality with high levels of overcrowding. In the larger or more rapidly growing urban centres it is common for a high proportion of the population to live in informal or illegal settlements, largely because of the gap between what low-income households can afford to pay for accommodation and land prices on the formal land market (which are often boosted by inappropriate regulations and complex procedures for purchasing and developing land for housing). Much of the housing in which lower-income groups live is made of non-permanent materials that are also flammable, and since a high proportion of such households also rely on open fires or kerosene stoves for cooking (and often heating) and kerosene lamps or candles for light, there are high levels of risk from accidental fires.
2 Much of the urban population lacks safe, regular, convenient supplies of water and provision for sanitation. As described in detail in UN-Habitat (2003b), official statistics on the extent of provision for 'improved' water and sanitation in urban areas in most low- and middle-income nations greatly understate the scale and extent of inadequate provision. It is common for large sections of the population in cities and smaller urban centres to use water that is contaminated and to have no provision for sanitation in their home. There is a growing literature on the extent to which open defecation is common in urban centres in low-income nations (UN-Habitat, 2003b). In many low-income nations, particularly in sub-Saharan Africa, it is also common for large sections of the middle-income population in urban

areas to have inadequate provision for water and sanitation, with these inadequacies relating more to the extreme weakness in local (public, private or community) service providers than to a lack of capacity to pay. Table 4.1 gives estimates for the number and proportion of the urban population lacking adequate provision for water and sanitation; the uncertainties in regard to the precise figures are partly due to poor quality data, and partly due to the lack of agreement as to how (and where) to draw the dividing line between what is and what it not 'adequate' (Hardoy et al, 2001; UN-Habitat, 2003b).

The extent of the problem with inadequate provision for water and sanitation in smaller urban centres is also likely to be underestimated, as most studies have concentrated on larger cities (Hardoy et al, 2001; UN-Habitat, 2003b), although most of the urban population in low-income nations and most middle-income nations live outside large cities.[7]

Box 4.1 shows the deficiencies in provision for Bangalore and who is most affected. Bangalore is chosen as an example because it is one of the most successful cities within low-income nations in developing a strong economic base and attracting foreign investment. This example shows that economic success and large-scale public investments in infrastructure do not necessarily bring environmental health benefits to low-income populations.

3 Much of the urban population in low- and middle-income nations lack regular services to collect household waste – indeed, many live in settlements that lack the paved roads needed to allow conventional garbage collection trucks to provide a door-to-door service. It is common for large sections of middle- or even upper-income groups to have inadequate or no provision in low-income nations. Again, it is likely that the extent of the problem in smaller urban centres is underestimated. The environmental health implications of a lack of garbage collection services in urban areas

Table 4.1 *Estimates for the proportion of people without 'adequate' provision for water and sanitation in urban areas*

Region	Number and proportion of urban dwellers without adequate provision	
	Water	Sanitation
Africa	100–150 million (~35–50%)	150–180 million (~50–60%)
Asia	500–700 million (~35–50%)	600–800 million (~45–60%)
Latin America and the Caribbean	80–120 million (~20–30%)	100–150 million (~25–40%)

Source: UN-Habitat (2003b)

Box 4.1 Bangalore's water and sanitation problems, despite its prosperity and economic success

In 2000, in this city with close to 6 million inhabitants, a baseline survey found that 73 per cent of the 2923 households surveyed in the municipal corporation area had access to water supply from the official network within the house or compound. But only 36 per cent had individual connections, with 36 per cent sharing the connection with others such as the landlord, other tenants or other users in an apartment and commercial complex. Twenty-seven per cent of the population did not have access to the piped water network.

A survey conducted among slum dwellers as part of the 'report card on urban services' revealed that irregular supply, the long distance to the tap and insufficient water were the problems frequently faced (Paul and Sekhar, 2000). Twenty-nine per cent of all households (and a large part of low-income households) draw water from some 18,000 water fountains (although a much smaller proportion rely only on these); it is common for women to spend two hours collecting water from these fountains. A study of public fountains found that many were located in unhygienic surroundings, with 45 per cent having stagnant wastewater in the surrounding area, 31 per cent having a solid waste dump in the immediate vicinity and 24 per cent having evidence of defecation in the surroundings. Wastewater drainage was found only in 48 per cent of standpipes, in two-thirds of which water was only available on alternate days, for an average of six hours a day.

The survey found that two-thirds of households in Bangalore reported the presence of a toilet within the premises, but less than half of these had a tap in the toilet and only 4 per cent had a flush tank. Twenty-eight per cent shared a toilet with other households, and a fifth of those reported problems with the arrangement – such as too many people per toilet, problems with blockages, poor maintenance and lack of cleaning. Four per cent used public toilets, and many users complained that they were dirty, not cleaned regularly and lacked lights. One per cent reported that they defecated in the open. Only a third of poor households in the city had access to satisfactory sanitation facilities. In a study of five slums, two had no water supply, one was supplied by boreholes and two had to depend on public fountains where between one and two wells and one tap served a population of between 800 and 900. Residents of the four slums had to walk from 20m to 1km to fetch water. With regard to sanitation, 113,000 households are reported to have no latrine at all. In a study of 22 slums, 9 with a total population of 35,400 had no latrine facilities. In another 10, there were 19 public latrines serving 102,000 people. Defecation in the open was common.

Source: Benjamin and Bhuvaneshari (1999); Benjamin (2000); Sinclair Knight Merz et al (2002). The data on the study of five slums comes from Achar et al, 2001. The data on sanitation in the 22 slums come from Schenk-Sandbergen (2001)

are obvious – most households dump their wastes on any available empty site, into nearby ditches or lakes, or simply along streets. The problems associated with this include smell, disease vectors and pests attracted by rubbish, and drainage channels blocked with waste. Where provision for sanitation is also inadequate, many households dispose of their toilet waste into drains or dispose of faecal matter within their garbage. Uncollected waste is obviously a serious hazard, especially for children playing in and around the home (and for many playing with items drawn from uncollected garbage), as well as for those who sort through rubbish looking for items that can be reused or recycled (Hardoy et al, 2001; Hunt, 1996).

4 Studies in many of the larger cities show that it is common for large sections of the population to live on sites at constant risk of disasters – for instance, on land that is often flooded, or slopes that are at high risk from landslides (Hardoy et al, 2001). This may also be found in smaller urban centres, although this will be most common in urban centres that grow rapidly and where (formal and informal) land markets exclude poorer groups from safer sites.

5 There are relatively few detailed studies of the health problems of populations in urban centres, but the studies that do exist suggest very large health burdens, relating primarily to infectious and parasitic diseases and accidents (McGranahan, 1991; Bradley et al, 1991; Hardoy et al, 1992, 2001; WHO, 1992, 1999; Pryer, 2003). They are 'environmental' in that they are caused by biological disease-causing agents in the air, water, food or soil, or transmitted by disease vectors, or caused by physical hazards in the living or working environment. This is consistent with the aforementioned studies that show the scale and depth of the deficiencies in provision for basic infrastructure and services and the quality of housing. It is likely that there are large health burdens arising from unsafe working conditions, and a considerable part of this occurs within the residential environment, since this is where a high proportion of low-income people generally work. There are also studies of health problems, or of the prevalence of certain diseases, in particular low-income settlements – see for instance, Pryer (1993) for a study that looks specifically at the health burden in a low-income settlement in Khulna, Bangladesh, and its economic consequences. See also Harpham et al, 1988; Bradley et al, 1991; Stephens, 1996, for summaries of available case studies; also APHRC, 2002 and Pryer, 2003.

There are many cities and smaller urban centres, or particular settlements within cities, where levels of air pollution considerably exceed WHO guidelines – for example, certain centres of heavy industry, mining or quarrying, or cities with high concentrations of motor vehicles with elevated levels of polluting emissions. Although this has a large health impact affecting large sections of the urban population (see McGranahan and Murray, 2003), in general, ambient air pollution is rarely considered among the most serious health risks compared to illnesses caused by infectious and parasitic diseases and accidents. The health burden linked to HIV/AIDS, for example, has grown rapidly in most nations and in many it now represents a large part of the total health burden in urban areas and much of the most serious 'non-environmental' health burden (van Donk, 2006).

Most of the work on health burdens, including those that arise from different diseases, is only for regions, or sometimes nations, not for urban populations or for the population of individual cities. But these give some idea of the scale of the health burden: for instance, figures for the health burden from diarrhoeal diseases (measured in terms of disability-adjusted life years lost to premature death and illness) show that the health burden per person in sub-Saharan Africa is 200 times that in Europe or North America (World Bank, 1993).

Available data on infant, child or under-five mortality for the urban populations of low- and middle-income nations is consistent with the above. Averages for entire nations' urban populations show infant, child or under-five mortality rates that are five to twenty times what they should be if populations had adequate nutrition, good environmental health and a competent health care service; also that in many low-income nations, these mortality rates increased during the 1990s (Montgomery et al, 2003). Table 4.2 shows infant and child mortality rates for urban and rural populations for a range of nations. What is perhaps surprising is how high these remain for urban populations in most nations, especially when considering the common assumption that urban populations benefit from 'bias' in government services. In many nations, the differences between the rural and the urban infant and child mortality rates are not very great. This is also surprising in that most urban areas have economies of scale and proximity in most of the measures that help reduce infant and child mortality rates (such as good provision for water and sanitation and for health care).

When viewing the figures in Table 4.2, it must be remembered that in all low- and middle-income nations, a high proportion of these nations' middle- and upper-income groups live in urban centres and such groups are likely to have relatively low infant and child mortality. So these 'averages' for nations' urban populations can hide the extent of the problem faced by low-income urban populations. In virtually all cities for which data are available in low-income nations, and for most in middle-income nations, there are also dramatic contrasts between different areas (districts, wards, municipalities) of the city regarding living conditions and health outcomes (Stephens, 1996; Hardoy et al, 2001). For example, for Nairobi, Kenya, under-five mortality rates were 151 per 1000 live births in its informal settlements (where over half the population live) and 62 for Nairobi as a whole (see Table 4.3). For informal settlements in one part of Nairobi, Embakasi, the under-five mortality rate was 254 – in other words, one child in four was dying before the age of five. It is likely that the under-five mortality rate in the wealthier parts of Nairobi are 10 to 20 per 1000 live births. The example is not unusual; many other studies have shown under-five mortality rates of 100–250 per 1000 live births in particular settlements. But it is unusual in that it was based on a representative sample of all Nairobi's informal settlements.

Table 4.2 *Infant and child mortality rates for urban and rural populations in selected nations, estimated mortality rates among infants (age less than 1) and children (ages 1–4 years)*

Country and year	Deaths per 1000 births*					
	Age <1			Age 1–4		
	Urban	Rural	Total	Urban	Rural	Total
SUB-SAHARAN AFRICA						
Benin (1996)	84	112	104	72	98	90
Burkina Faso (1998/99)	67	113	109	66	137	130
Cameroon (1998)	61	87	80	53	80	72
Central African Rep. (1994/95)	80	116	102	53	70	63
Chad (1997)	99	113	110	101	103	103
Comoros (1996)	64	90	84	18	36	32
Côte d'Ivoire (1994)	75	100	91	49	73	65
Eritrea (1995)	80	74	76	53	92	83
Ethiopia (2000)	97	115	113	58	88	85
Gabon (2000)	61	62	61	30	40	32
Ghana (1998)	43	68	61	36	58	52
Guinea (1999)	79	116	107	76	107	99
Kenya (1998)	55	74	71	35	38	37
Madagascar (1997)	78	105	99	53	77	72
Malawi (2000)	83	117	113	71	106	102
Mali (1996)	99	145	134	102	149	137
Mozambique (1997)	101	160	147	55	92	84
Namibia (1992)	63	61	62	25	36	32
Niger (1998)	80	147	136	107	212	193
Nigeria (1999)	59	75	71	52	73	67
Rwanda (1992)	88	90	90	74	80	80
Senegal (1997)	50	79	69	41	94	75
Sudan (1990)	74	79	77	46	71	63
Tanzania (1996)	82	97	94	42	59	56
Togo (1998)	65	85	80	38	79	69
Uganda (1995)	74	88	86	64	78	77
Zambia (1996)	92	118	108	90	98	95
Zimbabwe (1999)	47	65	60	23	37	33
NEAR EAST & NORTH AFRICA						
Egypt (2000)	43	62	55	10	19	15

Table 4.2 *continued*

Country and year	Deaths per 1000 births*					
	Age <1			Age 1–4		
	Urban	Rural	Total	Urban	Rural	Total
Jordan (1997)	27	39	29	5	7	5
Morocco (1992)	52	69	63	7	31	22
Turkey (1998)	42	59	48	10	16	12
Yemen (1997)	75	94	90	22	38	35
EUROPE & EURASIA						
Kazakhstan (1999)	44	64	55	7	10	9
Kyrgyz Republic (1997)	54	70	66	4	13	10
Uzbekistan (1996)	43	44	44	9	14	12
ASIA & PACIFIC						
Bangladesh (2000)	74	81	80	24	35	33
Cambodia (2000)	72	96	93	22	34	32
India (1999)	49	80	73	17	35	31
Indonesia (1997)	36	58	52	12	22	19
Nepal (1996)	61	95	93	23	53	51
Pakistan (1990/91)	75	102	94	21	33	29
Philippines (1998)	31	40	36	15	23	20
Vietnam (1997)	23	37	35	7	12	12
LATIN AMERICA & CARIBBEAN						
Bolivia (1998)	53	100	74	20	38	28
Brazil (1996)	42	65	48	7	15	9
Colombia (2000)	21	31	24	3	5	4
Dominican Republic (1996)	46	53	49	9	18	13
Guatemala (1998/99)	49	49	49	9	20	16
Haiti (2000)	87	91	89	27	65	53
Nicaragua (1997)	40	51	45	9	14	11
Paraguay (1990)	33	39	36	13	10	11
Peru (2000)	28	60	43	11	27	18

Note: *Infant and child mortality rates for the 10-year period preceding the survey.
Source: Demographic and Health Surveys (DHS); STAT compiler, www.measuredhs.com

It is obvious from Table 4.3 that there are very large differences among the population of Nairobi in regard to under-five mortality rates, and to the risk of contracting diarrhoeal diseases and the impacts that this has on health status and livelihoods. For high-income groups, diarrhoeal diseases are generally no more than a minor inconvenience; for low-income groups, they can have a serious impact on incomes (as income-earners are incapacitated), and are often life-threatening for infants or young children.

Pryer's (2003) study of the contribution of illness to poverty in the slums in Dhaka highlights another aspect of the scale of the health burden faced by low-income groups – in this instance, the extent to which ill health caused a deterioration in households' financial status. In her study of 850 households, ill-health was the single most important cause of such a deterioration, explaining 22 per cent of cases where households reported a deterioration in their financial status. Illness led to reductions in income and increased expenditures; often more loans taken out, assets sold and more adults resorting to begging (Pryer 2003). This link between illness, increasing indebtedness to cope with the drop in income and increase in health care expenditure and poverty was also described by Amis and Kumar (2000), drawing from research in Visakhapatnam, India. Although it is dangerous to draw general conclusions from one or two studies, the conditions described by Pryer (2003) and Amis and Kumar (2000) are similar to those in informal settlements or tenements in many other urban centres in low- and middle-income nations, so comparable links between high health burdens and poverty would be expected.

Table 4.3 *Infant and under-five mortality rates and diarrhoea prevalence in Kenya*

Location	Infant mortality	Under-five mortality rate	Prevalence of diarrhoea with blood in children under three in two weeks prior to interview
Kenya (rural and urban)	74	112	3.0
Rural	76	113	3.1
Nairobi	39	62	3.4
Other urban	57	84	1.7
Informal settlements in Nairobi	91	151	11.3
Kibera	106	187	9.8
Embakasi	164	254	9.1

Source: APHRC (2002)

Another indication of the scale of the health burden in urban areas in low-income nations is very low life expectancies. In many low-income nations there is evidence of average life expectancies not only being very low (between 35 and 50) but also falling, and this is known to relate in large part to mortality from AIDS (or to the lack of treatment available to those who are HIV-positive) or other diseases to which populations with HIV are vulnerable. But in many nations, this is also likely to be related to rising infant and child mortality rates that may be more linked to malnutrition and deterioration in health care services, housing conditions and provision for water and sanitation (which in some nations is, in turn, related to civil war). No data on life expectancies in urban areas were found for low-income nations.

Within this general picture of very limited achievements in reducing the health impact of environmental hazards in urban areas there are important exceptions. For instance, many Latin American nations that now have predominantly urbanized populations have managed to sustain long-term trends of falling infant and child mortality rates and increasing average life expectancies. This is also true for some Asian and African nations, although the scale of the improvements in urban areas is not clear, as the data are for nations or sub-national (state or provincial) units, not for urban populations or particular urban areas.

In the absence of data available in each city or smaller urban centre on what are the most serious environmental health problems and who is most at risk, it is obviously difficult to set priorities – both for action and for research. When this is combined with research and action agendas strongly influenced by external funding and external professionals, it can lead to surprising choices. For instance, it is very unlikely that small particle pollution or ground-level ozone in the outdoor city environment is among the most serious health risks in most cities in sub-Saharan Africa, but research is being funded on this topic. In part, this is because these are important topics in high-income nations (which leads to an assumption that it is important in all urban centres). But is also because the research is funded externally and supports professionals from high-income nations who have experience in this field – and no training or knowledge in the infectious and parasitic diseases that are likely to be far more significant health risks for most city populations.

Social and spatial aspects

Although it is common for assumptions to be made that there is urban bias in government expenditures and investments, the earlier section on the health burden in urban areas suggests that a large proportion of the urban population does not benefit much (if at all) from any such bias. For many nations, the differences in infant and child mortality rates between rural and urban areas are not as large as might have been expected if there was strong urban bias, especially since urban 'averages' will be improved by the concentration of middle- and upper-income groups in urban areas.

There are also factors that should help ensure an 'urban advantage' in regard to health and to the avoidance of environmental health risks. The first is because of the economies of scale and proximity arising from the concentration of people and production in urban areas, which reduces the unit costs of most forms of infrastructure and services that contribute to better environmental health (Hardoy et al, 2001). The second is because average incomes are generally higher in urban areas and this implies a greater capacity among the population to pay. The concentration of production units in urban areas that also require infrastructure and services, and which also have a capacity to pay for these, should also help fund such provision. One would also expect the concentration of voters in urban areas and the greater possibility of these becoming politically organized, and also the concentration of middle- and upper-income groups in urban areas (or particular cities) to contribute to a bias in government investments that favour urban areas (or particular cities). However, these potential urban advantages can only be realized if the government authorities responsible for governing these urban centres are capable of ensuring action on them either through public investment and management or through providing the framework supporting private or community provision. This, in turn, depends on national or provincial/state governments permitting urban governments the power, resources and legislative base they need to do so.

This section explores the links between environmental health and the quality of urban governance both from the point of view of urban populations and from that of possible social or spatial displacements/transfers of environmental health burdens. Given the range and complexity of environmental health risks concentrated in urban areas, this section focuses on one subset of actions to reduce environmental health problems through provision for piped water, sewers and drains (a critical part of what might be termed conventional urban infrastructure). This is an indicator of the extent to which urban populations have an important part of their environmental health needs addressed.

It is obviously not the whole story, since good environmental health depends on more than this (e.g. good quality homes and workplaces located in sites that are not at high risk from disasters, regular solid waste collection, controls on exposure to chemical pollutants and noise). Electricity might be considered an essential part of environmental health because of the reduction this should bring in fire risks, since people no longer need to rely on candles or kerosene lamps. Good quality health care and emergency services can also be considered a key part of good environmental health, not because these reduce environmental health risks, but because they greatly diminish their actual health impacts on populations who become ill or injured from environmental health risks.

In many urban locations, it is also possible to have good quality provision for water and sanitation without large-scale infrastructure – for instance, through tapping local water sources (usually groundwater) and stand-alone latrines – although in large population concentrations with high densities it is rare for these to be adequate in terms of minimizing risks from exposure to faecal-oral diseases or for the other diseases associated with inadequate water quality or quantity.

The extent to which an urban population has easy and convenient access to regular, good quality piped water (from both a health and convenience perspective, preferably piped into their home) and easy, convenient access to good quality, clean toilets (again preferably in their homes) is a measure of the quality of local (urban) government. This is not to imply that urban governments are the only possible providers of trunk infrastructure for water, sanitation and drainage but it is generally their responsibility to ensure provision, whether by public, private, community or cooperative organizations. Ensuring good quality provision for water, sanitation and drainage that reaches most or all the city population is a relatively severe test of the competence and capacity of local government, since it is more complex and expensive to plan, install and manage than most other interventions to address environmental health problems, such as better health care and garbage removal. Many case studies have demonstrated that it is possible to reach low-income groups with significant improvements in water, sanitation and drainage at costs that even relatively poor city authorities can afford (Hasan, 1993, 1997; Hanchett et al, 2003; Burra et al, 2003; Cain et al, 2002; Díaz et al, 2000). But ensuring good infrastructure provision for water, sanitation and drainage implies a cost and thus an allocation of resources – and consequently politics.

If a more detailed information base was available on the quality and extent of provision for water, sanitation and drainage for urban centres, it would be possible to place these centres in different categories according to the proportion of the population affected by the inadequacies of provision. Five categories are suggested here, with some ideas as to the urban centres or kinds of urban centre that would fall into these categories. Again, the weakness of the information base must be stressed – and most of the assessments of the cities given as illustrations below are not comprehensive assessments. For example, assessments of water supply are generally based only on the availability of a piped water supply, not on the quality of the water, the regularity of the supply and the extent to which it is affordable for lower-income groups. The categories below are used only to illustrate the wide spectrum in regard to the quality and extent of provision for different urban centres.

1   All, or virtually all, the population not served by piped water systems with treated water and drains because of little or no trunk infrastructure, in terms of water treatment, mains distributing water, surface drains and any provision for sanitation. In effect, a complete failure of local institutions to ensure provision for a water supply that is treated and distributed by pipes and provision for sanitation and drainage. Virtually all the population is affected – although a proportion of this population may be avoiding any serious problems through tapping local water sources safely, choosing sites that are not at risk of floods and defecation practices that minimize the risk of faecal-oral diseases. One suspects that most small urban centres in most nations in sub-Saharan Africa, and many low-income Asian nations, fall into this category (UN Habitat, 2003b; UN-Habitat, 2006). Some smaller urban centres in Latin America are also likely to fall into this category

– especially in the poorer nations or poorer regions of richer nations, or in new urban centres growing on agricultural or forest frontiers (Browder and Godfrey, 1997).

2   Similar to Category 1, except that the proportion of the population lacking provision has fallen to 75–90 per cent as a small proportion of the population has conventional infrastructure for piped water and drains and perhaps sewers — generally some central city areas and areas with a high concentration of middle- and upper-income groups. Again, a proportion of those under-served by conventional infrastructure may be avoiding environmental health risks through tapping local water sources and provisions made by households or communities. Many cities in Africa, some in Asia and a few in Latin America fall into this category in regard to provision for piped water, including some national capitals (e.g. Luanda and Ouagadougou in Africa, and Port-au-Prince in the Caribbean). Most major cities and virtually all other urban centres in sub-Saharan Africa, with the exception of cities in South Africa, would fall into this category in regard only to sanitation (with more than three-quarters of the proportion of the population lacking sewer connections to their home).

3   Around 40–74 per cent of the population still inadequately served, but the problem has become more spatially focused, with most of this population concentrated in particular geographic areas, much of it in informal or illegal settlements. Most major cities in sub-Saharan Africa, including national capitals, and many major cities in low-income nations in Asia fall into this category – that is, with 40–74 per cent of their population not served by piped water supplies or only with access to piped supplies through difficult to access public provision (UN-Habitat, 2003b). Many major cities in Latin America are also in this category (UN-Habitat, 2003b). In some cities, community-driven improvements or community-municipal partnerships have brought significant improvements in provision to a significant proportion of the low-income population – as in Karachi through the sewer construction programme supported by Orangi Pilot Project-Research and Training Institute (Hasan, 2006), and the community-driven toilet block construction and management programme in Mumbai (Burra et al, 2003).

4   Between 10 and 39 per cent of the population still inadequately served; for major cities the main difference between this and Category 3 is often that a considerable proportion of the population in informal settlements now has water supply and drainage systems that are integrated into city infrastructure systems. Provision for sanitation usually lags well behind provision for water. Available data suggest that many urban centres in Latin America and many major cities in North Africa and South Africa fall into this category.

5   Less than 10 per cent of the population not served by trunk infrastructure, although their actual numbers can still be quite high in large cities (e.g. 5 per cent of the population of a city with 15 million inhabitants lacking piped water still means 750,000 people). Most cities in Chile and many

of the larger and more prosperous Brazilian and Mexican cities are in this category (UN-Habitat, 2003b).

For large urban centres there are usually large differences in the quality and extent of provision for water and sanitation between different areas within the city. This most often shows up when there are data on provision for water and sanitation at municipal or district level within large cities with many separate local government areas. For instance, in Greater Buenos Aires, there are municipalities in which virtually all the population has piped water and sewers, but also municipalities in which only the minority have provision for piped water and sewers (Arrossi, 1996). These differences within cities can also be seen in the chapters on Accra (Chapter 6) and Delhi (Chapter 7).

What inhibits improvements in environmental health?

Various explanations can be put forward to explain deficiencies in water and sanitation and to explain why some cities have had improved provision and others have not. Some are briefly discussed in this section, but this is a tentative list, based on a weak and often undetailed evidence base and limited detailed literature. There is probably more detailed documentation and discussion of why provision for water and sanitation improved in Europe and North America during the second half of the 19th century and early part of the 20th century than there is today on the 2 billion people who live in urban areas in Africa, Asia and Latin America. The list is put forward in the hope that other researchers will critically examine it, both for its general relevance, and for its relevance in particular cities or city districts.

With regard to factors that explain the deficiencies in provision for water and sanitation (and thus prevent or slow down the diminution of environmental health problems), two focus on local issues that are largely linked to broader national or international contexts (the weakness of an urban centre's economy and of its local government). Two more centre on local governance limitations (the unwillingness or reluctance of local governments to prioritize water and sanitation for poorer groups or to develop models for improving provision that are more appropriate to local needs and resource availabilities), and two centre on the role of international agencies (the low priority allocated to water and sanitation in urban areas and international agency perceptions as to what role local governments should have in this).

1   *Weakness of the urban economy producing a limited capacity by most households to pay for water and sanitation.* This is perhaps the best explanation for the lack of provision for piped water and sanitation in most small, unprosperous urban centres in low- and middle-income nations. In addition, most of the cities with good quality provision reaching most of their populations are wealthy cities in middle-income nations. There are some case studies

of how poor provision is in particular small urban centres (Hardoy et al, 2001; UN-Habitat, 2003b; UN-Habitat, 2006) but given that there are tens of thousands of such urban centres, the documentation is sparse. An analysis of the results of demographic and health surveys from 43 low- and middle-income nations on water and sanitation provision showed that in general, provision was worse in urban centres with less than 100,000 inhabitants than for larger urban centres in all regions (North Africa, sub-Saharan Africa, Southeast Asia. South and West Asia and Latin America).[8] However, small urban centres are not always the worst served – and, for instance, in Brazil, smaller urban centres in prosperous regions are often better served than larger urban centres in poorer regions (Jacobi, 2002). Many small urban centres in Mexico are also relatively well served (UN-Habitat, 2006).

The weakness of the local economy and the lack of real demand for water and sanitation may also be a partial explanation for the inadequacies in provision for larger cities in many low-income nations – but as discussed later, other important competing explanations would be the weakness of local government, the unwillingness of local government and external funders (including national government and international agencies) to prioritize water and sanitation for low-income groups and the inability of all official agencies to support more appropriate models for provision.

2   *The weakness of local government* including its lack of funding for investment, weak revenue base (and limited possibility of enhancing this), and lack of staff with expertise in provision for water and sanitation. This certainly applies to most small and unprosperous urban centres. But this factor can act independently of the strength or weakness of the local economy. For instance, provision for water and sanitation is inadequate for much of the population in many cities in sub-Saharan Africa that have had periods of considerable prosperity, such as Port Louis, Gaborone, Dakar, Kinshasa, Kampala and Nairobi. The statistical base is too poor to see whether these periods of prosperity were accompanied by improved provision for water and sanitation that was extended to increasing proportions of the city population. But for much of sub-Saharan Africa, the scale of the failure in provision for water and sanitation even in the main cities is dramatic. There is also some evidence of deterioration in the quality and extent of provision over time (for instance, Thompson et al, 2000, for East Africa).

For most of Africa, and for many Asian nations, the weakness of local government generally has its roots in colonial rule. Most nations had a lack of qualified staff at independence, as well as inadequacies in their systems of urban government (and urban government revenue collection). The norms, codes and regulations in place for building, planning and land-use management were also largely inappropriate, being virtually all imported, mostly intended for use only within areas of the colonial city that were inhabited by 'non-natives', and never seen as measures to be implemented for entire city populations. Another reason for local government weakness is the speed with which many cities in sub-Saharan Africa grew at the time of independence. In many nations this was due to the removal of colonial controls on the rights of people from that nation to live in cities, while in all countries it was a result of building up the institutions of governance, education and health care that are part of an independent nation state. So perhaps part of the explanation for the inadequacies in provision is how weak and ill-equipped the city governments were for the very rapid growth they experienced around independence.

For some, it may also be the acceptance that government provision of conventional water and sanitation infrastructure was beyond their capabilities – for instance, accepting that households and residential areas would tap groundwater or surface water directly and that each would make their own provision for sanitation and drainage (or go without). Part of the explanation is also likely to be the reluctance of higher levels of government to allow local (city) governments the autonomy and strong fiscal base they need to become more effective in ensuring infrastructure and service provision. This, in turn, is linked to attempts to maintain national unity; the civil wars and regional conflicts that have plagued so many sub-Saharan African nations are a reminder of the reasons for this and the underlying cause (the boundaries created for nation states by European colonial powers that were not based on existing political structures).

3 *The unwillingness of local governments to prioritize provision for water and sanitation for poorer groups.* It is difficult to determine the extent to which the lack of attention to water and sanitation in a city with a prosperous economic base is due to the weakness of the local government or the politics that keep resources away from improving and extending provision. For example, provision for water and sanitation in Nairobi is very poor for the half of the population living in informal or illegal settlements (Alder, 1995; APHRC, 2002; Weru, 2004), and this cannot be blamed on Nairobi's lack of prosperity. But it is difficult to separate out whether this has been the result of political decisions at city level or at national level. For some prosperous cities, the explanation for the inadequacies in provision can be linked to political decisions – for instance, in Bangalore, one of India's most successful cities, as Box 4.1 highlighted, provision for water and sanitation for much of the population is inadequate. This can be linked to conscious decisions made by the development authority to prioritize other things, and the lack of accountability of this authority to the city's population. The municipal corporation is more accountable to city populations, but it is the development authority, not the municipal corporation, that controls most decisions with regard to public investments (Benjamin, 2000).

The inadequacies in provision for water and sanitation in Mumbai, India's financial capital, are also perhaps more plausibly explained by conscious political decisions not to give these a higher priority than by a lack of power and resources within city government. In part, this is because the middle- and upper-income groups in Mumbai and Bangalore do not have their health threatened by the diseases suffered by poorer groups – so the classic political mechanism that drove heavy investment in water, sanitation and drainage in most cities in Europe and North America in the late 19th and early 20th centuries are not operating (Chaplin, 1999). This can also be explained by élite attitudes to the poor that question why they came to the city, and feel that if provision for water and sanitation (and housing) improves it will attract more in-migration.[9] The fact that in many cities, large sections of the low-income population lacking provision live in illegal settlements where plot ownership is ambiguous and plot boundaries unclear, also helps explain the lack of provision, since this greatly complicates delivery of water and sanitation to individual plots and charging for services. But this problem has been resolved in many instances, where political circumstances allowed a pro-poor solution to be developed.

4 *The reluctance of local authorities to develop models for improving provision that are more appropriate to local needs and resource availabilities.* There is a long-established body of literature showing how innovations can reduce the unit costs of improving provision for water and sanitation, or increase poorer households' capacity to pay for improved provision (e.g. through credit that spreads the repayment of the cost of improved provision or of connection to existing networks). But most such innovations have not been taken up by local governments – or if they are, it is only after a longer struggle. See, for example, the time it took for local governments in Pakistan to recognize the relevance of the Orangi Pilot Project model for them (Hasan, 1997), or the time taken for Indian local governments to accept new models for community toilets pioneered by the National Slum Dwellers Federation, Mahila Milan and the Indian NGO The Society for the Promotion of Area Resource Centres (SPARC) (Burra et al, 2003). Although the extent of provision for piped water networks and sewers in urban areas is generally lowest in sub-Saharan Africa, most local governments in this region have opposed innovations that can reduce unit costs because these go against official regulations. For instance, improved provision for water and sanitation for low-income groups can be made much cheaper by permitting incremental development – that is, with communal provision of taps and toilets to sites on which poorer groups can develop their own housing with provision made to allow their upgrade to individual house provision when this can be afforded. There is generally a complex political economy underlying this – for example, the anti-poor attitudes of many of those within government and outside.

5 *What gets prioritized by international agencies.* Provision for water and sanitation in urban areas has never enjoyed a high priority from international agencies (Satterthwaite, 1997a, 1998, 2001; UN-Habitat, 2006). For some agencies, this is because provision for water and sanitation is not a priority; for others it is because they prioritized water provision in rural areas. And the limited amount of funding provided for water and sanitation by international agencies tends to concentrate in larger cities.

In some ways, this low priority by international agencies to water and sanitation is a puzzle since, in theory, the official bilateral and multilateral agencies are well-suited to funding water and sanitation in cities. This needs large investments (which make it easier for these agencies to spend their money and to cover their staff costs with a low proportion of total funding) in systems that should bring major benefits to low-income groups. It should also strengthen the city economy – something that came to be on the priority agenda of some agencies during the 1980s and 1990s as the importance of well-functioning city economies for successful national economies was recognized. There is also good potential (in theory) for cost recovery, as virtually all city inhabitants and businesses have some capacity to pay. Part of the reason for the low priority was the reluctance to invest in urban areas because these are believed to benefit from urban bias. There were also problems with maintenance for the new water and sanitation systems built in the 1970s, as no attention had been given to building the competence and capacity of local bodies to maintain the new systems – or what attention had been given to this proved unsuccessful. For the 1990s, the influence of the 'crisis narrative' of the world running out of water, and the assumption that privatization was one of the keys to improving provision, drew attention away from a conventional donor-funded programme of support for improved water and sanitation.

Part of the explanation is also the ignorance of senior international agency staff. For instance, one wonders why the Millennium Development Goals leave out diarrhoeal diseases from the list of diseases mentioned to which priority will be given. One wonders why sanitation was left out of the original Millennium Development Goals (did those who drafted them think that sanitation was unimportant?) and only squeezed in later, at the World Summit on Sustainable Development (WSSD) in 2002. One wonders why improved provision for water and sanitation is seen as part of 'improving environmental sustainability' within the Millennium Development Goals.

6 *Internationally driven perceptions as to what role governments (especially local governments) should have in infrastructure and service provision.* One reason for so little donor support for water and sanitation in urban areas during the 1990s was because of the assumption that privatization would greatly increase private sector funding for new investment. This can be illustrated by the policies of the World Bank. During the 1970s, the World Bank significantly increased its lending to support improved provision for water and sanitation in urban areas – both in terms of water and sanitation projects, and in terms of housing schemes that improved provision or water and sanitation to low-income populations (through upgrading and serviced site schemes). For this, it seems that there was no questioning that it was a legitimate role for government agencies to provide piped water supplies, sewers and drains. And, after all, this is a bank owned by governments whose central role is lending to governments for investment in development.

By the 1990s, this changed fundamentally, as the World Bank's official policy was much more to support the privatization of water and sanitation. One important research area that needs undertaking is the scale and nature of the bank's shift from funding governments to improve provision for water and sanitation to funding privatization and to applying pressure to governments to move down this route. The World Bank (and other donors) are now admitting that other solutions need to be sought (World Bank, 2004), since the amount of private capital attracted into improving and extending provision for urban water and sanitation has been disappointing, as has the performance of most privatized water and sanitation utilities (Budds and McGranahan, 2003).

What contributes to improvement in environmental health?

Having discussed factors that help explain why provision for water and sanitation is so poor, it is important to give some attention to factors that help explain why the deficiencies in provision are not worse.

There are two obvious factors: firstly, stronger city economies (so more can be spent on provision for water and sanitation); and secondly, stronger, more effective and more accountable city governments arising from decentralization, and the responsiveness of city governments to local democratic pressures and national governments to national pressures. These are certainly part of the explanation for improvements in the quality and extent of provision for many Latin American and Asian cities (see López Follegetti, 1999, and Velasquez, 1998, for case studies of particular cities), although their relative importance and the way they acted will differ much from nation to nation, and probably from city to city and within each city over time too. It is clear that part of the explanation for improved provision for water and sanitation for much of the urban population in Brazil is linked to the return to democracy (with greater priority to water and sanitation pre-dating the return to democracy, although it can be interpreted as a response to citizen demands). But again, there is a need for nation-specific and city-specific analyses as to what drove or underpinned improvements.

Also of importance are changes in governance processes that allowed or created more pro-poor outcomes. These include the many examples of 'participatory governance' where new mechanisms were introduced to encourage the involvement of poorer groups not just in specific projects but also in local governance. In part, these have been supported by a recognition within government that participatory governance is a necessary complement to representative democracy, to ensure greater attention to the interests of less-powerful groups, especially in situations of resource scarcity, where elections become a way of allocating limited state benefits rather than making political choices (Mitlin, 2004). They have also been partly the result of politicians or political parties seeking to increase their legitimacy and voter base; decentralization and democratization have created a new generation of politicians that are often more committed to participatory governance. More participatory governance models often include involving poorer groups (or citizens) in decisions about how public resources are allocated, as in participatory budgeting (Cabannes, 2004). More participatory governance systems mean the inclusion of those who do not find it easy to participate in state structures and processes because they are generally far removed from their own cultures and practices. Such groups generally face many forms of discrimination, including those based on gender, ethnicity and often simply poverty, when they try to engage with state agencies. Participatory governance processes open up new possibilities, by allowing poorer groups and their organizations greater 'space for negotiation' – although such processes are also politically contested and generally opposed by many elected politicians (Cabannes, 2004; Mitlin, 2004).

There are also an increasing number of examples of changes in local governance processes driven by organizations and federations formed by the urban poor that have importance for reducing environmental health burdens. These include the many upgrading programmes and new house development projects that federations of the urban poor or homeless have implemented (often with local government support, sometimes with central government funding) in many nations including India (Patel and Mitlin, 2004), South Africa (Bauman et al, 2001), Thailand (Boonyabancha, 2003, 2005), Kenya (Weru, 2004), Zimbabwe (Chitekwe and Mitlin, 2001) and the Philippines (Yu and Karaos, 2004). They also include the very large programme of community-designed, built and managed toilet blocks in cities in India, developed and promoted by the National Slum Dwellers Federation and savings groups formed by women slum and pavement dwellers (which are called *Mahila Milan*) and supported by local authorities, which has greatly improved provision for sanitation for millions of slum dwellers (Burra et al, 2003). These examples demonstrate how the costs of greatly reducing the environmental health burdens associated with poor quality, overcrowded housing and lack of basic infrastructure and services can be cut where governments (and other actors) work in partnership with low-income groups and their representative organizations.

Final comments

This chapter has described the very large environmental health burdens that large sections of the urban population in low- and middle-income nations suffer, and has discussed why these have received little attention in most low-income and many middle-income nations. One obvious reason is that most such environmental health burdens do not directly threaten middle- and upper-income groups. Most of the environmental health hazards suffered by the poorer groups linked to very poor quality housing and lack of infrastructure and services pose little threat to middle- and upper-income groups. The classic 19th century middle-class fear of 'contagion' was once a spur to heavy city-wide investment in piped water and sanitation that served everyone; as Chaplin's analysis of India shows (Chaplin, 1999), this now has much less influence because the middle- and upper-income groups know how to protect themselves from the diseases suffered by those without good provision for water and sanitation. This is unlike the middle- and upper-income groups for much of the 19th century, who did not know how to protect themselves from the 'miasmas' that were thought to spread disease, and suffered from (or fled from) cholera epidemics. This assumption that disease was spread by miasmas also helps explain why malaria is so misnamed as *mal aire* or bad air.

But fear of epidemics (and perhaps more to the point for city governments and vested interests, epidemics' potential economic consequences) and, to a lesser extent, a poor environmental health record, can act as a spur to investment. The city of Surat in India is reported to have had much improved provision for basic infrastructure after its plague epidemic in 1994. Perhaps some of the investment in improved water and sanitation in Latin America during the 1990s was influenced by the outbreak of cholera in Peru in 1993 and its spread. This factor can be linked to another contemporary influence that encourages investment in water and sanitation – the re-awakened recognition of the need for cleaner safer cities if they are to attract foreign investment. Again, this needs local analyses to see how this influences better provision for low-income groups; it does not necessarily do so, as the example of Bangalore given earlier demonstrates. But certainly, some cities have improved and extended provision for water and sanitation (and addressed other environmental health burdens suffered by poorer groups) as part of a conscious drive to attract foreign investment. This is also likely to have influenced investment patterns in cities that rely on tourism (although here it may only be the tourist sites that get improvements).

One important aspect that deserves more analysis is whether the environmental health problems in poor settlements that get priority from city authorities do so based on the extent to which they provide a risk for those living elsewhere. As noted previously, epidemics are a threat to city prosperity, even if they no longer pose much threat to middle- and upper-income groups. But many of the main causes of serious illness or injury and premature death among low-income populations provide little or no threat to middle- and upper-income groups – for instance, many of the diseases that take the greatest toll on their health, such as ARIs, pose little or no threat to those living elsewhere. This is also true for the physical hazards arising from poor quality, overcrowded homes and neighbourhoods lacking infrastructure. High risks from accidental fires used to be a city-wide threat, as can be seen by historical records in many cities; now it is largely concentrated in low-income areas with houses made of flammable materials with widespread use of open fires or kerosene stoves for cooking and heating, and candles or kerosene lamps for lighting. The risks from flooding or landslides are generally concentrated in specific settlements. To reiterate, then: where city authorities do address the environmental health problems suffered by lower-income groups, is there a tendency for them to prioritize those that do pose more threats to middle- and upper-income groups?

Another issue that deserves more analysis is the extent to which solutions to the most serious environmental health problems can cause displacements or transfers of environmental health burdens. Solutions to some of the most serious environmental health problems faced by poorer groups do not displace or transfer risk (e.g. better quality homes and neighbourhoods reducing risk from physical hazards, better health care reducing mortality and morbidity from ARIs), but most require some cost transfers (as capital for investment in solutions draws from revenues raised from entire city populations or economies), and many require partnerships with low-income groups and their community organizations. Some 'solutions' can transfer environmental burdens, but the scale of the risk from doing so and the number of people affected is generally much smaller than the risk reduction they cause – for instance, for garbage collection, drainage and sewage.[10]

Notes

1 The World Bank classifies all national economies by per capita income into low income, lower middle income, upper middle income and high income. In 2003, there were no high-income nations in Africa and very few in Latin America and the Caribbean (the exceptions are mostly small, wealthy Caribbean nations) or Asia (the exceptions being Japan; Taiwan, China; and Hong Kong, China; Singapore and a few small oil-rich or Pacific nations) (World Bank, 2003).
2 According to the Boston Consulting Group, in December 2003, China had 85,000 households with more than US$1 million in investible assets; the figures for some other low- and middle-income nations in Asia were 19,000 in India, 10,000 in Thailand, 5100 in Indonesia, 4600 in Malaysia, 1900 in Pakistan and 1500 in the Philippines (*The Economist*, 12 June 2004, p83).
3 Again, this is not to deny that there are groups within high-income nations that fall outside this, but at least for provision for water, sanitation and drainage, which is the main focus of this chapter, these proportions are very low.
4 Later to become the United Nations Human Settlements Programme: UN-Habitat.
5 See Table 2 in the back of World Bank, 2003; also Satterthwaite, 2004.
6 This draws on many city studies that are summarized in such books as Hardoy and Satterthwaite, 1989; UNCHS, 1996, 2001; Hardoy et al, 2001; UN-Habitat, 2003b, 2006. See also particular issues of *Environment and Urbanization*, such as October 1989, April and October 1995, and April and October 1997.
7 In Africa, three-fifths of the urban population lived in urban centres with less than 500,000 inhabitants in 2000; for Asia it was 49.6 per cent, for Latin America and the Caribbean it was 49.2 per cent (UN, 2004). This proportion does not necessarily increase if and when regions get wealthier; in Europe in 2000, for example, 63.5 per cent of the urban population lived in urban centres with less than 500,000 inhabitants (UN, 2004).
8 There were some exceptions. Also, the findings do not allow a judgement of the extent of public provision in different-sized urban centres – the measures were the percentage of households with piped or well water on the premises (the piped water could come from households tapping groundwater) and the percentage of households with flush toilets (which were not necessary connected to sewers and could be connected to septic tanks).
9 This is not to pretend that this is an accurate view; if 'better housing conditions' was an important influence on migration, why is it that very bad housing conditions do not appear to be a deterrent?

10 It is fashionable to demonize sewers as an inappropriate solution for sanitation in urban areas with little attempt to quantify the (very large environmental health) benefits these provide in comparison to the costs, and who bears these costs. For large and concentrated populations, well-functioning sewer systems enormously reduce the risk of human contamination by faecal matter, as well as providing cheap and convenient means to dispose of other household wastewater and to protect groundwater resources. They also take up relatively little room in space-constrained homes and serve multi-storey housing much better than any alternative. There are also adaptations of traditional sewer systems whose costs per households can come down to what most low-income households can afford (UN-Habitat, 2006). Most of their ecological disadvantages can be reduced or removed (e.g. high water consumption, large volumes of sewage concentrated at one point that needs expensive treatment, loss of nutrients for food production).

References

Abrams, C. (1964) *Man's Struggle for Shelter in an Urbanizing World*, MIT Press, Cambridge, MA

Achar, K. T. V., Bhaskara Rao, B. and de Bruijne, A. (2001) 'Organization and management of water needs in slums', in H. Schenk (ed), *Living in India's Slums: A Case Study of Bangalore*, The Indo-Dutch Programme on Alternatives in Development, Manohar, New Delhi, pp161–186

Alder, G. (1995) 'Tackling poverty in Nairobi's informal settlements: Developing an institutional strategy', *Environment and Urbanization*, vol 7, no 2, October, pp85–107

Amis, P. and Kumar, S. (2000) 'Urban economic growth, infrastructure and poverty in India: Lessons from Visakhapatnam', *Environment and Urbanization*, vol 12, no 1, pp185–197

APHRC (2002) *Population and Health Dynamics in Nairobi's Informal Settlements*, African Population and Health Research Center, Nairobi

Arrossi, S. (1996) 'Inequality and health in Metropolitan Buenos Aires', *Environment and Urbanization*, vol 8, no 2, October, pp43–70

Bartlett, S., Hart, R., Satterthwaite, D., de la Barra, X. and Missair, A. (1999) *Cities for Children: Children's Rights, Poverty and Urban Management*, Earthscan, London

Basta, S. S. (1977) 'Nutrition and health in low-income urban areas of the Third World', *Ecology of Food and Nutrition*, vol 6, pp113–124

Baumann, T. Bolnick, J. and Mitlin, D. (2001) *The Age of Cities and Organizations of the Urban Poor. The Work of the South African Homeless People's Federation and the People's Dialogue on Land and Shelter*, IIED Working Paper 2 on Poverty Reduction in Urban Areas, IIED, London

Benjamin, S. (2000) 'Governance, economic settings and poverty in Bangalore', *Environment and Urbanization*, vol 12, no 1, pp35–56

Benjamin, S. and Bhuvaneshari, R. (1999) *Urban Governance and Poverty: A Livelihood Perspective from Bangalore*, University of Birmingham, Birmingham

Bentinck, J. (2000) *Unruly Urbanization on Delhi's Fringe: Changing Patterns of Land Use and Livelihood*, Netherlands Geographical Studies 270, Utrecht/Groningen

Black, M. (1986) *The Children and the Nations: The Story of UNICEF*, United Nations Children's Fund (UNICEF), New York

Black, M. (1996) *Children First: The Story of UNICEF, Past and Present*, Oxford University Press, Oxford and New York

Boonyabancha, S. (2003) *A Decade of Change: From the Urban Community Development Office (UCDO) to the Community Organizations Development Institute (CODI) in Thailand*, Poverty Reduction in Urban Areas Working Paper 12, International Institute for Environment and Development (IIED), London

Boonyabancha, S. (2005) 'Baan Mankong; going to scale with "slum" and squatter upgrading in Thailand', *Environment and Urbanization*, vol 17, no 1, pp21–46

Bradley, D., Stephens, C., Cairncross, S. and Harpham, T. (1991) *A Review of Environmental Health Impacts in Developing Country Cities*, Urban Management Program Discussion Paper No. 6, The World Bank, United Nations Development Program (UNDP) and United Nations Centre for Human Settlements (Habitat), Washington, DC

Browder, J. D. and Godfrey, B. J. (1997) *Rainforest Cities: Urbanization, Development and Globalization of the Brazilian Amazon*, Columbia University Press, New York and Chichester

Budds, J. and McGranahan, G. (2003) 'Are the debates on water privatization missing the point? Experiences from Africa, Asia and Latin America', *Environment and Urbanization*, vol 15, no 2, pp87–114

Burra, S., Patel, S. and Kerr, T. (2003) 'Community-designed, built and managed toilet blocks in Indian cities', *Environment and Urbanization*, vol 15, no 2, pp11–32

Cabannes, Y. (2004) 'Participatory budgeting: A significant contribution to participatory democracy', *Environment and Urbanization*, vol 16, no 1, pp27–46

Cain, A., Daly, M. and Robson, P. (2002) 'Basic service provision for the urban poor; The experience of development workshop in Angola', Working Paper 8 on Poverty Reduction in Urban Areas, IIED, London

Chaplin, S. E. (1999) 'Cities, sewers and poverty: India's politics of sanitation', *Environment and Urbanization*, vol 11, no 1, April, pp145–158

Chitekwe, B. and Mitlin, D. (2001), 'The urban poor under threat and in struggle: Options for urban development in Zimbabwe, 1995–2000', *Environment and Urbanization*, vol 13, no 2, pp85–101

Clauson-Kaas, J., Surjadi, C., Hojlyng, N., Baare, A., Dzikus, A., Jensen, H., Aaby, P. and Stephens, C. (1999) *Crowding and Health in Low-Income Settlements: Kali Anyar, Jakarta*, United Nations Centre for Human Settlements/Avebury, Aldershot, p136

Connolly, P. (2004) 'The Mexican National Popular Housing Fund (FONHAPO)', in D. Mitlin and D. Satterthwaite (eds) *Empowering Squatter Citizens; Local Government, Civil Society and Urban Poverty Reduction*, Earthscan, London, pp82–111

Díaz, A. C., Grant, E., del Cid Vargas, P.I. and Velásquez, V. S. (2000) 'El Mezquital – A community's struggle for development in Guatemala City', *Environment and Urbanization*, vol 12, no 1, pp87–106

Dwyer, D.J. (1975) *People and Housing in Third World Cities*, Longman, London

Goldstein, G. (1990) 'Access to life-saving services in urban areas', in J. E. Hardoy, S. Cairncross and D. Satterthwaite (eds), *The Poor Die Young: Housing and Health in Third World Cities*, Earthscan, London, pp213–227

Hanchett, S., Akhter, S. and Khan, M. H., summarized by Mezulianik, S. and Blagbrough, V. (2003) 'Water, sanitation and hygiene in Bangladesh slums: A summary of WaterAid's Bangladesh Urban Programme Evaluation', *Environment and Urbanization*, vol 15, no 2, pp43–56

Hardoy, J. E. Cairncross, S. and Satterthwaite, D. (eds) (1990) *The Poor Die Young: Housing and Health in Third World Cities*, Earthscan, London
Hardoy, J. E., Mitlin, D. and Satterthwaite, D. (1992) *Environmental Problems in Third World Cities*, Earthscan, London
Hardoy, J. E., Mitlin, D. and Satterthwaite, D. (2001) *Environmental Problems in an Urbanizing World: Finding Solutions for Cities in Africa, Asia and Latin America*, Earthscan, London
Hardoy, J. E. and Satterthwaite, D. (1981) *Shelter: Need and Response; Housing, Land and Settlement Policies in Seventeen Third World Nations*, John Wiley & Sons, Chichester
Hardoy, J. E. and Satterthwaite, D. (1987) 'Housing and health in the Third World – Do architects and planners have a role?', *Cities*, vol 4, no 3, Butterworth Press, pp221–235
Hardoy, J. E. and Satterthwaite, D (1989) *Squatter Citizen: Life in the Urban Third World*, Earthscan, London
Harpham, T., Lusty, T. and Vaughan, P. (eds) (1988) *In the Shadow of the City: Community Health and the Urban Poor*, Oxford University Press, Oxford
Hasan, A.(1993) *Scaling-up of the Orangi Pilot Project's Low Cost Sanitation Programme*, Orangi Pilot Project-Research and Training Institute, Karachi
Hasan, A. (1997) *Working with Government: The Story of the Orangi Pilot Project's Collaboration with State Agencies for Replicating its Low Cost Sanitation Programme*, City Press, Karachi
Hasan, A. (2006) 'Orangi Pilot Project; the expansion of work beyond Orangi and the mapping of informal settlements and infrastructure', *Environment and Urbanization*, vol 18, no 2, pp451–480
Hunt, C. (1996) 'Child waste pickers in India: The occupation and its health risks', *Environment and Urbanization*, vol 8, no 2, October, pp111–118
Islam, N., Huda, N., Narayan, F. B. and Rana, P. B. (eds) (1997) *Addressing the Urban Poverty Agenda in Bangladesh, Critical issues and the 1995 Survey Findings*, The University Press Limited, Dhaka
Jacobi, P. (2002) 'Management of urban water and sanitation in Brazil', Background Paper prepared for United Nations Human Settlements Programme (UN-Habitat), New York
Kanji, N. (1995) 'Gender, poverty and structural adjustment in Harare, Zimbabwe', *Environment and Urbanization*, vol 7, no 1, April, pp37–55
Latapí, A. E. and de la Rocha, M. G. (1995) 'Crisis, restructuring and urban poverty in Mexico', *Environment and Urbanization*, vol 7, no 1, April, pp57–75
Leeds, A. (1974) 'Housing–settlement types, arrangements for living, proletarianization and the social structure of the city', in F. M. Trueblood and W. A. Cornelius (eds), *Latin American Urban Research*, Sage Publications, Thousand Oaks, CA
Leonard, H. J. (1989) 'Environment and the poor: Development strategies for a common agenda', in H. J. Leonard (ed), *Environment and the Poor: Development Strategies for a Common Agenda*, Overseas Development Council, Transaction Books, New Brunswick, NJ and Oxford, pp3–45
Lipton, M. (1977) *Why Poor People Stay Poor – Urban Bias in World Development*, Temple Smith, London
López Follegatti, J. L. (1999) 'Ilo: A city in transformation', *Environment and Urbanization*, vol 11, no 2, October, pp181–202
Mangin, W. (1967) 'Latin American squatter settlements; A problem and a solution', *Latin American Research Review*, vol 2, no 3, Summer, pp65–98

Maxwell, D., Levin, C., Armar-Klemesu, M., Ruel, M., Morris, S. and Ahiadeke, C. (1998) *Urban Livelihoods and Food and Nutrition Security in Greater Accra, Ghana*, International Food Policy Research Institute, Washington, DC

McGranahan, G. (1991) *Environmental Problems and the Urban Household in Third World Countries*, Stockholm Environment Institute, Stockholm

McGranahan, G. and Murray, F. (eds) (2003) *Air Pollution and Health in Rapidly Developing Countries*, Earthscan, London

Menegat, R. (2002) 'Participatory democracy and sustainable development: Integrated urban environmental management in Porto Alegre, Brazil', *Environment and Urbanization*, vol 14, no 2, pp181–206

Mitlin, D. (2004) 'Reshaping local democracy', *Environment and Urbanization*, vol 16, no 1, pp3–8

Mitlin, D. and Satterthwaite, D. (2001) 'Urban poverty: Some thoughts about its scale and nature and about responses to it', in S. Yusuf, S. Evenett and W. Wu (eds), *Facets of Globalization; International and Local Dimensions of Development*, World Bank, Washington, DC, pp193–220

Mitlin, D. and Satterthwaite, D. (eds) (2004) *Empowering Squatter Citizens; Local Government, Civil Society and Urban Poverty Reduction*, Earthscan, London

Montgomery, M. R., Stren, R., Cohen, B. and Reed, H. E. (eds) (2003) *Cities Transformed; Demographic Change and its Implications in the Developing World*, The National Academy Press/Earthscan, Washington, DC

Moser, C. O. N. (1996) *Confronting Crisis: A Summary of Household Responses to Poverty and Vulnerability in Four Poor Urban Communities*, Environmentally Sustainable Development Studies and Monographs Series No. 7, The World Bank, Washington, DC

Moser, C. O. N. (1998) 'The asset vulnerability framework: Reassessing urban poverty reduction strategies', *World Development*, vol 26, pp1–19

Moser, C. O. N., Herbert, A. J. and Makonnen, R. E. (1993) *Urban Poverty in the Context of Structural Adjustment; Recent Evidence and Policy Responses*, TWU Discussion Paper DP No. 4, the Urban Development Division, May, World Bank, Washington, DC

Patel, S. and Mitlin, D. (2004) 'Grassroots-driven development: The alliance of SPARC, the National Slum Dwellers' Federation and Matilda Milan', in D. Mitlin and D. Satterthwaite (eds), *Empowering Squatter Citizen: Local Government, Civil Society and Urban Poverty Reduction*, Earthscan, London, pp216–241

Paul, S. and Sekhar, S. (2000) *Benchmarking Urban Services: The Second Report Card on Bangalore*, Public Affairs Centre, Bangalore

Pryer, J. A. (1993) 'The impact of adult ill-health on household income and nutrition in Khulna, Bangladesh', *Environment and Urbanization*, vol 5, no 2, October, pp35–49

Pryer, J. A. (2003) *Poverty and Vulnerability in Dhaka Slums; the Urban Livelihoods Study*, Ashgate, Aldershot

Rakodi, C. I. (1996) 'Conceptualising poverty reduction: What can be done at the urban level?' Paper to UN Centre for Human Settlements Regional Workshop on Urban Poverty Reduction at City Level in Sub-Saharan Africa, Nairobi

Rodhe, J. E. (1983) 'Why the other half dies: The science and politics of child mortality in the Third World', *Assignment Children*, vols 61–62, pp35–67

Satterthwaite, D. (1990) 'La ayuda internactional', in N. Clichevsky, H. Herzer, P. Pírez, D. Satterthwaite, A. Azuela, S. Finquelievich, N. Marqués, A. Rofman,

M. Schteingart and H. Sretter (eds), *Construcción y Administración de la Ciudad Latinoamericana*, IIED–AL y Grupo Editor Latinoamericano, Buenos Aires, pp435–492

Satterthwaite, D. (1995) 'The underestimation of poverty and its health consequences', *Third World Planning Review*, vol 17, no 4, November, ppiii–xii

Satterthwaite, D. (1997a) *The Scale and Nature of International Donor Assistance to Housing, Basic Services and Other Human Settlements Related Projects*, World Institute for Development Economics Research, Helsinki

Satterthwaite, D. (1997b) 'Urban poverty: Reconsidering its scale and nature', *IDS Bulletin*, vol 28, no 2, April, pp9–23

Satterthwaite, D. (1998) *The Constraints on Aid and Development Assistance Agencies Giving A High Priority to Basic Needs*, PhD thesis, London School of Economics and Political Science, London

Satterthwaite, D. (2001) 'Reducing urban poverty: Constraints on the effectiveness of aid agencies and development banks and some suggestions for change', *Environment and Urbanization*, vol 13, no 1, pp137–157

Satterthwaite, D. (2004) *The Under-estimation of Urban Poverty in Low- and Middle-Income Nations*, IIED Working Paper 14 on Poverty Reduction in Urban Areas, IIED, London

Satterthwaite, D. (2005) *The Scale of Urban Change Worldwide 1950–2000 and its Underpinnings*, Human Settlements Discussion Paper, IIED, London

Satterthwaite, D., Hart, R., Levy, C., Mitlin, D., Ross, D., Smit, J. and Stephens, C. (1996), *The Environment for Children*, Earthscan and UNICEF, London and New York

Schenk–Sandbergen, L. (2001) 'Women, water and sanitation in the slums of Bangalore; a case study of action research', in H. Schenk (ed), *Living in India's Slums: A Case Study of Bangalore*, IDPAD, Manohar, New Delhi, pp187–216

Sinclair Knight Merz and Egis Consulting Australia, Brisbane City Enterprises and Feedback HSSI – STUP Consultants – Taru Leading Edge (2002), *Bangalore Water Supply and Environmental Sanitation Masterplan Project; Overview Report on Services to Urban Poor Stage 2*, AusAid, Canberra, March

Stephens, C. (1996) 'Healthy cities or unhealthy islands: The health and social implications of urban inequality', *Environment and Urbanization*, vol 8, no 2, October, pp9–30

The Economist, 12 June 2004, p33

Thompson, J., Porras, I. T., Wood, E., Tumwine, J. K., Mujwahuzi, M.R., Katui-Katua, M. and Johnstone, N. (2000) 'Waiting at the tap: Changes in urban water use in East Africa over three decades', *Environment and Urbanization*, vol 12 no 2, pp37–52

Turner, J. F. C. (1968) 'Housing priorities, settlement patterns and urban development in modernizing countries', *Journal of the American Institute of Planners*, vol 34, pp354–363

Turner, J. F. C. (1976) *Housing By People – Towards Autonomy in Building Environments*, Ideas in Progress, Marion Boyars, London

Turner, J. F. C. and Fichter, R. (eds) (1971) *Freedom to Build*, Macmillan, New York and London

UN (1976) *Report of Habitat: United Nations Conference on Human Settlements*, A/CONF.70/15, United Nations, New York

UN (2004) *World Urbanization Prospects: The 2003 Revision*, United Nations Population Division, Department of Economic and Social Affairs, ST/ESA/SER.A/237, New York, 323pp

UNCHS (United Nations Centre for Human Settlements) (Habitat) (1996) *An Urbanizing World: Global Report on Human Settlements, 1996*, Oxford University Press, Oxford and New York
UNCHS (Habitat) (2001) *Cities in a Globalizing World*, Earthscan, London
UN-Habitat (2003a) *The Challenge of Slums: Global Report on Human Settlements 2003*, Earthscan, London
UN-Habitat (United Nations Human Settlements Programme) (2003b) *Water and Sanitation in the World's Cities: Local Action for Global Goals*, Earthscan, London
UN-Habitat (2006) *Water and Sanitation in Small Urban Centres*, Earthscan, London
UNICEF (1988) Improving Environment for Child health and Survival, *Urban Examples* No 15, United Nations Children's Fund, New York
UNICEF and WHO (World Health Organization) (1984) *Primary Health Care in Urban Areas: Reaching the Urban Poor in Developing Countries – A State of the Art Report*, SHS/84.4, United Nations Children's Fund, New York
van den Berg, L., van Wijk, M. S. and Van Hoi, P. (2003) 'The transformation of agriculture and rural life downstream of Hanoi', *Environment and Urbanization*, vol 15, no 1, pp35–52
van Donk, M. (2006) '"Positive" urban futures in sub-Saharan Africa: HIV/AIDs and the need for ABC (a broader conceptualisation)', *Environment and Urbanization*, vol 18, no 1, pp155–176
Velasquez, L. S. (1998) 'Agenda 21; a form of joint environmental management in Manizales, Colombia', *Environment and Urbanization*, vol 10, no 2, pp9–36
Ward, B. (1976) *The Home of Man*, W. W. Norton, New York
Weru, J. (2004) 'Community federations and city upgrading: The work of Pamoja Trust and Muungano in Kenya', *Environment and Urbanization*, vol 16, no 1, pp47–62
WHO (1989) *Urbanization and its Implications for Child Health: Potential for Action*, Publications Division, World Health Organization, Geneva
WHO (1992) *Our Planet, Our Health*, Report of the WHO Commission on Health and Environment, World Health Organization, Geneva
WHO (1999), 'Creating healthy cities in the 21st century', in D. Satterthwaite (ed), *The Earthscan Reader on Sustainable Cities*, Earthscan, London, pp137–172
World Bank (1990) *World Development Report – 1990: Poverty*, Oxford University Press, Oxford
World Bank (1993) *World Development Report 1993; Investing in Health*, Oxford University Press, Oxford
World Bank (2001) *World Development Report 2000/2001: Attacking Poverty*, Oxford University Press, Oxford and New York
World Bank (2003) *Sustainable Development in a Dynamic World: Transforming Institutions, Growth and Quality of Life, World Development Report 2003*, World Bank and Oxford University Press, New York
World Bank (2004) *World Development Report 2004: Making Services Work for Poor People*, World Bank and Oxford University Press, Washington, DC
World Commission on Environment and Development (1987) *Our Common Future*, Oxford University Press, Oxford
Wratten, E. (1995) 'Conceptualizing urban poverty', *Environment and Urbanization*, vol 7, no 1, April, pp11–36
Yu, S. and Karaos, A. M. (2004) 'Establishing the role of communities in governance: The experience of the Homeless People's Federation Philippines', *Environment and Urbanization*, vol 16, no 1, pp107–120

5
Improving Urban Water and Sanitation Services: Health, Access and Boundaries

Kristof Bostoen, Pete Kolsky and Caroline Hunt

Introduction

Those who live in cities depend upon resources from outside city boundaries. Use of external water resources is one of the important ways a city affects, and is affected by, its surroundings. During rapid urban growth, these interactions become increasingly important for both the city and its environment.

Water flows back and forth between the natural environment and the urban community (Figure 5.1). Water supply brings water from the broader environment into the community, while drainage and sewerage returns it to the 'natural' environment. Water in such transfers is never pure H_2O, but is always mixed with other matter, as illustrated in Figure 5.1. Often this 'other matter' includes pathogens (disease-causing organisms).

Whatever water comes into a community has to be returned to the natural environment. Even with recycling and storage, the outflow must more or less equal the inflow, or else flooding will occur. Despite this fairly obvious fact, efforts are frequently made to improve community water supply without improving drainage. If water is returned to the natural environment with chemical or biological pollution, the contamination does not always disappear or die off, but can return to threaten the health of the polluting community or that of one downstream.

This chapter examines issues related to water supply and sanitation services, which are of particular relevance to low-income communities. The second section of this chapter looks at health issues relating to water, while the third section examines the targets set by the international community for water and sanitation and the challenges regarding the achievements of these targets.

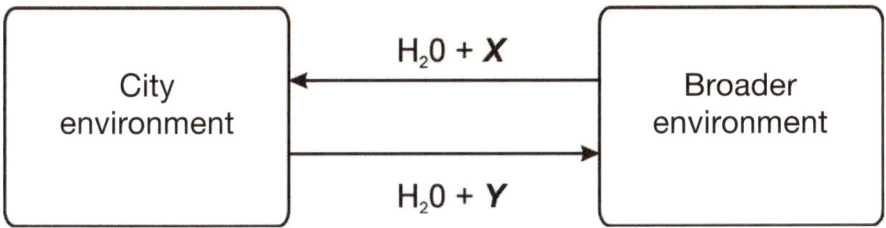

Figure 5.1 *The water balance*

As problems of water shortages and pollution are transferred from the local to the broader environment, the challenge shifts from one of maintaining human health to one of preserving the integrity of life-support systems for future generations (McGranahan et al, 2001). These transitions are well known and documented with regards to the water cycle.

The water cycle within the urban area, as illustrated in Figure 5.1, occurs at each spatial scale of the city; for a given neighbourhood, the rest of the city is seen as the broader environment. These different subdivisions, or boundaries, often create institutional issues which in turn have an impact upon service quality and health; these are explored in the fourth section of the chapter.

How water supply, sanitation and hygiene affect health

Below, in the following two sub-sections is a description of two common models to describe the relationships between water supply, sanitation, hygiene and human health, as they are understood at present. The third sub-section refers to recent and forthcoming research in this field.

Classifications of water-related infections

The first model has evolved from earlier work, grouping *water-related* infectious diseases by broad routes of transmission (Feachem et al, 1977; White et al, 1972). The categories are defined by the types of intervention that can control morbidity and mortality, rather than by the biological taxonomy of the organisms that cause them. As such, this model has helped engineers and public health professionals to work together on practical control strategies (Kolsky, 1993). A similar classification exists for *excreta-related diseases* (Feachem et al, 1983a) but has been less widely used. There are four categories in the Bradley-Feachem classification of water-related disease:

- *Faecal-oral* (waterborne *and* waterwashed)
  These include infections that are transmitted by swallowing faecally contaminated matter (food and water) containing pathogens. They can

be caused by lack of sufficient water to maintain personal and domestic hygiene *as well as* by drinking contaminated water. Diseases in this group include, among others, diarrhoeal diseases, typhoid, cholera and hepatitis A and E.
- *Strictly water-washed* (skin and eye infections)
  These are conditions that are exacerbated by lack of water for washing and hygiene, but are *not* faecal-oral. These diseases are largely related to skin and eyes, such as scabies, trachoma and conjunctivitis.
- *Water-based aquatic intermediate host*
  Aquatic organisms such as snails act as hosts to parasites, which then infect humans either by being swallowed or through contact in water (e.g. by piercing the skin of those wading in the water). Diseases in this group include guinea worm and schistosomiasis.
- *Water-related insect vector*
  These diseases depend on insect vectors, such as mosquitoes and flies, which breed in or near water. They transmit disease to humans, for example, through bites. The diseases involved include malaria, filariasis, yellow fever, dengue and onchocerciasis (river blindness).

From the four categories in the Bradley-Feachem classification it becomes clear that interventions focused on water quantity have broader impact than those focused on water quality. Water quality only affects faecal-oral diseases, whereas quantity affects both faecal-oral and water washed diseases. The relative importance of water quantity and its quality will be discussed later in this chapter.

Diarrhoeal diseases, which are faecal-oral, are responsible for the greatest number of episodes of illness (morbidity) and deaths (mortality) worldwide, compared to any other single classification of water and sanitation-related disease. This is shown in Table 5.1, based on data presented for World Health Organization (WHO) member states. It has been estimated that diarrhoeal disease represents 90 per cent of the health impact associated with water supply and sanitation (White et al, 1972). Diarrhoeal diseases are estimated to kill around 1.8 million people every year worldwide (WHO, 2004) of which the overwhelming majority is children. This toll is equivalent to 12 jumbo jet crashes every day or almost twice (1.9) the number of people who 'died in the World Trade Center on the 11th of September 2001' per day. There is some reason to believe that the number of deaths has fallen since the 1980s, possibly due to water and sanitation programmes and increased use of oral rehydration therapy (Bern et al, 1992). However, it appears that the number of episodes of diarrhoeal disease has remained constant.

Approximately 90 per cent of diarrhoeal disease cases are estimated to be attributable to environmental factors (Murray and Lopez, 1996). Apart from water supply, sanitation and hygiene, diarrhoeal disease is also associated with a number of other risk factors including age, malnutrition, lack of breast-feeding, and seasonality.

Table 5.1 *Health impacts of water- and sanitation-related diseases*

	Mortality estimates for 2000 (thousands)	DALYs* estimates for 2000 (thousands)
Faecal-oral		
Diarrhoeal disease	1798	61,966
Poliomyelitis	1	151
Water-washed		
Trachoma		2329
Water-based		
Schistosomiasis	15	1702
Water-related vector		
Malaria	1272	46,486
Lymphatic filariasis	19	5777
Dengue		616
Intestinal nematode infections	12	2951

Note: *DALYs or Disability-Adjusted Life Years is a indicator attempting to quantify 'time lived with a disability' and the 'time lost due to premature mortality' developed for the World Bank's 1993 World Development Report: Investing in Health.

Source: Adapted from WHO (2004)

The F-diagram

A second model is the F-diagram, depicted by Wagner and Lanoix (1958) (Figure 5.2), which has been widely used as a model of faecal-oral disease transmission. Unless faeces are isolated from potential contact with humans, animals and insects, pathogens may be carried on unwashed hands, in contaminated water or food, or via flies and other insects on to further human hosts. The first way to stop or reduce transmission is to ensure the safe disposal of faeces, through sanitation. Safe excreta disposal and washing hands following defecation is referred to as 'the first barrier' and considered the most important health intervention, as it keeps faecal pathogens out of the living environment. Children's faeces in particular are known to contain a high load of pathogenic organisms, such as Ascaris and Trichuris, but are also least likely to be safely disposed of (Cairncross, 1989; Kolsky, 1993). The secondary barriers to faecal-oral disease transmission protect people from whatever faecal contamination of the environment is present. These are based on hygienic practices, such as washing hands before handling food, fly control, safe food storage and the use of footwear.

The F-diagram graphically presents multiple routes of transmission. A single type of pathogen may be transmitted by several of these routes, and the population at risk may be vulnerable to many different pathogens, which may

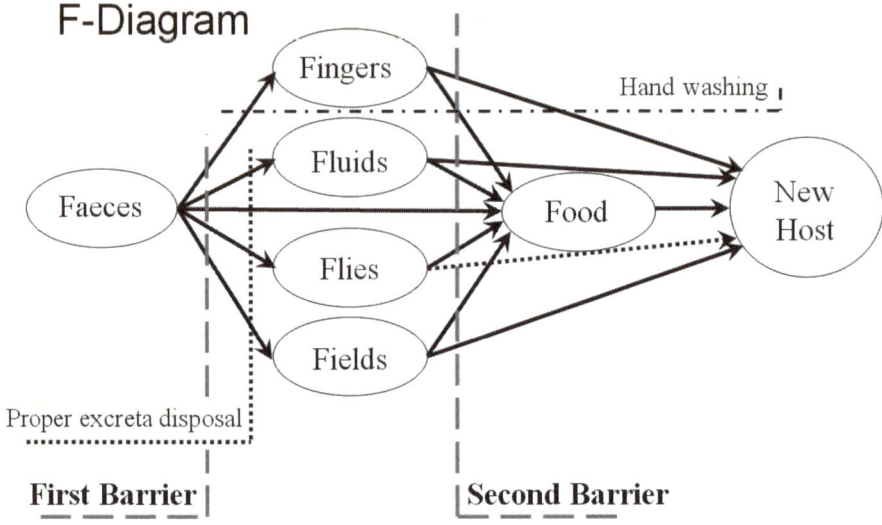

Source: After Wagner and Lanoix (1958)

Figure 5.2 *F-diagram*

favour different routes. Numerous commentators have advocated integrated measures to control diarrhoeal disease by combating multiple routes of transmission (Lewin et al, 1996; Curtis et al, 2000a). The greater the range of interventions, the greater the chance in successfully reducing diarrhoeal disease transmission. The F-diagram also shows that while water quality only affects one route, the quantity of water available for personal and domestic hygiene affects almost all routes.

Following the discovery in the 19th century of the undeniable role that water quality played in the epidemics of cholera and typhoid, there was a natural focus on the improvement of drinking water quality. This focus produced dramatic results in the reduction of waterborne epidemics. The F-diagram clearly shows that this would be the case where water contamination is the main route of transmission.

Where routes other than water consumption are more important for disease transmission, however, improving water quality will have far less effect. While waterborne epidemics are dramatic and alarming surprises, the sad truth is that the everyday endemic (non-epidemic) toll of faecal-oral disease is far, far higher, and most of the latter seems to be transmitted through routes other than water. While improving water quality does not necessarily affect endemic transmission, increasing the quantities of water available to improve personal and domestic hygiene *can* have a greater effect on this unacceptable toll (Cairncross, 1995). Most health benefits will be obtained from large amounts of water of a good quality. But if resources are scarce, public health professionals generally recognize the greater importance of access to water in quantity for hygiene, compared with the quality of that water (Esrey et al, 1985, 1991). Unfortunately, in practice, the main efforts in 'water and sanitation for low-income areas' are often still directed towards the improvement of the water quality of the public water supply, rather than improving access by poor households, and thus the quantity of water that those households can use. The beneficiaries of such efforts are more likely to be people who already have access to water than people with no access. Sanitation and hygiene promotion are even lower priorities in practice, although the principles of the F-diagram suggest that they should have equal or higher priority (Curtis et al, 2000b).

Recent and forthcoming research

Most studies of the impact of water quality have been based upon water quality measurements at the source or collection point. It is known however, that the degree of faecal contamination of water increases during transport to the household (Clasen and Bastable, 2003). There is also increasing evidence that improving water quality at the point of use has a positive health impact (Conroy et al, 2001; Iijima et al, 2001; Fewtrell et al, 2005; Clasen, 2006). This is regarded by some as an exception to the dominant paradigm (Clasen and Cairncross, 2004). While data support health benefits for people that have at least 15 litres of water per capita per day there are reasons to believe that benefits are reduced when access levels to water are lower (Clasen, 2006). However, at this time not enough data are available to substantiate this (Clasen, 2006). Systematic reviews and meta-analyses, such as those of the Cochrane Library infectious diseases group (Clasen et al, 2004) and field research will be needed to clarify these new findings and examine if these are in conflict with the current paradigm.

What does all this mean?

These models clarify the complex relationships between water, sanitation, hygiene behaviour and health. For example, good hygiene is more important in low-income areas where environmental exposure to pathogens is greater; residents of relatively clean areas can (and often do!) get away with lower standards of hygiene. Those who practise poor hygiene are certainly at greater risk than those who practise good hygiene, even in relatively clean environments. Water quantity is generally more important than water quality, because increased quantities of water promote good hygiene, and can prevent faecal-oral transmission by a number of different routes; increased quantities of water also reduce skin and eye infections. Only when drinking water is the main source of infection will water quality be more important than quantity. This is rarely the case where diarrhoea is endemic.

This means that in most cases within an urban setting, water distribution (access to water in quantity) is more important than public water treatment (its quality) until a certain relatively high level of environmental hygiene has been reached. Water treatment at the point of use seems to give health benefits, but it is not clear if a certain level of access to water is required to profit from such an approach. Finally, the quantity of water which people can actually use is vitally dependent upon access, as shown by Figure 5.4; the issue of access will be further explored in the following section.

Access to improved sanitation and water sources

'Water for life', the 2005 global water and sanitation assessment, contains the most up-to-date coverage data for most of the countries in the world (WHO/UNICEF, 2005). Since the Global Assessment 2000 Report (GA2000) (WHO/UNICEF, 2000) the United Nations (UN) Joint Monitoring Programme (JMP) does not report on 'safe' drinking water and 'adequate' sanitation. Instead, access to 'improved' water supply and sanitation technology types are now reported (see Table 5.2). This change in terminology reflects both past misrepresentation, and future uncertainty, in judging and defining services as *safe* in terms of human health. According to the report, over 2.6 billion people worldwide are without access to improved sanitation and over 1.1 billion do not have access to improved water supply. While many people have gained

Table 5.2 *Water supply and sanitation technologies considered to be improved and unimproved in WHO/UNICEF Global Assessment 2000*

The following technologies were considered to be improved:	
Water supply	**Sanitation**
Household connection	Connection to a public sewer
Public standpipe	Connection to septic tank
Borehole	Pour-flush latrine
Protected dug well	Simple pit latrine
Protected spring	Ventilated improved pit latrine
Rainwater collection	
The following technologies were considered not improved:	
Water supply	**Sanitation**
Unprotected well	Service or bucket latrines (where excreta manually removed)
Unprotected spring	Public latrines
Vendor-provided water	Open latrine
Bottled water*	
Tanker truck provision of water	

Notes: *Bottled water has been reclassified by the JMP as an '"improved" source of drinking only when there is a second source that is improved'.

Source: WHO/UNICEF (2000, 2005)

access since 2000, the number without access has remained the same (WHO/UNICEF, 2006).

Asia and Africa have the lowest levels of service coverage. In Asia, less than half the region's population have access to adequate sanitation. When comparing individual countries, the African region has the highest proportion of countries with less than 50 per cent water supply and sanitation coverage. In all regions, apart from North America, rural coverage is lower than urban coverage for both water supply and sanitation.

The Global Assessment 2000 presented the status of the sector using consumer-based data for the first time. These data were drawn from large nationally representative household sample surveys, such as the United States Agency for International Development's (USAID) Demographic and Health Survey (DHS) and United Nations Children's Fund's (UNICEF) Multiple Indicator Cluster Survey (MICS) and national census data. The GA 2000 thus presented a better baseline for future targets than previous reports. The WHO/UNICEF Joint Monitoring Programme for Water Supply and Sanitation (JMP), which published the GA 2000, continues to update these data to monitor progress to the Millennium Development Goal Target for water and sanitation. The latest report to date is *Meeting the MDG Drinking Water and Sanitation Target* (WHO/UNICEF, 2006), which is available together with current data at the JMP website (www.wssinfo.org).

Targets for the future

The UN Millennium Summit adopted the target of halving the proportion of people who are unable to reach, or to afford safe drinking water by the year 2015. The 2002 UN World Summit on Sustainable Development (WSSD) in Johannesburg has adopted the same target for access to sanitation facilities and the application of hygiene practices.

The compilation of the GA2000 has greatly improved data quality, using survey data. However, there is still a need to standardize survey outcomes to make results comparable. Figure 5.3 shows the typical scatter of results of the different surveys over the last 20 years for access to improved water in urban Niger. The variation in results is less for urban than for rural areas. Variations between different surveys are also less for household connections than for other improved access, which is probably a reflection of the interest of the water utility in keeping track of its customers and the ease of defining this way of delivering water to households.

Urban populations in Asia and Africa are predicted to almost double over the next 30 years. Against this trend, meeting the International Development Target of halving the proportion of those unserved by water by 2015 would mean providing water services to more than 300,000 additional people every day over the next 15 years. Halving the number unserved by sanitation requires provision of services to over 400,000 additional people per day.

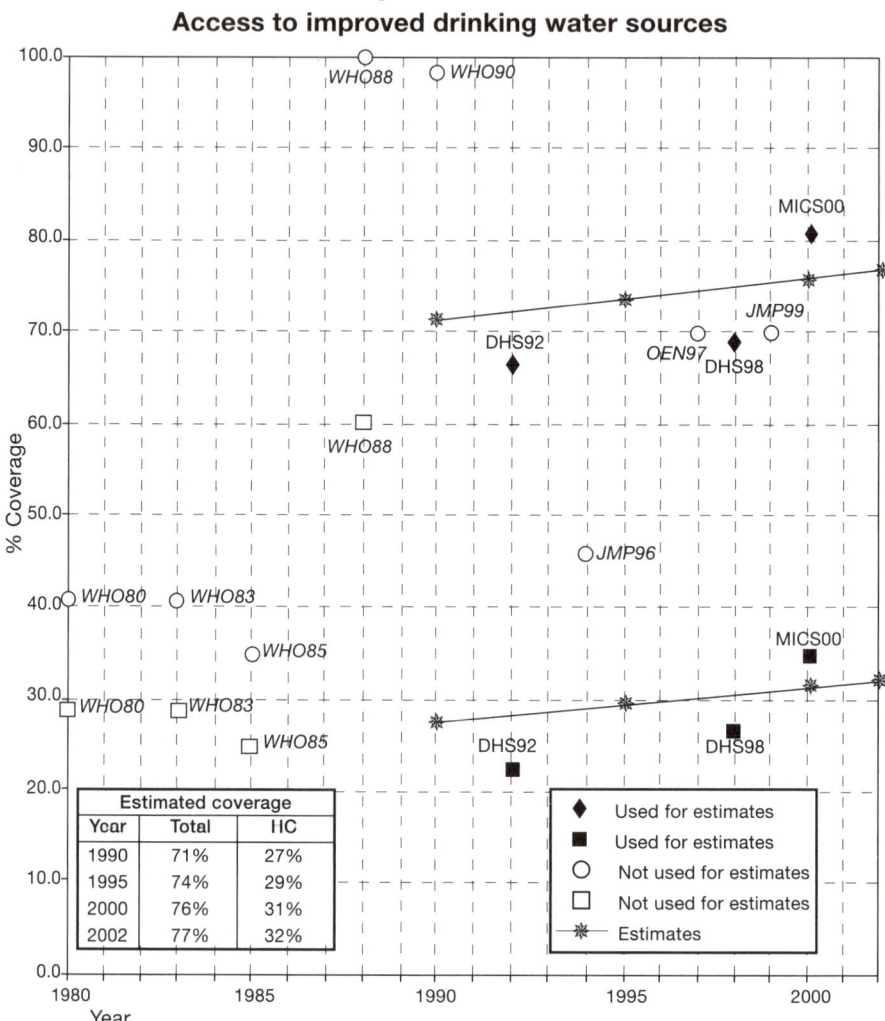

Note: HC = household connections.

Source: www.childinfo.org/areas/water/pdfs/niger_wat_02.pdf

Figure 5.3 *Variation on access in various surveys*

Limitations to the use of routine data sources

The use of existing surveys such as the DHS, MICS and national censuses, as in the GA 2000, has the advantage of being cost efficient, but it also has its drawbacks. These surveys have usually been designed to give a picture of 'the average household' on a national level and sometimes on a regional scale. Designed for other purposes, they cannot provide sector-specific (water, sanitation and hygiene) data at a local level that can be used for project implementation, evaluation and local decision making.

Large surveys such as the DHS require major administrative and organizational work, which means that they are unlikely to take place in countries or areas experiencing conflict or natural disasters.

To achieve better measurements in the field of water, sanitation and health practices there is a need for a simple standardized sample technique and a standardized set of indicators. Moreover, not only the collection techniques, but the interpretation and the type and extent of analyses need to be agreed upon if results are to be compared worldwide.

Access to improved services and its relation with health

The level and type of service both have the potential to influence health. However, numerous other factors, which influence the use and nature of the service, also affect health risk, in some cases to a greater extent than the level or type of service itself. These factors include: access to, and use of services; system maintenance; treatment; seasonality; water sources; and pathogen-specific factors. Poverty is very often a key variable behind many of the factors listed above, most notably access to services.

The improvement of water supply and sanitation has attracted particular interest in reducing diarrhoeal disease (Feachem et al, 1983b; Esrey et al, 1985). These environmental improvements, together with improvements in living standards, played a major role in reducing diarrhoea rates and controlling endemic typhoid and cholera in Europe and North America between 1860 and 1940 (Esrey et al, 1985). Similar effects were anticipated from equivalent improvements in low income countries, and these expectations contributed to the declaration of the Water Decade. In 1977, the UN Water Conference at Mar del Plata set up an 'International Drinking Water Supply and Sanitation Decade' for 1981–1990. Its aim was to make access to clean drinking water available across the world.

Although improving access to water and sanitation projects improves health (Feachem et al, 1983b; Esrey et al, 1985), it is difficult to link these achieved health benefits back to *specific* improvements. Many attempts have been made to measure the health impact from water supply, sanitation and, more recently, hygiene practices. Even attempts under the supervision of eminent specialists to measure the health impacts of water supplies and sanitation have produced almost useless or meaningless results (Cairncross, 1999). Health impact studies are, for that reason, not an operational tool for project evaluation or 'fine tuning' of interventions (Cairncross, 1990). This led various organizations like the WHO and the World Bank to adopt the Minimum Evaluation Procedure (WHO, 1983), which concentrates on measuring functioning and use of services rather than measuring their health impacts (World Bank, 1976; Briscoe et al, 1985; Esrey et al, 1985; Cairncross, 1999).

While the health benefits of increased water quantity are known, they are difficult to measure and to attribute back to increases in supply. There is, for example, a clear but counterintuitive relation between the time needed to

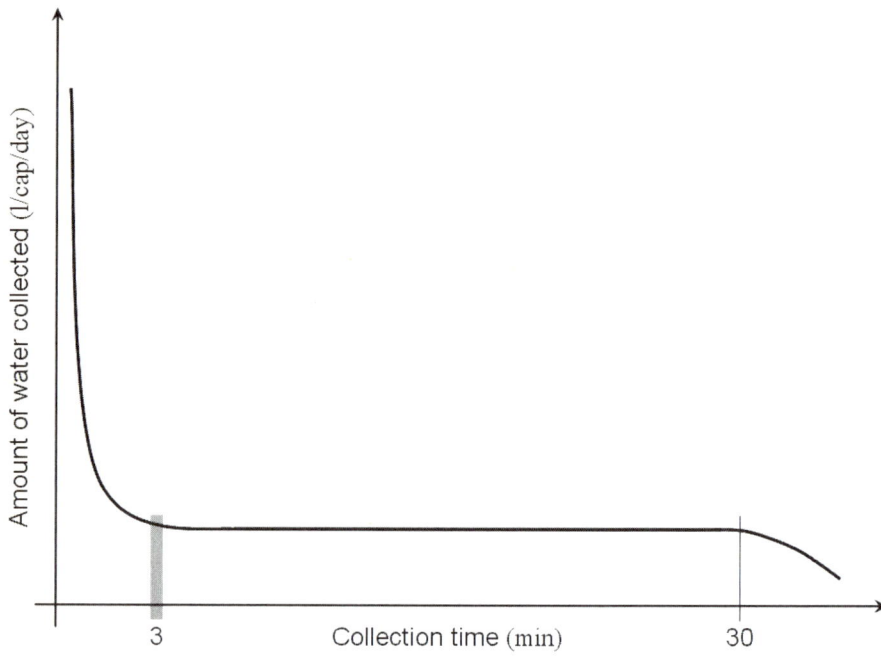

Note: Cap = Capita.

Source: Cairncross and Feachem (1993, p53)

Figure 5.4 *Relation between water consumption and time involved in water collection*

collect water and the amount of water collected (Figure 5.4). It is known that an increase in water consumption increases the water used for hygiene, which improves health. So, one might expect that reducing the time it takes to secure daily supplies to below 30 minutes would have a beneficial health impact. However, a reduction of the collection time of between 30 and 3 minutes will actually have little impact. Those who spend less than 3 minutes for water collection usually have a household connection. Note also that collection time includes queuing time, which can be significant in areas with relatively closely spaced taps with intermittent service, or in areas that are serving large populations. In various parts of Africa, reports show that while distance to source diminished or stayed the same, collection times for water increased (Thompson et al, 2002; UN-Habitat, 2003).

What does all this mean?

While the main ways water, sanitation and health practices relate to health are broadly understood in theory, their real-world interactions are far more complex. However, increasing access to water and sanitation is clearly recognized as leading to health benefits, and new international targets have been set for that reason. Although improved access might improve health, it is methodologically extremely difficult to attribute improvements in health exclusively to specific interventions, on a project-by-project basis. This makes health impact an unsuitable outcome measure for project evaluation. The current indicators based on level and type of service also have their limitations. Better sector specific indicators and survey tools need to be developed.

Boundary issues and the urban environment

Differing perspectives

Figure 5.5 shows the urban environment from the point of view of the householder. The home is at the centre, and is the householder's first priority for environmental management. If the householder is able to maintain the home in a relatively clean and pleasant condition, the next priority becomes the surrounding street (peri-domestic). Local and informal lane arrangements may, for example, develop among neighbourhoods to ensure that rubbish does

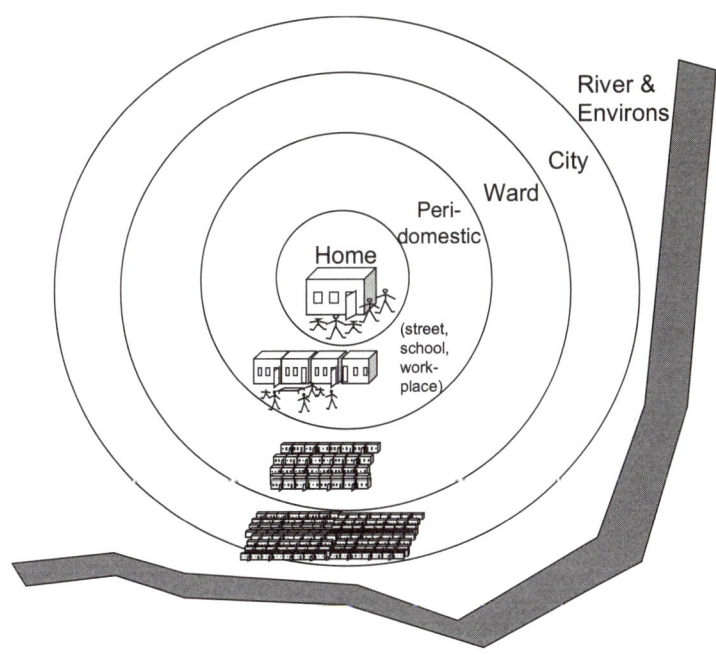

Source: Kolsky (1996) unpublished lecture notes, London School of Hygiene and Tropical Medicine

Figure 5.5 *Scales of the urban environment, as seen by a householder*

not pile up in the street, or clog drains. If the home and street are relatively clean, then some citizens will be concerned about the state of the environment in their larger neighbourhood (ward); when these problems are addressed, attention can then be focused on the rest of the city, and eventually, on the environment outside the city. It is natural that the focus of environmental concern spreads further outwards as the problems at each of the smaller (inner) scales are resolved.

It turns out that the householder's perspective and priorities are similar to those that emerge from a public health perspective. As most of the victims of poor environmental health are children under five, it makes sense to focus attention on where they spend the most time, which is at home. We have also seen that water access *at the household scale* is critical to increasing the quantity of water used to improve hygiene. The construction of public toilets half a kilometre from the house may offer limited improvement in convenience and dignity of some adults, but will not significantly improve the health of children in the community, who will rarely if ever use such services. Improving the quality of river water by controlling the quality of the wastewater discharge may be of ecological benefit, but it makes no difference to a household's health *unless* that improvement is translated into improved drinking water quality at the household scale. The public health priority for environmental improvement thus becomes the household, followed by its immediate neighbourhood.

These differing scales of the urban environment are reflected in the structure of environmental service provision. Water supply, sewerage, storm drainage and solid waste management all involve the flow of mass between the individual household and the larger environment. Figure 5.6 shows the superimposition of a water supply system upon the scales of the environment shown in Figure 5.5, and similar sketches can be prepared for the other environmental services such as sewage disposal and solid waste management. Table 5.3 also shows the physical infrastructure associated with each level of aggregation.

Table 5.3 *Scales of the urban environment and water, sewerage and drainage and solid waste-related infrastructure issues*

Service	Household	Street	Neighbourhood	Ward	City	Environs
Water	house connection or vendor	street main or public tap	secondary main	primary main	treatment works	intake
Sewerage and drainage	sewer connection	street sewer	secondary collector	primary collector	treatment works	outfall
Solid waste	collect or carry to bin	collective bin or skip	Recycling point	transfer station	Transfer/ recycling centre	landfill

Note: Recycling often occurs at each stage of the solid waste chain.
Source: Kolsky (1996) unpublished lecture notes, London School of Hygiene and Tropical Medicine

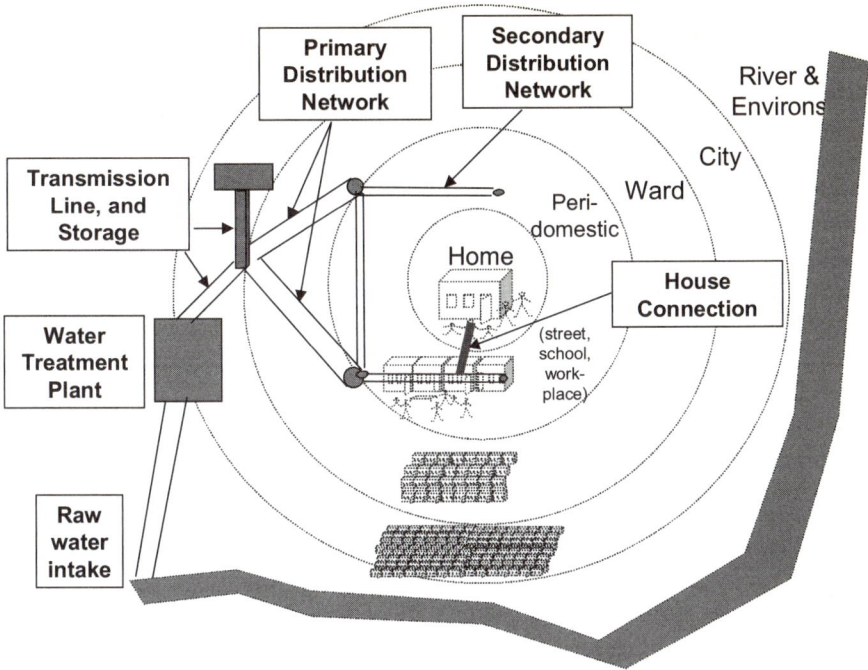

Source: Kolsky (2006) unpublished lecture notes, London School of Hygiene and Tropical Medicine

Figure 5.6 *Scales of water supply infrastructure matched to the urban environment*

The perspective of the service provider (e.g. the sewerage utility, the water supply engineer, and so on) is often different from that of the householder and public health specialist. They will look at the same system, but usually focus on different concerns, as shown in Figure 5.7. The highest priority of the water engineer is often the intake and central treatment works. Their attitude is, if this link of the chain fails, all the rest will fail.

Construction of centralized works often involves substantial amounts of financial capital and technical sophistication, both of which contribute to the professional standing of the individuals involved. Primary mains are the second priority after the central works, because of the relatively large impact of the failure or inadequacy of these links in the chain. While central pipes are more expensive per metre, the majority of the cost in the distribution system is tied up in the large number of small outlying lines. Individual street mains and house connections are often at the periphery of the technical professional's vision. Indeed, in many cases the 'outermost rings' of the technical professional are virtually ignored; water is provided to a public tap to serve a neighbourhood, and what happens to the water after that is not the practical concern of the

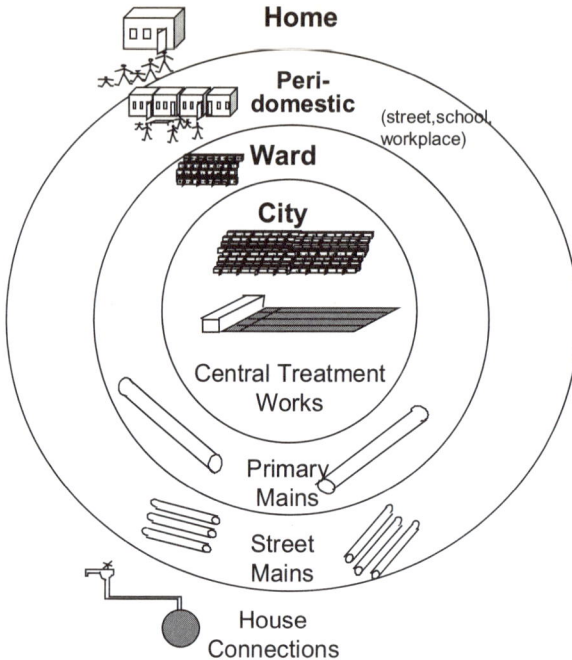

Source: Kolsky (2006) unpublished lecture notes, London School of Hygiene and Tropical Medicine

Figure 5.7 *Water Supply infrastructure and priorities, as seen by technical professionals*

utility. This focus on centralized works is even less appropriate for sanitation infrastructure, where large amounts of resources may easily be spent upon centralized waste treatment works of marginal public health benefit in comparison with the provision of basic household access.

Unfortunately, the poor live in the outermost 'marginal' rings. These 'rings' often represent very clearly definable neighbourhoods which are rarely laid out for service provision, have dubious tenure relationships, and are thus not a priority for service delivery. They are therefore easily excluded, yet it is here where the battle for urban public health is won or lost.

Boundaries and their impact on services

We define a boundary as the limit beyond which an individual or group feels no responsibility. Boundaries can be marked by physical, legal, bureaucratic, psychological or customary limits. They can include formal district boundaries in rural areas, and ward boundaries in urban areas; they may also be as informal, but also as strong, as the sense of community around a courtyard or square.

The use of a subjective word like 'feels' in the definition seems odd, but the reason for its use is a pragmatic one. The world is filled with legal or administrative boundaries without practical meaning because those within the boundary do not feel, directly or indirectly, the consequences of their actions.

Boundaries permit the breakdown of complex problems into simpler parts, and the corresponding delegation of responsibilities and tasks. By limiting responsibilities, they become manageable, both for the person delegating and for the person performing the task. It is difficult to be responsible for everything, but we can accept responsibility within given boundaries.

Classification of boundary problems

While boundaries are a useful political, social and administrative device, they are not without drawbacks. There are, at least, four related categories of problems arising from boundaries.

Complete externalities

When the damage of my actions to others does not affect me, I have no incentive to change my behaviour. This situation is known to economists as an 'externality', and in this case, explains why wastewater treatment always lags behind drinking water as a community priority. Drinking water affects the members of a community directly, so it is in their interest to take appropriate action to ensure its quality. Sewage effluent quality affects only downstream users; upstream community members causing the problem have no direct interest in resolving it for their downstream neighbours. Government action is often required to solve such problems. Although environmental pollution is the classic example of such a boundary problem, there are also administrative examples of externalities, as described below.

Partial externalities

In some cases, I *do* care about the damage I cause outside my boundaries, but if most of that damage will occur in any event, why should I change my behaviour? Hardin describes this in his classic paper *The Tragedy of the Commons* (Hardin, 1968). The paper examines the behaviour of individual shepherds responsible for grazing on common land. Collectively, it is clear that they would be better off with fewer sheep grazing on the commons, because in the long term, overgrazing will destroy the resource. Individually, however, each shepherd is better off grazing as many of his sheep on the land as possible; after all, if he restrains his behaviour, the commons will still be destroyed by others, and he will have sacrificed in vain. Some form of social arrangement needs to be worked out, or else the commons will be finished. The difficulty of working out such social arrangements, complete with effective sanctions against those who violate them, lies at the heart of many environmental and social problems. Rubbish dumping in urban slums often falls within this category.

Badly drawn boundaries

In some cases, things could work better if boundaries were simply redrawn. Engineers often refer to drainage networks and other urban infrastructure as trees. Such trees have large 'trunk' mains, and smaller 'branch' lines, and the image of a tree effectively conveys the notion of many smaller entities combining into a larger one. Most of the length of a drainage network, and most of the cost, lies in the many small branches rather than in the trunk line. While trunk lines are certainly more expensive per metre, the total cost of a network is dominated by the smaller branches at a smaller unit cost making up the outer branches. Such lines are often technically simple and individually not expensive; added up they represent the main cost of the network.

Traditionally, centralized authorities have taken responsibility for entire drainage networks on the grounds that all drains up to the individual property boundaries are public goods. This model was developed in the industrialized world and has served quite well there. Alternative approaches, however, have emerged in the cities of the developing world. Municipal authorities there are devolving the responsibility for these smaller lines to local community groups or non-governmental organizations (NGOs). This can be done because small branches are technically simple, and can be managed better by closer supervision within the community than by the distant municipal authorities. This does not mean that central authority should disengage totally in regards to compliance and appropriateness of these activities.

Redrawing boundaries in the water sector is not a new idea; the regional water authorities in Britain (Okun, 1978) are another practical demonstration of the benefits of more rational boundary definition. The environmental economics literature (Ruff, 1970) stresses the need to draw boundaries so that benefits and costs of activities are felt by those who take the action, thus reducing the boundary problem of externalities.

The category of badly drawn boundaries includes also cases where there are gaps between boundaries (so no group feels responsible), and cases where boundaries overlap (where more groups claim the same responsibility or authority).

Boundaries as barriers

Movement of ideas and resources across boundaries is always more difficult than *within* boundaries. For example, at the start of the International Drinking Water Supply and Sanitation Decade, a need was recognized to integrate the services of water supply and 'sanitation', used here as a euphemism for excreta disposal. This made sense both in conceptual public health terms (as both are involved in the spread of faecal-oral disease) and in practical engineering terms where water-based sewerage is the main means of excreta disposal, and thus depends directly on the water supply. Communications between these services improved as intended, under the new boundaries.

Redrawing the boundaries around water supply and excreta disposal, however, caused the nominal separation of sanitary sewerage and surface water drainage even where these systems are physically interconnected. In the same street separate crews clean these 'separate' drains, requiring separate transport and equipment at different times. Because of the new administrative boundaries, these crews no longer communicate, and no longer share access to the same resources.

Other types of boundary create other communications problems in water, sanitation and other sectors. Because external support agencies do not wish to become entangled in an open-yended commitment to paying recurrent costs, they have traditionally limited their involvement to capital investment. This means that both international and local resources are drawn to the investment sector, to the neglect of the operational aspects. Recently created municipal development authorities created in recent decades are responsible for municipal investment, but not for the day-to-day maintenance of the infrastructure they develop. There is often a strong feeling among those running the infrastructure that those planning and designing it don't understand basic operational reality.

Boundaries define the problem considered by those with responsibilities within these same boundaries. There is, for example, an administrative boundary between street sweeping and drain maintenance. The street sweeper's boundary extends only to keeping the street clean, so sweeping sediment and debris into a drain is seen as perfectly acceptable. Similarly, the drain cleaner's boundary extends only to keeping the drain clean, so that emptying debris onto the street for subsequent pick-up by the solid waste department is also seen as acceptable. If the street sweeper returns before the solid waste department picks up the debris, then it will be swept back into the drain, thus achieving the environmental goal of perfect recycling! In this case, the real issue is 'solid waste management' which cuts across the pre-existing boundaries of street sweepers and drain cleaners.

Public and private domains as a special example of boundary problems

For a long time public health engineering has been focused on *public* domains to bring health improvements to deprived populations. This has involved construction of large-scale urban water and sanitation infrastructure. A spatial model of disease transmission in public and private domains illustrates the move away from this traditional, engineering approach to public health (Cairncross et al, 1996). Diseases transmitted on the household level have to be dealt with via interventions that reach the household level. This puts more emphasis on *private* health at household level and the need to understand decisions made and actions taken at household level and their relation to environmental health. The private domain is distinctly different from the public domain in which the intervention of a public authority is required to prevent disease transmission. Some of these spatial problems relating to this model have been discussed in the paragraphs above. This model acknowledges the importance of household practices and behaviour without ignoring the public domain.

Many studies have investigated the links between public water supplies and household contamination of stored water (Kirchhoff et al, 1985; Deb et al, 1986; Jonnalagadda and Bhat, 1995; Mintz et al, 1995; Jagals et al, 1997; Quick et al, 1999). Findings have been rather mixed. There is a growing consensus that diarrhoeal disease pathogens originating within the home, as found in household water storage vessels, are less of a threat to household health than pathogens found in source water supplies (e.g. from public wells) (VanDerslice and Briscoe, 1993). There appears to be degrees of immunity to pathogens commonly found within the household. These complexities are acknowledged and explored within the public-private domain model. The recent literature on point-of-use treatment of drinking water (Conroy et al, 2001; Hellard et al, 2001; Iijima 2001; Fewtrell et al, 2005; Clasen, 2006) seems to indicate that there are health benefits from improving drinking water quality at the household level. Further research will be required to determine if this is true for households with limited access to water.

Several studies have looked at the impact of interventions at household scale when the neighbourhood is contaminated with faeces (Cairncross et al, 1996). Others have logically argued that interventions should be targeted at those most in need. It has been suggested that children who are not breast-fed are more susceptible to diarrhoeal disease and as such may benefit more from water supply and sanitation interventions than breast-fed infants (Esrey et al, 1985, 1991). The logistical difficulties in carrying out this approach may, however, make it impracticable.

Water stress at global and household scales, a boundary point of view

At the International Conference on Freshwater in Bonn in 2001, water scarcity was attributed to growing demand and increasing pollution and waste of freshwater sources. There is confusion between water stress at the household and regional scales. Regional water stress is sometimes portrayed as the major determinant of households' access to adequate water and sanitation, as well as the prevalence of water-related diseases (UN-Habitat, 2003). The amounts of water required to meet basic needs are relatively modest. It is estimated that on a worldwide basis, agriculture accounts for about 69 per cent of annual water withdrawals; industry about 23 per cent, and domestic use about 8 per cent (Hinrichsen et al, 1997). In Africa the percentage of domestic water use is estimated to be 7 per cent, while in Asia only 6 per cent (Hinrichsen et al, 1997).

The most common measure for water stress at a national level is the Falkenmark indicator (Falkenmark et al, 1989) which estimates the amount of freshwater available per capita per year. Benchmark values for the Falkenmark indicator are less than 1700m^3 per capita per year, which indicates water stress, while less than 1000m^3 per capita per year indicates severe water stress. In Figure 5.8 the relationship between urban water access, national water stress and national gross domestic product (GDP) per capita is shown based on data from the United Nations Environment Programme (UNEP) Data Compendium (UNEP, 2002). The figure shows the counter-intuitive relationship that water stressed nations have a larger proportion of their populations with access to

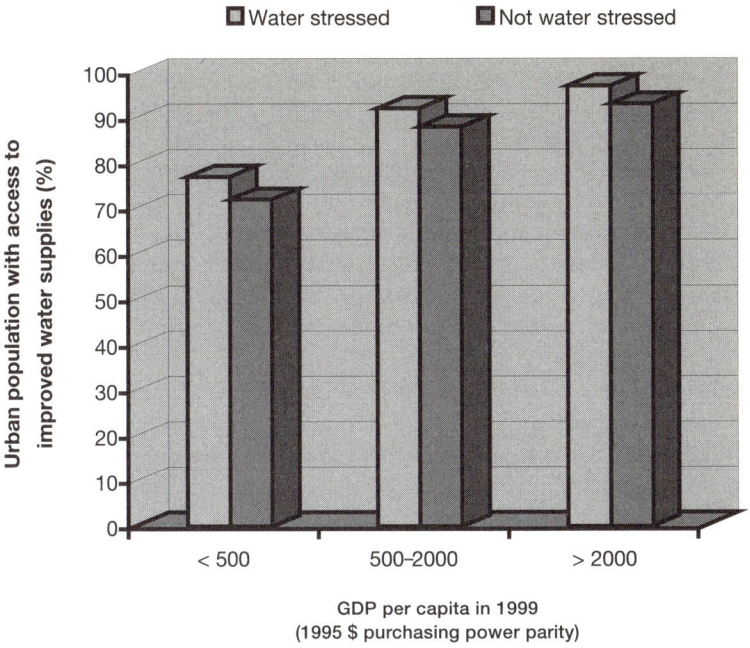

Source: UN-Habitat (2003)

Figure 5.8 *Relationship between urban water access, national water stress and national GDP per capita*

water than those nations not considered 'water stressed'. Figure 5.8 shows that it is an erroneous over-simplification to extrapolate freshwater stress at a national level to imply similar reduced access to water at the household level, regardless of the GDP per capita. Apart from a variety of other possible explanations, it is crucial to bear in mind that domestic drinking water supply and sanitation make up a very small fraction of any nation's demand for water resources.

Solutions to boundary problems?

Boundaries at different levels can be helpful in identifying problems within environmental services. Water provision by a utility to the neighbourhood level, but not to the individual household, invites the creation of an informal sector that may or may not operate as an extortionate cartel. The interface between differing levels can also be critical where the volumes in and out of the interface do not match. In the case of solid waste management, waste may be carried to a neighbourhood bin, but not picked up from that bin by the municipal service. Local drains that serve one neighbourhood by flooding their neighbour can also reflect a failure to consider the whole system. So service provision must extend to the true 'end user' in the outer boundaries to ensure that the goals of access are really achieved.

Solutions to boundary problems are rarely obvious, or else sound trite. Nevertheless, two strategies emerge:

Careful definition of boundaries, with flexible mechanisms for redefining them. Redrawing boundaries along hydrological rather than political lines has worked well in some European watersheds (Okun, 1978). This has resulted in the advocacy for an integral holistic and transboundary management system of river basins (BHS, 1998). Unforeseen problems will, however, still arise, and there must be a reasonable way to refine or redraw the boundaries in the light of experience.

Good mechanisms to identify and confront cross-boundary problems. Identification of cross-boundary problems is not so difficult, especially if a group is given the task of finding out, for example, why government services don't perform properly. *Resolution* of inter-departmental conflicts, however, is a timeless management problem (Handy, 1985). 'Interdepartmental Task Forces' can be effective, or simply a sop offered in order to appear to be doing something about the problem.

Conclusion

The problems of the poor, are suffered by the poor, and dealt with by the poor, The problems of the rich, are suffered by the public, and dealt with by the Government. (Marianne Kjellen)

Scale of the problem

At present, an estimated 2.6 billion people lack access to improved sanitation and 1.1 billion people lack access to improved water supply. While there is substantial uncertainty about the definition or measurement of these terms, there is no doubt that a very large proportion of the human race does not have what public health authorities could accept as reasonable access to water and sanitation, and thus to the prerequisites for better hygiene. The toll of this inadequate access is high in terms of health, time, money, comfort and dignity. More than 2 million people, mainly children, die every year from diarrhoeal disease, equivalent to a jumbo jet crash every two hours. Unfortunately, this burden falls almost exclusively upon the poor, and most of it could be avoided with promotion of better water, sanitation and hygiene services.

Water, sanitation, hygiene and health

It has been well established for nearly 30 years that the quantity of water that people use for personal and domestic hygiene is more important in maintaining health than the quality of the water they drink. While recent epidemiology on household water treatment justifies greater attention to this intervention, the implications for the public utility remain the same: that physical access to nearby waterpoints is a key aspect of service provision. The importance of excreta disposal and hygiene practices such as handwashing are obvious from the models mentioned and clearly visualized in the F-diagram.

Access to services

The practical interrelationship between water, sanitation and hygiene in relation to health are far more complex than the theoretical links. This makes health impacts difficult to attribute to improved services, and the measurement of health impacts unsuitable for routine project evaluation. Increasing the access to water and sanitation is widely recognized as leading to health improvements, and new targets have been set by the international community to improve worldwide morbidity and mortality.

So far, the type of technology used at household level has been used as the main proxy for access and health impact. More careful consideration shows that this idea has severe limitations, and that better indicators of access to environmental health services need to be developed. Reasonable access to services such as 'improved' water is a complex idea with many facets, encompassing not just water quality, but also its quantity, cost, operational reliability, seasonal availability and collection time and effort. Measuring access to improved services, although essential, is not straightforward. To measure progress on these targets, more powerful and sector specific survey tools will need to be developed, along with the will to use them.

Domestic water consumption increases greatly with convenient and affordable water delivery to the household, reflecting the importance of collection time as a determinant in the amount of water used. These types of proxy indicators are easier to measure than the health impacts obtained by these improved services.

Boundaries, scales of environmental challenges and access

Before they can be expanded, services have to become more efficient. Institutional boundaries are central to many environmental problems, and are closely linked to the economist's idea of 'externalities', which occur when individuals or agencies are unaffected by the consequences of their actions. In many cases, these boundaries are regarded as immutable, because of institutional resistance to change. Identifying, acknowledging and addressing the various problems associated with institutional boundaries requires time, energy and goodwill. Successful models can evolve only after trial and error, careful monitoring and evaluation and honest documentation of both success and failure.

Environmental health operates on a variety of levels. One simple spatial division is between public and domestic domains of infection. The most vulnerable groups in society in terms of mortality and morbidity (children and the elderly) spend most of their time in the domestic domain. Recent epidemiology of disease transmission and hygiene has stressed the need to create conditions in which households can manage the domestic domain more effectively, through increased water use, household sanitation and improved personal hygiene.

Regional water stress is sometimes portrayed as the major determinant of households' access to adequate water and sanitation. Household consumption in the world is below 8 per cent of the global water use, and this chapter shows that household access to water in urban areas is, on average, higher in water-stressed nations than in countries where there is no water stress, according to the Falkenmark indicator.

This chapter has also presented a range of scales in which to examine environmental challenges, varying from that of the household up through street, ward and city levels. These scales often reflect their own boundary problems, as when the capacity to collect waste at one point is not equal to the capacity to remove it to the next stage. Householders, public health experts, and infrastructure professionals view these environmental scales differently. Both householders and public health experts see the household as the natural focus for environmental service provision, while infrastructure professionals tend to focus, for good and bad reasons, upon the centralized infrastructure (treatment works, centralized pumping, and so on). Environmental service provision is a chain, and, like all chains, is only as strong as its weakest link. Improvements in environmental health infrastructure will only be significant if they lead to changes at the household level. This public health reality underscores the need for improved access to environmental services at the household level. We need reliable measures of such access if we are to improve it, and the need for better practical indicators of access to environmental services is thus critical for the sector at this time.

Disclaimer

The findings, interpretations and conclusions expressed in this article are entirely those of the authors, and should not be attributed to the World Bank.

References

Bern, C., Martines, J., Zoysa, I. D. and Glass, R. I. (1992) 'The magnitude of the global problem of diarrhoeal disease: A 10-year update', *Bulletin of WHO*, vol 70, no 6, pp705–714

BHS (1998) 'International Conference on Hydrology in a Changing Environment', University of Exeter, British Hydrological Society, www.hydrology.org.uk/exeter.html

Briscoe, J., Feachem, R. G. and Rahman, M. (1985) *Measuring the Impact of Water Supply and Sanitation Facilities on Diarrhoea Morbidity; Prospects for Case-control Methods*, World Health Organization (WHO), Environmental Health Division, Geneva

Cairncross, S. (1989) 'Water supply and sanitation: An agenda for research', *Journal of Tropical Medicine and Hygiene*, vol 92, pp301–314

Cairncross, S. (1990) 'Health impacts in developing countries: New evidence and new prospects', *Journal of the Institution of Water and Environmental Management*, vol 4, no 6, pp571–577

Cairncross, S. (1995) *Water Quality, Quantity and Health. Safe Water Environments*, Swedish International Development Cooperation Agency (Sida), Eldoret

Cairncross, S. (1999) *Measuring the Health Impact of Water and Sanitation*, WEDC, London School of Hygiene and Tropical Medicine (LSHTM), London, p2, www.worldbank.org/watsan/pdf/tn02.pdf, www.lboro.ac.uk/well/resources/fact-sheets/fact-sheets-htm/mthiws.htm

Cairncross, S., Blumenthal, U., Kolsky, P., Moraes, L. and Tayeh, A. (1996) 'The public and domestic domains in the transmission of disease', *Tropical Medicine and International Health*, vol 1, no 1, pp27–34

Cairncross, S. and Feachem, R. G. (1993) *Environmental Health Engineering in the Tropics*, John Wiley & Sons, Chichester

Clasen, T. (2006) 'Household water treatment for the prevention of diarrhoeal disease. Department of Infectious Diseases', PhD thesis, LSHTM, London, p271

Clasen, T. F. and Bastable, A. (2003) 'Faecal contamination of drinking water during collection and household storage: The need to extend protection to the point of use', *Journal of Water and Health*, vol 1, no 3, pp109–115

Clasen, T. F. and Cairncross, S. (2004) 'Editorial: Household water management: Refine the dominant paradigm', *Tropical Medicine and International Health*, vol 9, no 2, pp187–191

Clasen, T., Roberts, I., Rabie, T. and Cairncross, S. (2004) *Interventions to Improve Water Quality for Preventing Infectious Diarrhoea*, Cochrane Library, Infectious Diseases Group, John Wiley & Sons, Chichester

Conroy, R. M., Meegan, M. E., Joyce, T., McGuigan, K. and Barnes, J. (2001) 'Solar disinfection of drinking water protects against cholera in children under 6 years of age', *Archives for Diseases in Childhood*, vol 85, no 4, pp293–295

Curtis, V., Cairncross, S. and Yonli, R. (2000a) 'Domestic hygiene and diarrhoea – pinpointing the problem', *Tropical Medicine and International Health*, vol 5, no 1, pp22–23

Curtis, V., Cairncross, S. and Yonli, R. (2000b) 'Domestic hygiene and diarrhoea – pinpointing the problem', *Tropical Medicine and International Health*, vol 5, no 1, pp22–32

Deb, B. C., Sircar, B. K., Sengupta, P. G., De, S. P., Mondal, S. K., Gupta, D. N., Saha, N. C., Ghosh, S., Mitra, U. and Pal, S. C. (1986) 'Studies on interventions to prevent el tor cholera transmission in urban slums', *Bulletin of the World Health Organization*, vol 64, no 1, pp127–131

Esrey, S. A., Feachem, R. G. and Hughes, J. M. (1985) 'Interventions for the control of diarrhoeal diseases among young children: Improving water supplies and excreta disposal facilities', *Bulletin of the World Health Organization*, vol 63, no 4, pp757–772

Esrey, S. A., Potash, J. B., Roberts, L. and Shiff, C. (1991) 'Effects of improved water supply and sanitation on ascariasis, diarrhoea, dracunculiasis, hookworm infection, schistosomiasis and trachoma', *Bulletin of the World Health Organization*, vol 69, no 5, pp609–621

Falkenmark, M., Lundqvist, J. and Widstrand, C. (1989) 'Macro-scale water scarcity requires micro-scale approaches: Aspects of vulnerability in semi-arid development', *Natural Resources Forum*, vol 13, no 4, pp258–267

Feachem, R. G., Bradley, D. J., Garelick, H. and Mara, D. D. (1983a) *Sanitation and Disease; Health Aspects of Excreta and Wastewater Management*, John Wiley & Sons,

Chichester
Feachem, R. G., McGarry, M. and Mara, D. (eds) (1977) *Water, Waste and Health in Hot Climates*, John Wiley & Sons, Chichester
Feachem, R. G., Hogan, R. C. and Merson, M. H. (1983b) 'Diarrhoeal disease control: Reviews of potential interventions', *Bulletin of the World Health Organization*, vol 61, no 4, pp637–640
Fewtrell, L., Kaufmann, R. B., Kay, D., Enanoria, W., Haller, L. and Colford, J. M., Jr (2005) 'Water, sanitation, and hygiene interventions to reduce diarrhoea in less developed countries: A systematic review and meta-analysis', *Lancet Infectious Diseases*, vol 5, no 1, pp42–52
Handy, C. B. (1985) *Understanding Organizations*, Penguin Books, London
Hardin, G. (1968) 'The tragedy of the commons', *Science*, vol 162, pp1243–1248
Hellard, M. E., Sinclair, M. I., Forbes, A. B. and Fairley, C. K. (2001) 'A randomized, blinded, controlled trial investigating the gastrointestinal health effects of drinking water quality', *Environmental Health Perspectives*, vol 109, no 8, pp773–778
Hinrichsen, D., Robey, B. and Upadhyay, U. D. (1997) *Solution for a Water-Short World*, Johns Hopkins School of Public Health, Population Program, Baltimore, MA, www.infoforhealth.org/pr/m14edsum.shtml
Iijima, Y., Karama, M., Oundo, J. O. and Honda, T. (2001) 'Prevention of bacterial diarrhea by pasteurization of drinking water in Kenya', *Microbiology and Immunology*, vol 45, no 6, pp413–416
Jagals, P., Grabow, W. O. K. and Williams, E. (1997) 'The effects of supplied water quality on human health in an urban development with limited basic subsistence facilities', *Water South Africa*, vol 23, no 4, pp373–378
Jonnalagadda, P. R. and Bhat, R. V. (1995) 'Parasitic contamination of stored water used for drinking/cooking in Hyderbad', *South-East Asian Journal of Tropical Medicine and Public Health*, vol 26, no 4, pp789–794
Kirchhoff, L. V., McClelland, K. E., Do Carmo Pinho, M., Araujo, J. G., De Sousa, M. A. and Guerrant, R. L. (1985) 'Feasibility and efficacy of in-home water chlorination in rural North-eastern Brazil', *Journal of Hygiene*, vol 94, no 2, pp173–180
Kolsky, P. J. (1993) 'Diarrhoeal disease: Current concepts and future challenges. Water', *Transactions of the Royal Society of Tropical Medicine and Hygiene*, vol 87 (supplement no 3), pp43–46
Kolsky, P. (2002) 'Water, health and cities: Concepts and examples', Paper presented at an international Workshop on Planning for Sustainable Development: Cities and Natural Resource Systems in Developing Countries, University of Wales, Cardiff, 13–17 July, 1992
Lewin, S., Stephens, C. and Cairncross, S. (1996) 'Health impacts of environmental improvements in Cuttack and Cochin, India', Review prepared for the Overseas Development Administration by LSHTM, London
McGranahan, G., Jacobi, P., Songsore, J., Surjadi, C. and Kjellen, M. (eds) (2001) *The Citizens at Risk: From Urban Sanitation to Sustainable Cities*, Earthscan, London
Mintz, E. D., Reiff, F. M. and Tauxe, R. V. (1995) 'Safe water treatment and storage in the home. A practical new strategy to prevent waterborne disease', *Journal of the American Medical Association*, vol 273, no 12, pp948–953
Murray, C. J. L. and Lopez, A. D. (eds) (1996) *The Global Burden of Disease: A Comprehensive Assessment of Mortality and Disability from Diseases, Injuries, and Risk Factors in 1990 and Projected to 2020*, Harvard University Press, Harvard, MA
Okun, D. A. (1978) *The Regionalization of Water Management: A Revolution in England and Wales*, Applied Science Publishers, London

Quick, R. E., Venczel, L. V., Mintz, E. D., Soleto, L., Aparicio, J., Gironaz, M., Hutwagnar, L., Greene, K., Bopp, C., Maloney, K., Chavez, D., Sobsey, M. and Tauxe, R. (1999) 'Diarrhoea prevention in Bolivia through point of use water treatment and safe storage: A promising new strategy', *Epidemiology and Infection*, vol 122, pp83–90

Ruff, L. E. (1970) 'The economic common sense of pollution', *The Public Interest*, vol 19, Spring, pp69–85

Thompson, J., Porras, I. T., Tumwine, J. K., Mujwahuzi, M. R., Katui-Katua, M., Johnstone, N. and Wood, L. (2002) *Drawers of Water: 30 Years of Change in Domestic Water Use and Environmental Health – Summary*, Earthprint, www.iied.org/sarl/pubs/drofwater.html#9049IIED?to=9049IIED; www.iied.org/sarl/dow

UNEP (United Nations Environment Programme) (2002) *Global Environmental Outlook 3*, Earthscan, London

UN-HABITAT (United Nations Human Settlements Programme) (2003) *Water and Sanitation in the World's Cities*, Earthscan, London

VanDerslice, J. and Briscoe, J. (1993) 'All coliforms are not created equal: A comparison of the effects of water source and in-house water contamination on infantile diarrhoeal disease', *Water Resources Research*, vol 29, no 7, pp1983–1993

Wagner, E. G. and Lanoix, J. N. (1958) 'Excreta disposal for rural areas and small communities', *WHO monograph series*, Geneva, p39

White, G. F., Bradley, D. J. and White, A. U. (1972) *Drawers of Water*, University of Chicago Press, Chicago, IL

WHO (1983) *Minimum Evaluation Procedure (MEP) for Water Sanitation Projects*, World Health Organization, Geneva, p52

WHO (2001) *The World Health Report 2001; Mental Health: New Understanding, New Hope*, World Health Organization, Geneva, www.who.int/whr/

WHO (2004) The World Health Report 2004 – Changing History, WHO, Geneva, www.who.int/whr/2004/en/index.html

WHO/UNICEF (United Nations Children's Fund) (2000) *Global Water Supply and Sanitation Assessment 2000 Report*, www.unicef.org/programme/wes/pubs/global/global.htm

WHO/UNICEF (2005) *Water for Life. Making it Happen*, World Health Organization, Geneva, www.who.int/water_sanitation_health/waterforlife.pdf

WHO/UNICEF (2006) *Meeting the MDG Drinking Water and Sanitation Target: The Urban and Rural Challenge of the Decade*, www.childinfo.org/areas/water/pdfs/jmp06final.pdf

World Bank (1976) *Measurement of the Health Benefits of Investments in Water Supply*, The World Bank, Washington, DC

6
Poverty and the Environmental Health Agenda in a Low-income City: The Case of the Greater Accra Metropolitan Area (GAMA), Ghana

Jacob Songsore and Gordon McGranahan

Background

Introduction

> Urban environmental problems can be divided into two sets of issues, or two agendas. First, there are the items on the conventional 'sanitary' or environmental health agenda (often termed the 'brown' agenda), which have long been familiar to urbanists (Bartone et al, 1994; Leitmann 1994). These include unsanitary living conditions, hazardous pollutants in the urban air and waterways, and accumulations of solid waste. Such problems have many immediate environmental health impacts and tend to fall especially heavily on low-income groups (see, for instance, Bradley et al, 1991; McGranahan, 1991; Hardoy et al, 1992). Secondly, there are the items within the more recent 'green' agenda promoted by environmentalists (mostly from high-income countries): the contribution of urban-based production, consumption and waste generation to ecosystem disruptions, resource depletion and global climate change. Most such problems have impacts that are more dispersed and delayed, and often threaten long-term ecological sustainability. (McGranahan and Satterthwaite, 2000)

A conceptual model that seeks to integrate these polar opposite perspectives is the Urban Environmental Transition (UET) model (McGranahan and Songsore, 1994; McGranahan et al, 1996; and McGranahan et al, 2001). The UET model argues that the nature of environmental problems in cities changes with levels of economic development. Cities have two general categories of environmental risk to human well-being. There are those that directly affect health and those that operate indirectly by impairing ecosystems that humanity depends on (Smith and Lee, 1993). As a general rule, the urban environmental hazards that are immediately health threatening are those found in poor homes, neighbourhoods and workplaces of cities in Africa and other developing countries. Among these are inadequate water supply and sanitation facilities, poor and crowded housing, smoky kitchens, insect infestation, contaminated food, piles of uncollected garbage and bad drainage (McGranahan and Songsore, 1994).

By contrast, as the home and neighbourhood problems decline in importance as one moves from low- to middle-income cities, the most extreme examples of city-wide problems, such as ambient air pollution and polluted rivers, become dominant. The wealthiest cities of the northern hemisphere have taken measures to reduce the home, neighbourhood and city-wide pollution problems. However, through over-consumption, the wealthy cities of the northern hemisphere draw more heavily on the global resource base and generate a disproportionate share of worldwide pollution, accounting for a large share of global warming, acid rain and depletion of the ozone layer. These tend to have less direct and delayed effects on human health. 'The logic behind the transition is that the wealthy use more resources and create more waste, but also use part of their wealth to avoid personal exposure to unpleasant and hazardous pollutants' (Kjellen and McGranahan, 1998, pp67–68; Songsore, 2000, pp4–5).

Closely related to the above model is the concept of environmental risk transition developed by Smith and Lee (1993). Economic development is associated with a major reduction of some kinds of ill health and increases in others. 'The historically high "traditional" sources of ill health associated with rural poverty, such as infectious and parasitic diseases, trend downward with economic development, although at varying rates. As traditional diseases decline as causes of death, "modern" diseases like cancer and heart diseases take over' (Smith and Lee, 1993). These two frameworks provide the logical basis for Omran's model of epidemiological transition (Omran, 1971).

City context

The Greater Accra Metropolitan Area (GAMA), which serves as both Ghana's national capital and major industrial hub as currently defined, includes the Accra Metropolitan Area (AMA), Tema Municipal Area and Ga District. These three urbanized districts have grown into a major urban agglomeration, not only in a physical sense but also economically and functionally, even if not yet administratively. This greater metropolitan region, which had a combined population of 1.3 million in 1984, had a projected 1990 population of 1.7 million. Its population, according to the more recent 2000 census count, now stands at over 2.7 million people. The estimated population for 2005 is 3.2 million which is projected to increase to 4 million in 2010. It ranks as the single largest urban agglomeration in Ghana, with a growing tendency to being a primate city, as its growth is much faster than the next largest city of Kumasi. GAMA covers a land area of 420km^2. It has just over 10 per cent of the total population of Ghana and yet it has the most diversified economy in the country contributing between 15 and 20 per cent of gross domestic product (GDP). About 32 per cent of the country's manufacturing industries are located in the metropolis. Apart from the concentration of modern manufacturing industries, it also serves as the commercial, business, educational and cultural centre of the country (Benneh et al, 1993).

The rapid population growth and physical expansion of GAMA have been occurring in an uncertain economic environment, characterized by a near decade of economic decline (in the mid-1970s and early 1980s) which has been followed by nearly two decades of structural adjustment. While reversing the economic decline, the era of structural adjustment has so far failed to restore economic welfare to its pre-crisis levels (Songsore and McGranahan, 1993). A large number of households within the metropolis are living in sub-standard housing and overcrowded conditions without the resources for decent shelter. Increasing economic activity is also generating greater industrial, commercial and municipal wastes. The above socio-economic and demographic context poses a great environmental health challenge which negatively impacts on the quality of life of its population even though at the other end of the spectrum, its impact on global sinks is negligible.

Objectives

This chapter locates GAMA, a poor city in a low-income developing country, within the UET debate by presenting the environmental profile of the city together with existing intra-urban differentials by wealth group. This is followed by an evaluation of the local health impacts of these environmental conditions for the various wealth groups with respect to the monitored health conditions. These monitored health problems include prevalence of diarrhoea and acute respiratory infection (ARI) for children under six and cough symptoms for principal women homemakers. A companion survey on mortality within the city provides further supportive evidence on health differentials within the city. The chapter also attempts to assess the regional ecological impacts of city production, consumption and waste generation. The chapter argues that for low-income cities such as GAMA:

> *The immediate environmental threats for the residents of these cities are not long-term global warming, cumulative exposure to carcinogens, or even decade-long desertification, but rather the life and death immediacy of malaria, respiratory illness, and diarrhoea. Their threats are derived in part from household environments characterized by indoor air pollution, a bug-filled outdoors, near-the-door faeces, and far-from-the-door water. There are also the dangers connected with the use of insect sprays, uncontrolled sewage, and ambient air pollution.* (Kates, 1994, p1; also see Songsore and McGranahan, 1996)

Evidence in support of the UET model is visible not only between cities at different levels of economic development, but also within cities, and between different wealth groups, as will be demonstrated for GAMA.

Irrespective of the gender of the head of household, women play a dominant role in environmental management within the household. Overall, 95 per cent of all households in GAMA have women as the principal homemaker, with a mere 5 per cent of all households surveyed having a male as the principal homemaker. Women as a group therefore bear the main responsibility of managing the household environment. 'But on account of their class location, they may be confronted with an entirely different array of environmental hazards, and rich women may be in a position to transfer these burdens to poor working class women and men hired by the household' (Songsore and McGranahan, 1996; see also Songsore and McGranahan, 1998). The environmental burdens women endure are therefore mediated by wealth of the household, as subsequent sections illustrate.

This study draws its results from a city-wide survey of households in GAMA whose detailed methodologies have been presented elsewhere (Benneh et al, 1993; Songsore and McGranahan, 1993; Songsore et al, 1997). It was part of a larger comparative study of three cities, Accra, Jakarta and São Paolo with Accra being the poorest and São Paolo the wealthiest. In 1990, just before the survey, Ghana had a gross national product (GNP) per capita of about US$390, while Brazil's GNP per capita was US$2680. Indonesia's GNP was closer to the boundary between low-income and middle-income countries, with a per capita GNP of US$570. The results reflected a shift from local, household level problems for Accra to city-level problems for São Paolo, thus providing the initial empirical basis for the development of the UET model (McGranahan and Songsore, 1994).

Intra-urban differentials in access to environmental services and exposure to environmental hazards

Statistics summarizing environmental conditions are presented below for five wealth groups of roughly equal size – about 20 per cent of the sample households each. These wealth categories were constructed on the basis of an index computed from household ownership of consumer durables (Benneh et al, 1993). 'While this grouping is intended to be relevant to discussions on class and poverty, there is no suggestion that the groups constitute classes, or that poverty is defined by the absence of consumer durables' (Songsore and McGranahan, 1996, p17; see also Songsore and McGranahan, 1998).

Although average urban incomes in GAMA are higher than those in other urban centres in Ghana, and although most poor people live in rural areas, by international standards, poverty in GAMA is widespread and more recent evidence from national statistics shows that it is increasing (Ghana Statistical Service, 1995). For example, poverty in the metropolis more than doubled between 1988 and 1992, up from 9 per cent to 23 per cent, while the depth of poverty increased from 2 per cent to 6 per cent. This is based on arbitrary poverty lines set by the World Bank (World Bank, 1995, p44). More recent data suggest that the metropolis has, however, made the greatest gains in poverty reduction between 1991 and 1992 and 1998 and 1999 (Ghana Statistical Service, 2000).

Although structural adjustment in Ghana was initiated in response to economic crisis and has been accompanied by a moderate recovery, at least as measured by standard economic indicators, these benefits have not been equally shared by all strata of urban society. Many poor households have suffered as a result of retrenchment and jobless growth, while the economic benefits have accrued primarily to the relatively well-off (Songsore and McGranahan, 2000, p2). For example, recorded formal sector employment experienced a drastic decline from 464,300 in 1985 to 186,300 in 1992 (UNDP, 1997; Songsore, 2003b). In 1980, the ratio of workers in the informal sector of the Ghanaian economy to workers in the formal sector was 2:1. By 1990, the ratio was 5:1 and it has apparently grown since 1990, along with growing poverty in the informal sector, notwithstanding the rather rosy picture painted about poverty incidence in GAMA with rather contrived poverty lines (Maxwell, 1999; Songsore, 2003b). Poverty and marginality within the city has become structural with a growing number of youth who have nothing to offer to the globalizing and liberalized economy of GAMA 'except to add to the growing problem of street children, child prostitution, child labour, urban violence and the drug/criminal economy' (Songsore, 2003b, p25). In this respect, the environmental conditions have grown worse for the bulk of GAMA's population residing in the low-income areas. Even the relatively well-off in the peri-urban zone have not been adequately served as housing development goes ahead of planning and the delivery of services.

This section reviews intra-urban differentials in access to environmental services or amenities that promote good health and also intra-urban differentials in exposure to environmental hazards that induce ill-health.

Inequalities in access to environmental services

This sub-section reviews some of the most relevant intra-urban inequalities with regards to access to environmental services such as potable water and sanitation and to a lesser extent sullage and solid waste disposal facilities.

The importance of an adequate supply of potable water and good sanitation for health is now well established (Lindskog and Lundqvist, 1989; Cairncross, 1990; Hardoy et al, 1992). Table 6.1 shows a close relationship between wealth and access to potable water and sanitary services. Most households surveyed rely on the piped system for their water supply but the distribution pattern is highly uneven and service delivery erratic throughout the system, especially in low-income areas and new developments in the peri-urban zone. About 35 per cent of households had access to in-house piping, 24 per cent private standpipe, with another 8 per cent relying on communal standpipe as their drinking water source. As much as 28 per cent of households depended on

Table 6.1 *Access to water and sanitary services by wealth quintile of household (%)*

A Water source	Wealth quintile				
	1-Poorest col. %	2 col. %	3 col. %	4 col. %	5-Wealthiest col. %
In-house piping	6	17	26	49	78
Private standpipe	16	28	35	30	12
Communal standpipe	21	8	6	4	3
Vendor	49	41	29	15	7
Other	7	6	4	3	1
Total	100	100	100	100	100
(Sub-sample size)	(205)	(187)	(210)	(200)	(198)
B Toilet facility	1-Poorest col. %	2 col. %	3 col. %	4 col. %	5-Wealthiest col. %
Flush toilet	7	17	30	47	77
Pit/KVIP latrine	72	54	45	24	12
Pan latrine	15	25	22	28	8
Other/none	6	4	2	2	3
Total	100	100	100	100	100
(Sub-sample size)	(205)	(187)	(210)	(200)	(198)
C Toilet sharing	1-Poorest col. %	2 col. %	3 col. %	4 col. %	5-Wealthiest col. %
Not shared	6	14	18	31	65
Share with ≤10 hh	20	21	28	39	19
Share with >10 hh	69	62	51	28	12
No response	5	3	2	3	5
Total	100	100	100	100	100
(Sub-sample size)	(205)	(187)	(210)	(200)	(198)
D Outdoor defecation*	1-Poorest col. %	2 col. %	3 col. %	4 col. %	5-Wealthiest col. %
No	53	74	72	77	88
Yes	48	26	28	13	12
Total	100	100	100	100	100
(Sub-sample size)	(205)	(187)	(210)	(200)	(198)

Note: *Refers to reported outdoor defecation by neighbourhood children. Data are from 1991/92.

Source: Songsore and McGranahan (1993)

the informal water vendor for their daily water supply. Perhaps the most disadvantaged households were the 4 per cent of GAMA residents whose main source of water supply was from open waterways, rainwater collection, wells and other private sources.

Given the widespread practice of unhygienic water handling and storage in deprived low-income areas, it is not enough to focus on bringing 'water to the tap' as what is happening 'between the tap and the mouth' is also critical in determining health outcomes from water use (Lindskog and Lundqvist, 1989, p16; Benneh et al, 1993, p20).

While most rich households have in-house piping, typically connected to overhead storage containers, the poorest and most deprived households rely mostly on water vendors, communal standpipes and other less efficient water supply sources (Table 6.1A). These require in-house storage of water in drums and other containers that easily become contaminated as shown by the results of the physical tests of water quality (Benneh et al, 1993, pp20–26). The use of communal dip cups for drinking water also encourages drinking water contamination.

Sanitation management which is closely linked to access to water supply also has very critical health implications. About 36 per cent of households in GAMA have access to flush toilets; 41 per cent use pit latrines (some of which are Kumasi Ventilated Pit Latrines); 20 per cent use bucket or pan latrines; and 4 per cent have access to no toilet (Songsore and McGranahan, 1993, p20).

As Table 6.1B, C, and D show, inequalities in access to toilet facilities is related to household wealth as pit latrines are popular with low-income households, while flush toilets (both sewered and septic) dominate higher wealth groups.

Inadequate sullage and solid waste disposal is a common feature in the metropolitan area, although inequalities still exist between the wealthy few and the bulk of the urban population. These inadequacies provide the ecological niches for the breeding of vectors and parasites, and the transmission of disease. About 46 per cent of all household sullage is discharged through open separate drains, with yard and street dumping each accounting for 25 and 13 per cent, respectively. Only 3 per cent of all households dispose of their sullage through the sewerage system or septic tanks, with another 7 per cent served by closed separate drains. Most of the limited cases of best practice with regards to sullage disposal occur in wealthy households.

In terms of visibility, garbage or solid waste management remains one of the most intractable problems within GAMA although its health implications may not be as serious as that of the water sanitation complex. Residential domestic waste is estimated to form a greater proportion of all sources of solid wastes produced in Ghanaian urban settlements. Currently, the Waste Management Department of AMA, for example, is only capable of collecting 60 per cent of the more than 900 tonnes per diem of refuse and about 300 tonnes of night soil. The position is marginally better for Tema District but may be a lot worse in Ga District (Benneh et al, 1993). Garbage or solid waste disposal is dominated by open dumping of garbage, which is practised by 83 per cent of all households; only 11 per cent of the metropolitan population, often the wealthy, benefit from home collection service (Songsore and McGranahan, 1993, p31). Of late, garbage collection service has improved somewhat in AMA and Tema District although less so in Ga District within the peri-urban zone.

Inequalities in exposure to environmental hazards and pests

The inequalities in access to environmental services indicated above tend to correspond to other types of intra-urban differentials in exposure to insect pests and the way these pests are controlled. Apart from malarial mosquitoes, which are cosmopolitan in distribution with slightly higher concentrations in the peri-urban zone and in marshy areas, all other pests investigated such as flies, cockroaches and rats have a higher concentration in poor households and yet it is in these poor households that insect control methods are less than adequate (Table 6.2). For example, the mosquito coil, which is very popular in low-income households, has been associated with ARI in children under six years from the results of the survey.

Other hazards that exhibit similar patterns by their concentration in poor households include indoor air pollution from the use of inefficient fuels such as woodfuel and charcoal; the use of a cooking hut; crowding in sleeping rooms; and leaking roofs (Table 6.3). All these factors are implicated in ARI prevalence in children under six or are related to high doses of carbon monoxide and respirable particulate inhalation for women when cooking with charcoal and woodfuel respectively.

The results from our analysis of environmental risk factors also showed ecological or area-based variations in these environmental conditions, more especially at the level of the eight major residential sectors (identified for sampling purposes) which were subsequently collapsed to six for analysis. Poorer residential zones had the worst environmental conditions even though the patterns were less sharp as these residential sectors were not always clearly segregated by class or wealth (Benneh et al, 1993; Songsore and McGranahan, 1993; Songsore, 1999). Age of buildings and level of planning were other confounding variables.

These inequalities reflect in part the class character of structural adjustment policies as the poor have carried a disproportionate share of the burden. The reliance the structural adjustment programme has placed on market mechanisms, especially cost recovery through public service user fees, has tended to inhibit environmental improvement for the poor (Songsore and McGranahan, 2000). Results of a more recent study on the state of environmental health of GAMA show that in a number of respects, especially with regards to water supply, sanitation, hygiene, and access to shelter, conditions have grown worse since the time of the initial survey as the overall population increased from about 1.7 million in 1990 at about the time of the survey to 2.7 million at the last census count in 2000 (Songsore et al, 2005). This, along with physical expansion of the built-up area of GAMA, has not been accompanied by commensurate expansion of services. A World Bank loan for water supply expansion, which

Table 6.2 *Relationship between wealth and pests, and pest control methods (%)*

A Flies in kitchen at time of interview	Wealth quintile				
	1-Poorest col. %	2 col. %	3 col. %	4 col. %	5-Wealthiest col. %
None	4	7	11	21	49
Few	43	60	65	64	47
Many	32	22	14	12	4
Very many	21	11	10	4	2
Total	100	100	100	100	100
(Sub-sample size)	(205)	(187)	(210)	(200)	(198)
B Cockroaches in house	1-Poorest col. %	2 col. %	3 col. %	4 col. %	5-Wealthiest col. %
Never	5	10	9	8	9
Occasionally	12	16	20	15	31
Often a few	28	31	26	34	30
Often many	66	54	46	44	30
Total	100	100	100	100	100
(Sub-sample size)	(205)	(187)	(210)	(200)	(198)
C Rats enter house	1-Poorest col. %	2 col. %	3 col. %	4 col. %	5-Wealthiest col. %
Never	10	16	23	25	40
Occasionally	19	31	21	26	33
Often	16	16	22	19	12
Every night	56	38	34	30	15
Total	100	100	100	100	100
(Sub-sample size)	(205)	(187)	(210)	(200)	(198)
D Percentage using selected insect control methods	1-Poorest sub-sample %	2 sub-sample %	3 sub-sample %	4 sub-sample %	5-Wealthiest sub-sample %
Full screening	22	37	52	70	82
Mosquito nets	6	8	10	7	9
Mosquito coils	58	47	51	44	26
Aerosol pesticides	14	27	43	51	68
Pump pesticides	6	9	12	12	16
Traditional methods	7	3	4	2	2
At least one Total does not add	81	84	93	97	96
(Sub-sample size)	(205)	(187)	(210)	(200)	(198)

Source: Songsore and McGranahan (1996)

Table 6.3 *Relationship between wealth and indoor air and housing problems*

A Principal cooking fuel	Wealth quintile				
	1-Poorest col. %	2 col. %	3 col. %	4 col. %	5-Wealthiest col. %
Do not cook	1	2	1	0	2
Woodfuel	22	7	5	2	3
Charcoal	74	81	83	66	35
Kerosene	3	8	8	8	4
LPG	0	1	1	21	47
Electricity	0	2	2	3	11
Total	100	100	100	100	100
(Sub-sample size)	(205)	(187)	(210)	(200)	(198)
B Cooking locations include	1-Poorest sub-sample %	2 sub-sample %	3 sub-sample %	4 sub-sample %	5-Wealthiest sub-sample %
Separate kitchen	10	14	23	51	77
Multipurpose room	12	11	12	7	4
Communal room	2	4	4	4	1
Cooking hut	21	10	12	5	4
Verandah	27	33	29	20	17
Other/open air	59	53	55	45	27
Total does not add					
(Sub-sample size)	(205)	(187)	(210)	(200)	(198)
C M^2/person in most crowded sleeping room	1-Poorest col. %	2 col. %	3 col. %	4 col. %	5-Wealthiest col. %
<2	16	13	9	4	7
2–4	55	56	54	51	27
4–6	16	17	22	25	27
6+	12	14	15	20	39
Total	100	100	100	100	100
(Sub-sample size)	(205)	(187)	(210)	(200)	(198)
D Problems with building	1-Poorest sub-sample %	2 sub-sample %	3 sub-sample %	4 sub-sample %	5-Wealthiest sub-sample %
Dampness	8	6	7	8	5
Leaking roof	67	49	40	32	23
Mildew/mould	8	6	4	4	3
Total does not add					
(Sub-sample size)	(205)	(187)	(210)	(200)	(198)

Source: Songsore and McGranahan (1996)

could have brought some relief to the water sector, has stalled because of donor pressure on government to privatize the urban water system and civil society agitation against privatization.

Environment and health with regards to monitored ill-health problems

The results of the monitored children's diarrhoea and respiratory problems and the respiratory problems of the principal homemaker (often women) suggested the following classification of environmental risk factors:

- socio-economic variables;
- access variables;
- service efficiency variables (related to access variables);
- ecological/vector prevalence variables;
- behavioural variables (Songsore and McGranahan, 1993).

The prevalence of diarrhoea and the respiratory infection among children under age six and respiratory problem symptoms in the principal female homemakers showed variations across these variables.

Environmental risk factors and diarrhoea among children under six

According to classification of risk factors or health indicators, the socio-economic variables implicated in children's diarrhoeal prevalence in a bivariate analysis include wealth status of household and education of principal homemaker (often women) (Table 6.4).

The greatest number of environmental risk factors implicated in diarrhoea prevalence of children related to access to environmental services such as adequate water supply and sanitation facilities (Table 6.5).

Tables 6.6 and 6.7 also related diarrhoeal prevalence to efficiency of environmental services and crowding in the use of these services respectively.

Table 6.8 summarizes the prevalence of childhood diarrhoea among households with different behaviour, such as closure of water storage container, outdoor defecation of neighbourhood children, handwashing practice, covering of prepared food and cleaning of toilet.

One could describe the diarrhoeal problem as reflecting a complex of environmental risks in those households with a high prevalence of the disease (Table 6.9), while ecology or residential sectors based on the survey strata effectively structure those risks (Table 6.10). There is a clear difference in diarrhoeal morbidity between middle- and high-class areas on the one hand and most of the poor areas on the other.

Even though most risk factors are correlated, results of a logistic regression analysis indicate that a large number of variables remain significant. Table 6.11

Table 6.4 *Socio-economic characteristics and children's diarrhoea prevalence (%)*

Category	Size of sub-sample	Two-week prevalence of diarrhoea (%)
A Wealth quintile		
1 Poorest	130	22.3
2	95	11.6
3	107	12.2
4	114	9.7
5 Wealthiest	91	8.8
B Education of principal homemaker		
No formal education	150	16.0
Some primary	279	15.1
Some secondary	108	5.6

Source: Songsore and McGranahan (1993)

shows the approximate relative risks of the high risk environmental factors monitored for diarrhoeal prevalence.

Environmental risk factors and respiratory problems among children under six and principal homemakers

The survey collected information from the principal female homemaker on respiratory problems among children under six and on the respondent herself. For children, the questions were designed to identify ARI. For the principal homemakers, tracer symptoms were employed (e.g. sore throat, persistent cough) which would be expected to identify a broad range of respiratory problems, including possible chronic conditions. Tables 6.12 and 6.13 summarize some major risk factors from bivariate analysis for children's ARI and adult women's respiratory problem symptoms respectively. For both children and women the range of risk factors is not as diversified as those of childhood diarrhoea. These results are to a large extent consistent with other studies which are more epidemiological in design.

Table 6.14 shows the approximate relative risk of environmental factors monitored for children's ARI and respiratory problem symptoms among adult women. These results are generally consistent with existing data on morbidity in Ghana, which suggest that the pattern of morbidity differs slightly between rural and urban communities, with environment-related diseases, especially fever presumed to be malaria topping the list in both rural and urban communities. This is not surprising, as about 90 per cent of the world's

Table 6.5 *Access to environmental services and children's diarrhoea prevalence (%)*

Category	Size of sub-sample	Two-week prevalence of diarrhoea (%)
A Principal drinking water source		
In-house piping	162	6.8
Private-standpipe	139	10.1
Communal standpipe	50	42.0
Vendor	165	13.9
Other	21	14.3
B Number of taps in home		
None	375	16.3
One	53	5.7
More than one	108	7.4
C How water is carried home		
Hand carried	85	24.7
Head carried	170	17.7
Other (e.g. piped)	280	7.5
D Location of water source		
In-house compound	266	7.1
Out of house	270	19.6
E Access to toilet		
Flush	164	6.7
Pit latrine	241	17.0
Pan latrine	113	15.9
Other/none	19	10.5
F Pay for use of toilet		
No	159	6.9
Yes	344	16.9

Source: Songsore and McGranahan (1993)

malaria burden is estimated to occur in Africa. The 10 most common health problems reported at outpatient facilities, in descending order of importance, include malaria, upper respiratory tract infections, diarrhoea, skin diseases, accidents and intestinal worms. In Ghana, as in other developing countries in the Tropics, malaria and respiratory and diarrhoeal diseases are all major childhood killer diseases both in urban and rural settlements. Urban slum and squatter settlements are above all endemic areas of infectious hepatitis, cholera and typhoid (Songsore, 2003, p283–284).

Table 6.6 *Efficiency of environmental service and children's diarrhoeal prevalence (%)*

Category	Size of sub-sample	Two-week prevalence of diarrhoea (%)
A Water storage containers		
None	24	0.0
Pot	37	29.7
Bucket	82	13.4
Pigfoot container	73	16.4
Jerrycan	48	18.8
Barrel	245	11.4
Overhead tank	28	3.6
B Regular water interruptions		
No	99	6.1
Yes	438	15.1

Source: Songsore and McGranahan (1993)

Table 6.7 *Crowding and children's diarrhoeal prevalence (%)*

Category	Size of sub-sample	Two-week prevalence of diarrhoea (%)
A Sharing a toilet		
Not shared	118	7.6
Share with <10 households	137	8.8
Share with >10 households	261	18.2
B There are times when it is inconvenient to use toilet		
No	319	7.2
Yes	199	23.6

Source: Songsore and McGranahan (1993)

The disease burden has hardly changed over the years. Infectious diseases are still dominant, although chronic diseases have also become ascendant, thereby conflating the disease burden, especially for the urban poor.

Chronic non-infectious disease burden

Available studies would seem to suggest that the urban populations in developing countries are suffering the 'worst of both worlds' in their mortality profile as

Table 6.8 *Hygiene behaviour and children's diarrhoeal prevalence (%)*

Category	Size of sub-sample	Two-week prevalence of diarrhoea (%)
A Water storage container closure		
No container	17	0.0
Open	94	27.7
Closed	426	10.8
B Neighbourhood children defecate outdoors		
No	386	8.8
Yes	145	26.2
C Handwashing prior to food preparation		
Not mentioned	132	22.0
Mentioned	394	10.7
D Uncovered prepared food observed		
No	465	11.8
Yes	72	23.6
E Clean toilet		
No	209	19.6
Yes	311	9.3

Source: Songsore and McGranahan (1993)

Table 6.9 *Number of risk factors a household faces, and children's diarrhoea prevalence*

Number of high risk conditions	Size of sub-sample	Two-week prevalence of diarrhoea (%)
None	19	0.0
One	50	0.0
Two	133	3.0
Three	119	14.3
Four	77	14.3
Five	68	23.5
Six	21	57.1
Seven	13	69.2

Source: Songsore and McGranahan (1993)

Table 6.10 *Residential sector and children's diarrhoeal prevalence (%)*

Residential sector	Number in sub-sample	Two-week prevalence of diarrhoea (%)
RF	29	24.1
HDLCS	274	13.5
HDIS	81	25.9
MDIS	61	4.9
MCS	70	4.3
HCS	22	4.5
Total sub-sample	537	13.4

Note: RF = Rural Fringe; HDLCS = High Density Low Class Sector; HDIS High Density Indigenous Sector; MDIS = Medium Density Indigenous Sector; MCS = Middle Class Sector; HCS = High Class Sector.

Source: Songsore and McGranahan (1993)

Table 6.11 *Approximate relative risk of environmental factors with respect to diarrhoea among children under age six*

Environmental factor	Approximate relative risk
Pot used for storing water	4.3
Water interruptions occur regularly	3.1
Toilet shared with more than five households	2.7
Prepared food purchased from vendor	2.6
Water stored in open container	2.2
Neighbourhood children defecate outdoors	2.1
Many flies in kitchen during interview	2.1
Hands not always washed before food is prepared	2.0

Source: McGranahan and Songsore (1994); also see Benneh et al (1993)

they are confronted by infectious diseases arising from their own poverty and the chronic diseases associated with industrialization and urbanization (Smith and Lee, 1993, p179; Stephens, 1996; Songsore, 2000). These risk overlaps are particularly common in urban slum areas and deprived communities in GAMA where modern risks have begun to rise at a time when traditional risks are still significant although declining. As shown in Table 6.15, a recent study of Accra that examined mortality differentials by age for infectious, respiratory and circulatory diseases suggests that those living in socio-economic and environmental deprivation are at a consistently higher risk of death, for each cause group and for adults as well as children (Stephens et al, 1994; Stephens, 1996, p123).

Of the many factors that combine to degrade health, poverty stands out for its overwhelming role. It is therefore not surprising that the World Health

Table 6.12 *Environmental risk factors and children's acute respiratory infection prevalence*

Category	Size of sub-sample	Two-week prevalence of ARI symptoms (%)
A Household uses cooking hut		
No	481	10.6
Yes	63	20.6
B M^2/person in the most crowded sleeping room		
>4	388	13.9
4+	139	5.8
C Many flies in food preparation area		
No	379	9.0
Yes	165	18.2
D Use mosquito coils		
No	284	8.8
Yes	260	15.0
E Children present during cooking		
Rarely present	402	9.0
Usually present	137	20.4
F Principal homemaker has respiratory symptoms		
No	443	8.1
Yes	101	27.7

Source: Songsore and McGranahan (1993)

Organization (WHO) has called poverty the world's biggest killer (WHO, 1995; World Resources Institute, UNEP, UNDP, WB, 1998). Statistically, not only does being poor increase one's risk of being ill, but poor people are also more likely to live in unhealthy environments and work under poor environmental and hazardous conditions that also induce ill-health. Ill-health in turn lowers productivity and income, which in turn drives the poor into greater poverty and increased vulnerability to disease (Songsore and Iddrisu, 1999, p17).

Table 6.13 *Environmental risk factors and prevalence of respiratory problem symptoms among (female) principal homemakers*

Category	Size of sub-sample	Two-week prevalence of respiratory problem symptoms (%)
A Roof leaks		
No	559	16.1
Yes	352	21.3
B Use cooking unit		
No	816	17.3
Yes	94	25.6
C Open air cooking		
No	476	22.1
Yes	434	13.8
D Time stove is lit		
0–3 hours	506	15.4
3+ hours	403	21.3
E Smokes cigarettes (smokers excluded from rest of table)		
No	914	18.0
Yes	41	51.2
F pump pesticides		
No	817	15.8
Yes	93	37.6
G Child has respiratory illness		
No children	405	19.5
None of children ill	445	13.7
Has ill child	61	41.0

Source: Songsore and McGranahan (1993)

The city production and consumption systems and sustainability issues

GAMA's production, consumption and waste generation systems have both impacts on health and ecology. Industrial pollution from large industries is still restricted to a few sites in the metropolis. Because heavy and chemical

Table 6.14 *Approximate relative risk of environmental factors for respiratory disease*

A Risk factors with respect to symptoms of acute respiratory problems among children under age six	
Children often present during cooking	2.6
Many flies in kitchen during interview	2.4
Less than 4m per person in most crowded sleeping room	2.3
Water supply interruptions occur regularly	2.2
Mosquito coils used	1.8
Cooking never done outdoors	1.8
Roof leaks during rains	1.7
B Risk factors with respect to symptoms of respiratory problems among principal women of the households[a]	
Pump-spray insecticide used	3.5
Water supply interruptions occur regularly	1.6
Roof leaks during rains	1.5
Cooking never done outdoors	1.4
Cigarettes smoked per day[b]	1.1

Note: All these factors were statistically significant (greater than 95 per cent confidence) in a logical regression. Control variables (for example, wealth, age of principal women, number of children under age six) were included but are not presented. The approximate relative risk, or odds ration, is the odds of having the symptoms if the factor is present divided by the odds of having the symptoms if the factor is absent.
[a] Data for only one woman per household are included. That is, the woman who was interviewed and for whom more detailed information is available, on morbidity and education, for example.
[b] The relative risk factor for this environmental factor is determined per cigarette.

Source: McGranahan and Songsore (1994); also see Benneh et al (1993)

industries which are the major polluters are few, it is not appreciated that the bulk of the workplace pollution arises from dispersed medium- and small-scale industries that have a more dispersed distribution in the city (Yankson, 1999). Uncontrolled discharge of hazardous contaminants from these industries results in build-up of toxic constituents in surface water supplies and contamination of groundwater. In addition, open dumping of solid hazardous residues including possibly biomedical wastes and emission of toxic gases pose serious environmental health risks (ECA, 1997).

The existence of clusters of obsolete industrial establishments and poorly maintained motor vehicles generates localized air pollution. On account of its small size, low level of industrialization and poverty, most African cities do not pose a threat to the global sinks as cities in the northern hemisphere do. Industrial, water and air pollution have not yet become major problems, although there is growing concern about these city-wide problems, largely

Table 6.15 *Summary age-adjusted mortality differentials between socio-environmental zones in Accra (Ghana) 1991; mortality rates (per 10,000) and Relative risks (RR)*

Zone	Circulatory diseases Rate (RR)	Infectious and parasitic diseases Rate (RR)	Respiratory diseases Rate (RR)
1 (worst)	16.4 (2.3)	9.2 (2.0)	7.6 (1.9)
2	14.6 (2.1)	14.4 (3.0)	7.5 (1.9)
3	13.5 (1.9)	10.1 (2.1)	6.5 (1.6)
4 (best)	7.0 (1.0)	4.7 (1.0)	4.0 (1.0)

Source: Stephens (1996)

arising from the spread of the agenda of the global environmentalist movement. The regional ecological footprints (EFs) of GAMA are quite serious for, as we have indicated, most urban households in GAMA are large-scale consumers of biofuels – especially charcoal but also to some extent woodfuel – as their principal sources of domestic energy for cooking and for commercial activities such as food processing and other micro-enterprises in the large and growing informal economy of the city. Apart from its health effects already monitored this is one major hidden cause of ecological degradation (Songsore, 2003a).

Almost all the charcoal and some of the woodfuel consumed in the cities and towns in Ghana derives from the selective felling of suitable trees from bio regions sometimes hundreds of kilometres away from the main urban centres of consumption. Due to low conversion rates, it has been argued that urban demand for charcoal is very wasteful of wood. It is estimated that 100kg of dry wood yields only 34kg of charcoal. This implies a conversion rate of 34 per cent (Ahwere, 1994).

Overall, about 79 per cent of the country's charcoal supply comes from the ecologically more fragile savannah zone, 15 per cent from the semi-deciduous zone and only 6 per cent from the rainforest zone. The savannah areas therefore constitute the main charcoal-producing area for the major urban settlements (HDR(G), 1999; Songsore, 2003a). Most of the species found in Northern Ghana Guinea savannah ecosystem are also used as medicines, food sources for both humans and animals and as biofuels. 'Out of the 18 categories of medical uses according to the International Classification of Health Problems in Primary Health Care (ICHPPC-2), only four categories, viz: Neoplasms, Blood Diseases, Congenital Anomalies and Perinatal Morbidity and Mortality, are not shown to have treatment from plants in the Northern Guinea Savannah ecosystem' (HDR(G), 1999, p34).Unfortunately, because of fuelwood foraging, bush burning and extensive field-shifting agriculture there is a very high possibility of losing most of these medicinal herbs and plants in this zone even before they can be fully studied by researchers (HDR(G), 1999, p35; Songsore, 2003a).

Conclusion

As the analysis presented shows, environmental health issues and the unfinished brown agenda are of overriding concern to the city dwellers of GAMA, although the regional EFs of GAMA are causing damage to sensitive ecosystems in regions far removed from the city boundary itself. Even this has been due to the relative poverty of urban dwellers, which limits their capacity to move up the energy ladder to the use of cleaner fuels such as liquefied petroleum gas (LPG) and electricity. These legitimate concerns are being lost in the preoccupation with green issues in the northern hemisphere. The relative weight of water and sanitation issues, and environmental issues in general, in the burden of disease in GAMA and other developing world cities should determine the priorities for environmental interventions, while trying to avoid the development paradigm in the northern hemisphere based on over-consumption that is not sustainable in the long run. As Harvey puts it:

> *It is vital, when encountering a serious problem, not merely to try to solve the problem in itself but to confront and transform the processes that gave rise to the problem in the first place. Then, as now, the fundamental problem is that of unrelenting capital accumulation and the extraordinary asymmetries of money and political power that are embedded in that process. Alternative modes of production, consumption and distribution as well as alternative modes of environmental transformation have to be explored if the discursive spaces of the environmental justice movement and the theses of ecological modernization are to be conjoined in a programme of radical political action.* (Harvey, 1996, p97)

Acknowledgements

The research summarized in this chapter is part of a collaborative programme involving the Stockholm Environment Institute and the University of Ghana, funded by Sida.

References

Ahwere, S. K. (1994) 'Charcoal production in forest and savannah zones of Ghana', *Technical Report, No. 6,* Forest Resources Research Institute, Kumasi, pp8–12

Bartone, C., Bernstein, J., Leitmann, J. and Eigen, J. (1994) *Towards Environmental Strategies for Cities: Policy Considerations for Urban Environmental Management in Developing Countries,* United Nations Development Programme (UNDP), United Nations Centre for Human Settlements (UNCHS) and World Bank Urban Management Program No. 18, World Bank, Washington, DC

Benneh, G., Songsore, J., Nabila, J. S., Amuzu, A. T., Tutu, K. A., Yangyuoru, Y. and McGranahan, G. (1993) *Environmental Problems and the Urban Household in the Greater Accra Metropolitan Area (GAMA) – Ghana,* Stockholm Environment Institute, Stockholm

Bradley, D. Stephens, C., Cairncross, S. and Harpham, T. (1991) *A Review of Environmental Health Impacts in Developing Countries*, Urban Management Program Discussion Paper No. 6, World Bank, UNDP and UNCHS (Habitat), Washington, DC

Cairncross, S. (1990) 'Water supply and the urban poor', in J. E. Hardoy, S. Cairncross and D. Satterthwaite (eds), *The Poor Die Young: Housing and Health in the Third World*, Earthscan, London, pp109–126

ECA (Economic Commission for Africa) (1997) 'Ad hoc expert group meeting on measures pertaining to urban environmental health problems in Africa', Addis Ababa, Ethiopia, 26–30 May, ECA/FSSDD/ADHOC/EXP/UEHP/7

Ghana Statistical Service (1995) 'Ghana living standards survey: Report on the third round: GLSS3 September 1991 – September 1992', Ghana Statistical Service, Accra

Ghana Statistical Service (2000) *Poverty Trends in Ghana in the 1990s*, Ghana Statistical Service, Accra

Hardoy, J. E., Mitlin, D. and Satterthwaite, D. (1992) *Environmental Problems in Third World Cities*, Earthscan, London

Harvey, D. (1996) 'The environment of justice', in A. Merrifield and E. Swyngedouw (eds), *The Urbanization of Injustice*, Lawrence & Wishart, London, pp65–99

HDR(G) (Human Development Report (Ghana)) (1999) 'Ghana human development report 1999, sustaining the environment for poverty reduction', (Draft), UNDP, Accra

Kates, R.W. (1994) 'A tale of three cities (editorial)', *Environment*, vol 36, no 6

Kjellen, M. and McGranahan, G. (1998) 'Multi-city study on household environmental problems in third world countries', *Urban Health and Development Bulletin*, vol 1, no 2, pp64–70

Leitmann, J. (1994) 'The World Bank and the brown agenda: Evolution of a revolution', *Third World Planning Review*, vol 16, no 2, pp117–127

Lindskog, P. and Lundqvist, J. (1989) 'Why poor children stay sick: The human ecology of child health and welfare in rural malawi', Research Report No. 85, Scandinavian Institute of African Studies, Uppsala

Maxwell, D. (1999) 'The political economy of urban food security in sub-Saharan Africa', *World Development*, vol 27, no 11, pp1939–1953

McGranahan, G. (1991) *Environmental Problems and the Urban Household in Third World Countries*, Stockholm Environment Institute, Stockholm

McGranahan, G., Jacobi, P., Songsore, J., Surjadi, C. and Kjellen, M. (2001) *The Citizens at Risk: From Urban Sanitation to Sustainable Cities*, Earthscan, London

McGranahan, G. and Satterthwaite, D. (2000) 'Environmental health or ecological sustainability? Reconciling the brown and green agendas in urban development', in C. Pugh (ed) *Sustainable Cities in Developing Countries: Theory and Practice at the Millennium*, Earthscan, London, pp73–90

McGranahan, G. and Songsore, J. (1994) 'Wealth, health, and the urban household: Weighing environmental burdens in Accra, Jakarta, and São Paulo', *Environment*, vol 36, no 6, pp4–11, 40–45

McGranahan, G., Songsore, J. and Kjellén, M. (1996) 'Sustainability, poverty and urban environmental transitions', in C. Pugh (ed), *Sustainability, the Environment and Urbanization*, Earthscan, London, pp103–133

Omran, A. R. (1971) 'The epidemiologic transition: A theory of the epidemiology of population change', *Milbank Memorial Fund Quarterly*, vol 49, no 4, pp509–538

Smith, K. R. and Lee, Y. F. (1993) 'Urbanization and the environmental risk transition', in J. D. Kasarda and A. M. Parnell (eds), *Third World Cities: Problems, Policies and Prospects*, Sage Publications, London, pp161–179

Songsore, J. (1999) 'Urban environment and human well-being: A case of the Greater Accra Metropolitan Area', *Bulletin of the Ghana Geographical Association*, no 21, pp1–10

Songsore, J. (2000) 'Urbanization and health in Africa: Exploring the interconnections between poverty, inequality and the burden of disease', IWSG Working Papers, 12–2000, University of Frankfurt, Frankfurt

Songsore, J. (2003a) *Regional Development in Ghana: The Theory and the Reality*, Woeli Publishing Services, Accra

Songsore, J. (2003b) *Towards a Better Understanding of Urban Change: Urbanization, National Development and Inequality in Ghana*, Ghana Universities Press, Accra

Songsore, J. and Iddrisu, A. (1999) 'Housing and workplace environments affecting human development', Draft Report submitted for the *1999 Ghana Human Development Report*, UNDP, Accra

Songsore, J. and McGranahan, G. (1993) 'Environment, wealth and health: Towards an analysis of intra-urban differentials within the Greater Accra Metropolitan Area, Ghana', *Environment and Urbanization*, vol 5, no 2, pp10–34

Songsore, J. and McGranahan, G. (1996) *Women and Household Environmental Care in the Greater Accra Metropolitan Area (GAMA), Ghana*, Urban Environment Series No. 2, Stockholm Environment Institute, Stockholm

Songsore, J. and McGranahan, G. (1998) 'The political economy of household environmental management: Gender, environment and epidemiology in the Greater Accra Metropolitan Area', *World Development*, vol 26, no 3, pp395–412

Songsore, J. and McGranahan, G. (2000) 'Structural adjustment, the urban poor and environmental management in the Greater Accra Metropolitan Area (GAMA) – Ghana', *Bulletin of the Ghana Geographical Association*, no 22, pp1–14

Songsore, J., McGranahan, G. and Kjellen, M. (1997) 'An environmental health indicator profile for the Greater Accra Metropolitan Area (GAMA), Ghana', Prepared for Workshop on Public Participation and Urban Environmental Health Services, A Concerted Action Financed by the European Commission, 23–26 September, Bamako, Mali

Songsore, J., Nabila, J. S., Yangyuoru, Y., Amuah, E., Bosque-Hamilton, E. K., Etsibah, K. K., Gustafsson, J. and Jacks, G. (2005) *State of Environmental Health Report of the Greater Accra Metropolitan Area (GAMA) 2001*, Ghana Universities Press, Accra

Stephens, C. (1996) 'Research on urban environmental health', in S. Atkinson, J. Songsore and E. Werna (eds), *Urban Health Research in Developing Countries: Implications for Policy*, CAB International, Wallingford, pp115–134

Stephens, C., Timaeus, I., Akerman, M., Avle, S., Maia, P. B., Campanario, P., Doe, B., Lush, L., Tetteh, D. and Harpham, T. (1994) *Environment and Health in Developing Countries: An Analysis of Intra-Urban Differentials Using Existing Data*, London School of Hygiene and Tropical Medicine (LSHTM), London

UNDP (1997) *Ghana Human Development Report*, United Nationas Development Programme, Accra

WHO (World Health Organization) (1995) *The World Health Report 1995; Bridging the Gaps*, Geneva

World Bank (1995) *Ghana Poverty Past, Present and Future*, Report No 14504-GH, World Bank, Washington, DC

World Resources Institute, UNEP (United Nations Environment Programme), UNDP, World Bank (1998) *World Resources 1998–99: A Guide to the Global Environment*, Oxford University Press, Oxford

Yankson, P. W. K. (1999) 'Health and environmental impact of informal production and services units in Accra, Ghana', *Bulletin of the Ghana Geographical Association*, no 21, pp89–100

7
Dynamics of Growth and Process of Degenerated Peripheralization in Delhi: An Analysis of Socio–economic Segmentation and Differentiation in Micro–environment

Amitabh Kundu

Introduction

This chapter analyses the process of demographic and economic growth and the quality of the micro-environment in and around the city of Delhi, focusing on the differences between the core and periphery, based on both secondary as well as primary data. The specific question raised is whether there is a process of socio-economic segmentation, resulting in disparity in terms of access to basic amenities and physical quality of living wherein the environmental costs fall unevenly on the rich and poor and on the people living in the core and periphery. The larger objective is to see if an environmental transition in the context of urban India can be meaningfully analysed within the framework of shifting of costs from 'core' to 'periphery' wherein the two concepts have socio-political connotations and are not defined simply in terms of geographical location.

The second section in this chapter looks at demographic aspects, analysing the trend and pattern of growth within and around the metropolis. It shows how the uneven growth in population and differential absorption of migrants have accentuated the disparity in population density, having a direct bearing on the quality of life for different groups. The third section presents the changing profile of the metropolitan economy by probing into aspects of income growth, industrialization, employment, poverty, and so on, explaining the deceleration of migration and their diversion to peripheral towns and villages. The industrial policy, schemes and programmes launched in the national capital territory (NCT) of Delhi, which have led to the relocation of pollutant industries and of slum population to the periphery, are reviewed in the fourth section. The fifth section analyses the rehabilitation policies with regard to poor migrants and slum settlements. The availability of basic amenities, namely drinking water, electricity and toilets, and trends in intra-urban and rural-urban disparity have been analysed in the sixth section to highlight growing inequality across space. An attempt has been made here to see the extent to which market-based trends and state policies are responsible for the shifting of environmental burdens from the core to the periphery. Quality of life of slum dwellers and investments made by them in improving their living standards have been analysed, in relation to tenurial status and other socio-economic characteristics, in the seventh section, using survey data. The last section summarizes the conclusions and argues that transferring of environment burdens in space and the process of degenerated peripheralization have serious long-term cost implications for the regional economy.

Changing demographic profile in Delhi

An analysis of demographic data for the NCT of Delhi and its surrounding region over recent decades reveals several interesting features. Delhi urban agglomeration[1] (UA) has grown by over 4 per cent per annum in every decade since 1931. None among the other metropolitan cities – cities with more than a million people – in the country has experienced such a demographic growth. Based on these high growth rates, it has been argued that the city is absorbing a large number of people from the hinterland and thus has to bear the cost of augmenting its infrastructural facilities and suffer deterioration in quality of life. It is further observed that in the 1990s, when there was a significant deceleration in urban growth in the country and in the large majority of metro-cities, the UA maintained a high growth and even reported acceleration (Government of NCT of Delhi, 2006).

There is a certain amount of statistical noise underlying these propositions that needs to be removed. Population growth of 7.25 per cent per annum in urban Delhi during the 1940s (Table 7.1) can be considered as unusual as it is due to an influx of migrants from across the national boundaries at the time of partitioning of the country in 1947. Importantly, there is considerable fluctuation in the growth rate for urban Delhi but in the case of rural Delhi, it is more violent. These fluctuations can be attributed to a large number of villages in Delhi becoming part of the urban segment, resulting in an increase in urban area. Thus, the high urban growth is partly due to administrative decisions and does not reflect a process of rural-urban transformation. It would, therefore, make sense to consider the demographic growth in the entire NCT of Delhi as a whole for temporal analysis.

Table 7.1 *Demographic profile of the national capital territory of Delhi: 1951–2001*

	1951–61	1961–71	1971–81	1981–91	1991–2001
Total	52.44	52.93	53.00	51.45	46.53
Rural	−2.52	39.93	8.01	109.96	1.69
Urban	64.17	54.57	58.16	46.87	51.55
	1951–61	1961–71	1971–81	1981–91	1991–01
Delhi Urb. Agglo.	64.17	54.57	57.09	46.95	51.93
i DMC	84.11	59.47	48.55	47.55	36.21
ii NDMC	−5.35	15.39	−9.53	10.35	−1.99
iii Delhi Cantt.	−11.83	58.81	48.53	10.83	31.91

Source: Population Census for different years from 1951 to 2001

The negative impact of absorbing in-migrants within the city has not been as high as one may infer from the figures of population growth, as urban Delhi has expanded enormously over geographical space. The area of the Delhi Municipal Corporation (DMC) increased from 240.8km^2 in 1961 by about 20km^2 in 1971. It went up further by a similar amount during 1971–1981, and registered a record increase of about 46.0km^2 during 1981–1991. The census of 2001 reported an addition of 33 new towns that were rural settlements in 1991. Understandably, this growth dynamic has brought large chunks of village land within the urban fold.

The growth profile of settlements is highly disparate within the urban agglomeration as well as within the NCT. The New Delhi Municipal Corporation (NDMC) and Cantonment that had low population densities can be seen as successfully diverting the incremental migrant population to other segments and thereby experiencing a sluggish demographic growth. The growth rate for NDMC was negative for several decades while that for Cantonment was much less than that of the DMC. Understandably, the different growth rates have increased the disparities in density within the agglomeration. This has had an adverse effect on the quality of the micro-environment in many of the localities and wards in DMC area, as the pressure on its limited amenities increased tremendously over the years. The same is true for a large number of neighbouring rural settlements, as well, where population growth has been phenomenal.

Importantly, the census towns within the agglomeration, particularly those in the southern and eastern part of Delhi have experienced a very high population growth. The new towns added in the UA in the 1980s and 1990s having populations of more than 5000 have also recorded rapid growth. Among the 23 small towns in 1991, as many as 8 exhibit growth rates of over 7 per cent per annum, while the average growth rate of these urban centres is much above that of the DMC. An analysis of the district level data for Delhi NCT reveals that, excepting the three centrally located districts, namely central, northern and New Delhi, the demographic growth in the other six districts was about or above 4 per cent per annum during 1991–2001. One can, therefore, argue that the dynamics of growth in Delhi have unleashed a strong process of sub-urbanization in the hinterland. The growth has taken place both within as well as outside the urbanizable limits, as determined by the Delhi Master Plan (1960–1980).

There is, however, no underplaying the fact that several colonies within the core of DMC have experienced a high growth of population, absorbing a significant portion of rural migrants. A disaggregated analysis reveals that many low-lying areas, strips on the sides of railway tracks, vacant plots where development projects could not be launched in time, and so on, have attracted a large number of migrants, right in the heart of the city. These, of course, constitute a part of the 'periphery', defined not in geographical but in socio-political terms. Importantly, the geographical periphery, too, has absorbed a large majority of the migrants who wanted to make Delhi their destination. This has helped in decelerating population growth in several inner colonies, particularly those located in New Delhi and Cantonment.

The percentage of interstate intercensal migrants to the total population in the NCT decreased from 20.01 per cent in 1971 to 19.77 per cent in 1981 and further to 16.39 per cent in 1991. This has gone down further to 15.74 per cent (by place of last residence) in 2001. The decline of migration has, however, taken place only in urban Delhi. Indeed, there has been a dramatic rise in the percentage of interstate intercensal migrants in rural Delhi. The percentage of in-migrants to the rural population has gone up from a meagre 1.51 per cent in 1971 to 16.56 in 1981. The figure soared as high as 19.24 per cent in 1991. Taking the NCT as a whole, a decline in demographic growth (a low migration rate for the NCT in the 1990s can also be deduced from the fact that the sex ratio (*females per thousand males*) has remained stable) during 1991–2001, which is in line with the deceleration in population growth and the rate of migration in other urban centres of the country (Kundu, 2006). One can, of course, argue that the increase in population growth rate of urban Delhi is basically due to a large number of villages in Delhi being given urban status.

The sex ratio in urban Delhi has been increasing continuously in recent decades. This improvement could be explained in terms of betterment of housing conditions, which in turn has enabled the migrants to bring in their families. One would infer that it is the better-off sections among the prospective migrants who are in a position to find a foothold in the city. Also, there has been a slowing down in the stream of in-migrants who have a low sex ratio. This can be attributed to single male migrants, driven out of the rural hinterland through poverty and social distress, becoming fewer in number in recent decades. This seems to be the major reason for the improvement in the sex ratio.

In the case of rural Delhi, however, the trend is the opposite. Here, the sex ratio, both for the total, as well as the migrant, population, has gone down systematically which suggests that a substantial proportion of single male migrants are being absorbed in the periphery. Understandably, the people, forced out of the rural economy in the hinterland through crop failure or other types of calamity, could find shelter more easily in peripheral villages than in urban areas of the NCT.

Trends and patterns of income and employment generation and its implications for the environment in the metropolis

The NCT of Delhi is the most prosperous among the states and union territories of India. It has emerged as a major node in northwestern India with phenomenal growth in manufacturing, trading, communications and transport activities. Per capita State Domestic Product (SDP) in the year 2000–2001 was about 2.4 times that of the average for the country. Also, the growth rate of income during 1980s and 1990s was the same as that of the country, and consequently the gap between the NCT and national figures in per capita income has been maintained. The figures for the past couple of years, however, reflect differential growth rates and widening of the gap. The strong economic position of the NCT is reflected also in its high employment rate,[2] compared with the all India (urban) average, as also for other metro-cities. It is, therefore, not a surprise that the NCT has attracted a large number of migrants, although there has been a deceleration in its demographic growth during the 1990s, as is the case with most of the metro-cities in the country.

One would wonder why the growth in per capita income of Delhi has not been significantly higher than that of the country, as the metropolis has witnessed a high incidence of information technology- and management-linked jobs in the service sector. This can be explained in terms of a stagnating – even a declining – real income of a large section of informal sector workers. The worsening employment situation for men may be inferred from the decline in the percentage of workers during the 1990s. Both the Population Census and National Sample Survey (NSS) report a sluggish employment growth[3] and a corresponding increase in unemployment (Table 7.2).

The employment scenario for women in urban Delhi is worse compared with other metro-cities or smaller towns. This is due to the low incidence of primary sector-based activities, that are more likely to employ women. Also, Delhi's economy has a more formal component than other cities, offering fewer jobs to women. Despite marginal improvement in women's employment rates during the 1990s, it continues to be low and a source of anxiety for urban planners.

The distribution of workers in casual, regular and self-employed categories, as given by the NSS, gives an interesting picture for urban Delhi

Table 7.2 *Workers (work participation rates) by principal and principal+subsidiary status in Delhi and India in 15+ age groups in various National Sample Survey rounds (%)*

		Principal		Principal + subsidiary	
		Male	Female	Male	Female
RURAL					
Delhi	1993–94	90.9	12.5	90.9	16.2
	1999–00	79.1	3.3	79.2	4.4
All India	1993–94	84.6	34.6	86.4	48.6
	1999–00	82.9	35.0	84.1	45.2
URBAN					
Delhi	1993–94	79.6	12.7	79.6	13.2
	1999–00	74.3	11.9	74.5	14.1
All India	1993–94	75.8	17.5	76.8	22.3
	1999–00	74.5	16.6	75.2	19.7

Source: The 50th and the 55th rounds of the National Sample Survey publications, Government of India

(NSS Organisation, 2001). The percentage of casual employment decreased significantly during the 1990s, which was a reflection of changing working relationships in urban Delhi. Businesses, as well as households, generally prefer to employ people on a somewhat regular basis as that is supposed to ensure the safety and security of the enterprises, households and the neighbourhood. Arguing differently, urban Delhi has become increasingly hostile to in-migrants, who generally get absorbed in informal activities. The occasional drives to clear up the slum settlements and close down non-conforming industries have resulted in greater formalization in the labour market, less casualization and a reduction in employment.

The rural areas within the NCT present a sharply contrasting picture from the urban areas in terms of employment, both for men and women. As access to land and basic services within the central areas of the city has become increasingly difficult, a large percentage of the migrants have flocked into the rural periphery within and beyond the NCT, as noted above. Unfortunately, economic activities have not come up in the way it was envisaged in the Master Plan. Also, the process of agricultural land being put to urban uses or being kept vacant for speculative purposes has adversely affected the job opportunities in the primary sector. There has not been any compensatory growth in urban-linked activities, which could have absorbed the workforce displaced in agriculture, except in a few pockets. As a result, the rate of employment in the last decade has been low and the rate of unemployment high both for men and women in rural Delhi. More importantly, the situation worsened during the 1990s. Also, there was a high growth of workers in construction activities, basically because of the housing and infrastructural development activities in certain parts of the periphery. Employment in trade, too, grew rapidly. Both unfortunately, have a high incidence of informal employment, and their high growth confirms the process of degenerated peripheralization.

The percentage of workers among women in rural Delhi is extremely low – less than 5 per cent – which is several times below the all India rural or urban figure. This may be attributed to women being thrown out of agriculture due to rural land being used for urban purposes. Unfortunately, they could not be absorbed into the alternate employment opportunities that became available over time, either due to the nature of employment, their skill requirements or the inability to travel to places outside their villages.

It may be mentioned here that there has been deterioration in the urban employment situation in the country as a whole, attributed to the launching of the programmes of globalization, which resulted in significant deceleration in growth of the urban population in the 1990s.

The above overview suggests that decline in employment is much more serious in the case of Delhi. Interestingly, despite the worsening employment situation, the percentage of poor in the city in 1999–2000 works out to be as low as 7 per cent, compared to the national average for urban areas, which is 22 per cent. In the early 1970s, the former was as high as 55 per cent, which was above the national figure. Understandably, the decline in poverty can be attributed to the slowing down of in-migration of the poor. Delhi has, thus, grown through exclusion – exclusion of poor migrants as well as exclusion of the existing poor in the city. In fact, there is resistance on the part of the national capital, as in other cities, to provide employment, shelter and physical amenities to in-migrants who are uprooted from the hinterland and seek absorption in the metropolis (Premi, 1985). This is manifested clearly in Delhi, which is able to maintain, through this, a low density and a reasonably high quality of the micro-environment in select colonies.

Changes in the spatial structure of manufacturing industries: The battle for retention of income benefits and transference of environmental costs

The trend and pattern of industrial development in and around the NCT has been determined by competing economic and environmental agendas backed up by different pressure groups. On the one hand there is a demand from the upper and middle-class population for the generation of a larger number of regular skilled or semi-skilled jobs within the growing modern industrial and tertiary sectors. The local government, politicians and bureaucrats have, by and large, supported the demand as it enabled mobilization of greater financial resources through formal and informal channels. These prompt the establishment of policies and programmes that attract industries in Delhi. The established élite class, on the other hand, wants better environmental conditions and, therefore, demands restrictions on growth in pollutant industries, slums and poor migrants. There is also a commitment at the national level to disperse industries so that balanced development can take place in the region. These are behind the legal and administrative controls and restrictions that have been sought to be imposed on the location and functioning of industrial units from time to time.

Environmental concerns within the market-driven framework of growth

The Master Plan (1960–1980) was designed primarily with the objective of decongesting the national capital by removing the large scale and nuisance industries, including slum settlements, from the core to the peripheral region, to identified sites outside the urban limits. These sites were to accommodate much of the future growth of industries and industrial workers, as the plan imposed restrictions in various ways on their location in formal residential colonies. Slums that accommodated a large percentage of unskilled workers engaged in these industries were considered unhealthy for the sustained growth of the city. The perspective implicit in the Plan was to shift these from the city core; that can be seen clearly from pronouncements such as, 'The decentralisation of manufacturing is the most feasible mean of redistributing population... Much of the management and technical personnel, together perhaps with some of the skilled employees of decentralized manufacturing centres, may reside in Delhi and commute outward to work' (Government of Delhi, 1962).

Despite this perspective, Delhi experienced a rapid growth of manufacturing activities until the late 1970s. According to statistics from the Delhi Pollution Control Board (DPCB), and those generated through surveys conducted by the Delhi government's Department of Industries, the number of units, which was as low as 8000 in 1951, reached 40,000 in 1978. Investment and production in industries increased over 30 times. The growth of manufacturing employment was somewhat less, but it works out to be as high as 6 per cent per annum. Availability of infrastructure, wholesale markets, trade and other commercial services were the factors responsible for this growth.

The period of National Emergency, 1975–1977, when the perspective of 'environmental sanitization of the city had the upper hand' (see Kundu and Kundu, 2006), saw a large-scale eviction of slums and their economic activities. Resettlement of slum dwellers was sought in selected locations in the periphery where certain infrastructural facilities were provided. Sheds were provided by the Delhi government at specified locations to entrepreneurs whose units were dislocated, or to those seeking to establish new units. However, despite official claims, it is difficult to establish that the re-settlers had better infrastructural facilities to carry on their businesses. The programmes designed for their economic rehabilitation were mostly non-functional and ineffective. As a consequence, the occupancy rate of the sheds was very low and the growth of manufacturing activities in Delhi came down substantially. Employment in industries during 1978–1986 grew by 3 per cent per year, which was less than what was experienced since independence.

Importantly, this period of restrictions on entry to industries and migrants did not stimulate growth of manufacturing employment in the rural periphery of Delhi. It is indeed true that a number of medium-scale units came up here, but that did not push up the income level of the workers, since the high-value jobs were always taken up by skilled workers coming from the central city. The employment available here was mostly for a short duration and part-time in nature, where the wage rates were low. As a consequence, many in the periphery preferred to commute to the central city for better livelihoods.

Changed perspective favouring high industrial growth since the mid-1980s

The pro-industrial lobby seemed to get an upper hand in the early 1980s as the Delhi government promoted industrial growth, focusing on small-scale industries. The government brought out an industrial policy document in 1982 that envisaged rapid growth of modern small-scale units (with investment in fixed assets not exceeding Rs.10 million) that are non-polluting and non-hazardous in nature. The idea was to generate demand for skilled workers in the city without putting pressure on land and electricity supplies in the metropolis. Importantly, while different types of manufacturing were encouraged in industrial zones, certain types of unit were permitted even within residential areas; namely household industries. As many as 73 household industries, falling under Group A of Master Plan of Delhi (Government of NCT of Delhi, 1962),[4] could be set up in residential areas with permission of a High Powered Committee. Besides these, industries listed under A1 of MPD-2001 could be set up in rural areas.[5]

The Department of Industries came out with a concessional policy package and launched a number of programmes for providing training, marketing, quality control, and export facilities and so on, for promoting small-scale industries. It helped the latter in procuring raw materials and importing machinery by issuing necessary certificates. The department managed a number of industrial estates with basic infrastructural facilities, attracting a large number of entrepreneurs. Besides, it provided plots and flatted factories to small-scale entrepreneurs.

As a consequence of this changed policy perspective, Delhi saw a rapid growth of industries following the late 1980s – several of these being located in central or core areas – taking the pace of industrial growth much above that of the previous decades. Many of these units were labour intensive, and hence the annual growth rate of employment during 1988–1996 touched an all time record of 10 per cent. In real terms, the capital intensity (investment per worker) came down to a third of the figure of the early 1980s.

This high industrial growth is only partially due to the conscious government policy, as discussed above. One can clearly see the market successfully thwarting governmental efforts to restrict the scale of the units or take these out of the inner city. A large number of informal activities (many would hardly qualify as small-scale units) both in manufacturing and in trading rose up in the core and absorbed a sizable number of poor and migrant labour. Although the wage differential between formal and informal activities was high, this process enabled a trickling down of benefits from rich to poor within the city. The new rural migrants were absorbed in the slums, largely in the peripheries, but they mostly aspired to go closer to the core to take advantage of better economic conditions. This development has been viewed as a failure of the environment lobby, or those who advocated balanced regional development by taking industries out of the NCT to backward regions in neighbouring states, falling within the national capital region (NCR).

Assertion of the lobby for environmental cleansing in the inner city and creation of a differentiated and degenerated periphery

The forces behind the environment lobby became active once again by the late 1980s. The Delhi government had set up a working group in 1984 whose major concern was location of noxious and nuisance industries and warehouses in the rural periphery that aggravated environmental problems. The group had recommended putting a stop to industrial growth in the rural periphery and permitting only shopping complexes and group housing colonies. At the same time, it had recommended improving the level of basic amenities, including the means of transport and communications in the peripheries. Further, the Master Plan (1980–2000) had come forward with proposals to decongest the central city by taking industrial activities to the distant areas within the NCR. Construction of a large number of flyovers in the city was proposed to improve the accessibility of the élite colonies to the peripheries and establish better linkage among the colonies (Expert Group on the Commercialisation of Infrastructure Projects, 1996).

Notwithstanding all these, Delhi saw a rapid growth of industries through the late 1980s. It was only in 1996 that the DPCB came out with a survey, reporting as many as 92 per cent of 137,000 units operating in Delhi to be in residential and non-conforming areas, thereby highlighting the negative impact of this high growth on the morphology of the metropolis. The survey drew the attention of the Supreme Court of India, which observed that many of these industries, including those operating under the guise of small-scale units, violated the norms laid down in the Master Plan (1980–2000). It passed a directive for immediate closure of these industries and shifting of permissible small-scale units from non-conforming to conforming areas, namely 28 industrial estates within the NCT. In pursuance of that, no small-scale industry was to be registered in non-conforming areas. Even in the case of conforming areas, the units had to get clearance from the DPCB. To ensure compliance of the provisions of the Master Plan, a High Power Committee has established to examine the permissibility of certain (so-called household) units in non-conforming areas. Units failing to obtain necessary permission from the committee were to stop operating immediately and be rehabilitated in conforming zones.[6]

A compromise seems to be working out in the Master Plan 2021, which envisages participation of the private sector, as never before, for providing housing, infrastructure facilities and civic services, and large-scale regularization of industrial and commercial activities in non-conforming areas. However, it promises to protect environmental quality in core residential areas by shifting a large number of industrial units to the peripheral areas.[7] The guidelines in the Master Plan stipulate permitting mixed land use and regularization of industries in zones where their incidence is high (inhabited largely by the poor), because the real efficiency and cost advantage of Delhi lies in that. This would allow the entrepreneurs and executives to live in better-off colonies and commute to the 'periphery', which includes marginalized areas within the city besides the peripheral zone, defined in terms of geographic location, using the six- or eight-lane flyovers. Moreover, they would benefit from high industrial growth. A few noxious and large-scale units have been moved out to industrial estates, along with evicted slum populations in the neighbouring states within the NCR. These are being accommodated in designated pockets in the periphery, which would also reserve select areas for high-tech activities as well as high-income residences. The regulatory framework is being designed to save heritage zones, parts of Yamuna and the ridge, and maintain 'plenty of green areas' around the central localities in the city. No serious attempt has however, been made to improve the quality of life in the low-income pockets in the highly segmented periphery or tackle the problems of the intended and unintended industrial growth. This 'oversight' seems be a marginal cost the city has to pay to benefit from its growth potential, as it must also ensure quality living to a powerful élite in the metropolis, who would otherwise move to better destinations outside the country, along with their capital.

Given this clear rationale for maintaining a 'differentiated and degenerated periphery', it is no surprise that the Master Plan has failed in one of its key objectives: to disperse industries to distant growth centres within the NCR. There was lack of seriousness in the policy and no concrete measures were taken for provisioning basic services and infrastructural facilities of a certain quality in the towns outside the capital city. The guidelines of the Industrial Policy Document for dispersal were not observed while licences were issued during the 1980s and 1990s. Several administrators and political leaders in Delhi openly subverted decentralization efforts and worked against the shifting of industries out of Delhi, for individual or party gains. In the absence of a comprehensive framework and interstate coordination, industries grew rapidly in the immediate hinterland within and around the city. Importantly, the Master Plan 2001 would mark no significant departure from the earlier plans in terms of its focus of improving the environment in the core city. The strategy for balanced regional development, which appeared much like window-dressing, may now be discarded altogether.

The people in the periphery of Delhi, unlike in Delhi and other large cities, are generally unaware of the environmental implications of policies and are less effective in stalling the location of hazardous industries. The local bodies are financially weak and therefore cannot provide the necessary infrastructural support for attracting modern industries or control the production processes of hazardous units. The industries find these settlements around the metros convenient because the legislation linked to physical planning is not strictly implemented and the environmental lobby is almost non-existent. Furthermore, the policy of structural reform and industrial deregulation in the country has removed the restrictions on setting up units within or near the metropolises and helped informal relaxation in the Master Plans. This is certainly a major factor for degenerated peripheralization of Delhi.

Urban development and land policies in Delhi and the relocation of the slum population

The Delhi government has followed the policy of discouraging in-migration, particularly into its central areas, since the launching of the first Master Plan in the 1960s. From time to time, it took up programmes for shifting slum dwellers to identified colonies in the periphery. The Plan had maintained that a large number of migrants, because of their 'impoverished rural background' and dependence on agri-based activities, often create serious environmental problems in the heart of the city. With the twin objective of decongesting and improving the environment in the inner city and providing better public amenities to the slum dwellers, a resettlement scheme was launched for relocating these people and diverting fresh rural migrants towards 'urban villages' outside the city limits, where they can 'engage themselves in their traditional activities like metal work, handloom weaving, etc.' It was argued that these villages would act as a buffer zone for the migrants who make 'Delhi their first destination because of superior economic opportunities in the city' (Government of Delhi, 1962).

In the first phase, resettlement colonies were developed near the city core and the places of work, giving legal title to the people, who were residing in the city of Delhi before 1960. These were planned around the centre of high- and middle-income residential areas, with a view to making them an integral part of the neighbourhood. In all, 3700 households were given plots of $67m^2$ on lease for a period of 99 years. About 3600 tenements were constructed and given on leasehold basis. Understandably, there has been large-scale selling of property by the re-settlers as the plot size was large, which attracted high-income households. Indeed, these resettlement colonies have now been occupied by high- and middle-income households, making these an integral part of the central city.

This approach of providing land in the core of the city could not be sustained for long, as the cost of land worked out as prohibitive. By the late 1960s, the Delhi government had all but abandoned the scheme of giving such large-sized plots or built-up tenements. Resettlement of the slum population was, however, taken up on a massive scale in the second phase during 1975–1977, the period of the National Emergency. Taking advantage of the 'emergency-type' political situation, the government could make a gigantic effort to relocate 150,000 squatter families (almost 500,000 people) from within the Walled City and adjacent areas. This led to the emergence of 26 new colonies, while the number of slum households came down from 100,000 in 1973 to 20,000 in 1977. Each household in the new colony was given a plot measuring 21m^2 on a rental basis. Many of these came up outside the urban limits; a few were located even in agricultural green land. These did not conform to the minimum standards laid down in the Master Plan.[8]

Due to considerable public resentment, the government took the decision in the early 1980s to increase the plot size to 26m^2 and confer leasehold rights to all re-settlers at their 'camping sites', on payment of a small fee. For a short time after 1983, built-up tenements were also provided in the resettlement colonies. Further, Delhi Development Authority (DDA) recognized the deficiency in basic amenities and proposed to provide additional facilities in the Master Plan (1980–2000). Unfortunately, a large majority of the households did not come forward to pay the fees for conversion of their plots from rental to leasehold property.[9] The peripheral areas, thus, continued to have serious deficiencies in basic amenities.

In the 1980s the emphasis in slum development programmes shifted from resettlement to *in-situ* upgrades in the country. A number of schemes were launched, such as Environmental Improvement of Urban Slums, Environmental Improvement of Jhuggi Clusters, Urban Basic Services, and Conversion of Dry Latrines to Water Borne Latrines. In the case of Delhi, however, these remained as mere experiments and could not be scaled up at the city level due to the unwillingness of the land-owning agencies to transfer their land. In all, about 5000 households benefited from these development projects. The poor in-migrants, nonetheless, took advantage of the relaxed policy environment in finding shelter within the city (Planning Commission, 1983). This led to the steady growth of slums during the 1980s, so much so that the number went up to 250,000 in 1991.

Given the magnitude of the problem, the Slum and Jhuggi Jhompri Department of the DMC launched a three-pronged strategy which continues today. The first strategy involves shifting squatter families from areas that are considered physically hazardous and relocating them at alternative sites, mostly at faraway locations. The agency owning the land that is being cleared has to bear 75 per cent of the cost of procuring the alternate land. Unfortunately, many among the relocated families are 'selling out' the plots and coming back to their original place of squatting, as employment opportunities near the relocation sites are low. Also, providing leasehold titles, which is a part of this strategy, was stalled in 1992, following a High Court order urging the government to provide only licences for the use of land, which, it was hoped, would bring down the transfer of properties.

The second strategy envisages organizing on-site upgradation of slum clusters, wherever this is feasible, keeping in view the willingness of the land-owning agency. The third strategy involves provision of select basic amenities and training programmes in the slum clusters without necessarily improving the dwelling units or interfering with the physical layout of the area. A number of research studies, however, reveal that the cases of on-site development are only a few, and improvements in the quality of the micro-environment in the existing slums is at best marginal, as the Delhi government has maintained its major thrust on resettlement of the slum population in specified locations outside the core city.

The new slum strategy notwithstanding, the state continues to evict slum dwellers from the central city, basically to satisfy the upper- and middle- class lobby, using environmental, aesthetic and health arguments. Unlike many other metropolises, Delhi has not gone in for site and service schemes wherein land tenure could be provided at the site of occupation. The slum improvement schemes implemented in Delhi, such as Environment Improvement for Urban Slums, Urban Community Development, and so on, did not guarantee the households even informal land tenure.

Importantly, DDA is acquiring agricultural land within the urbanizable limits and creating new colonies for the rehabilitation of slum dwellers. DDA generally leaves some land vacant in the villages to enable the community to start alternate economic activities.[10] This has, indeed, facilitated a few landowners to start a business or build an additional apartment/room for renting purposes with the money received as compensation for their land. The landless class, however, has been forced to seek employment outside their traditional economy; unfortunately, only a few have succeeded. The impact for women deserves special mention, since not many of them could get absorbed in the industries that have come up in the hinterland. Consequently, female unemployment rates are very high. This has been accentuated by the liquidation of most of the traditional household industries due to the penetration of the city market and the changes in consumer preferences with modernization.

It is important that there has been high growth of residential *cum* commercial complexes in the southern and eastern part of the periphery, restricted to certain pockets. The development in the rest of the periphery is abysmally low, such that thousands of houses constructed under public schemes in the north and west lie vacant in the absence of demand. Unfortunately, in the villages in the southern periphery, which have been transformed into 'European cities', decent jobs have mostly been grabbed by migrants or those who commute from Delhi. This, too, has contributed to the worsening of the employment situation among the locals and has created tensions between them and the in-migrants in the fringe villages, as the former have lost their traditional jobs and have been unable to capture the newly created opportunities.

An overview of the present strategy, including land policies and related programmes launched during the past four decades, thus suggests that there is a conscious effort to divert the poor migrants towards the peripheral areas, barring a few pockets. The public agencies have a clear perspective of pushing out most of the existing slum colonies and informal manufacturing activities to the peripheral townships or villages. The upper- and middle-income sections of the city population, who could buy a plot or build a house legally or illegally, have thereby enjoyed maximum benefits out of the land policy. This has come in terms of subsidies and massive unearned income accruing through encroachment of extra space or transfer of property. Those going in for unauthorized construction for residential, commercial and industrial purposes in non-conforming areas or green belts, too, have benefited through regularization measures. The state has also been generous to those who have built without authorization in formal settlements wherein additions and alterations have been made, barring occasional demolition drives that have generally been taken as tokenism by the builders' lobby.

Unfortunately, similar benefits have rarely come the way of squatters, who have purposely been kept in a state of flux, to serve as a vote bank. Promises and even 'decisions' of *in situ* upgrading have been announced from time to time, but these proclamations were mostly for political gains, and have invariably been turned down by administrative order or judicial injunction. The squatters have been forcefully evicted as and when land has been needed for 'public use' and been shifted to resettlement sites. In doing this, the state has been cautious, apprehensive of political fallouts. In moving slums out of the centre of the city, the government has consequently taken three steps forward and one step back. Importantly, the courts have taken a critical view of not merely the growth of 'illegal squatters' but also of the lukewarm measures adopted by public agencies to evict them (Kundu, 2003). Public interest litigations that stalled slum evictions during the 1960s and 1970s are currently being filed, demanding clearance of public places and pushing the slums out of the city limits. All these have strengthened the process of segmentation of the city, particularly peripheralization and marginalization of the poor.

Access to basic amenities in Delhi and the process of spatial segregation

In 2001, the percentage of urban households having safe drinking water facility in Delhi was as high as 98 per cent, compared with 81 per cent for urban India, according to the Population Census. However, the city recorded much slower improvement in the coverage during 1981–2001, which is less than recorded at the national level. This is understandable because the figure for Delhi was already very high. What is surprising, however, is that the coverage of households by tap water has decreased. There are reasons to believe that poor and new migrants are being left out of the formal water supply system, which is due to unwillingness of public agencies to bear the cost of extending the pipelines to the periphery. The programmes of globalization and structural reform launched in the country in the early 1990s resulted in reduction in budgetary allocations (Planning Commission, 2002) and availability of subsidized funds for the provision of basic amenities and their increasing dependence on capital markets (Kundu et al, 1999). This has slowed down the rate of coverage of basic amenities in households of urban India, a sharper manifestation of which can be seen in Delhi.

It may be pointed out that about 60 per cent of the towns in Delhi UA had a percentage of households with safe drinking water in 2001, much below that for urban Delhi[11] and most of these towns are located in peripheral areas. On the contrary, the NDMC and DMC and a few nearby towns had about 99 per cent of their households covered by this facility (Table 7.3). Unfortunately, however, the figure for the Cantonment Board is low and decreased further by about 4 percentage points during 1981–1991. This can be explained in terms of the policy of the army establishment not to provide or permit continuation of the existing facilities to informal settlements.

Importantly, census definition of safe drinking water includes sources such as hand pumps and tube wells, as well as taps. The National Ground Water Commission has, however, shown through a survey conducted during 1999–2000 that a large percentage of groundwater in Delhi is not safe for drinking as it carries toxic chemicals and other metal particles. It would, therefore, be more important to look at the data on the percentage of households having access to tap water only. Delhi is one among the very few states where this indicator has reported a decline.

Spatial disparity in terms of coverage through tap water is appalling. While the average figure was only 68 per cent for the Delhi UA in 1981, in centrally located areas like the NDMC and the Cantonment, this is above 85 per cent. In contrast, many of the fringe towns had less than 30 per cent of their households with access to the facility.[12] Understandably, a large proportion of the households in the peripheral areas depend on hand pumps and tube wells that are not safe sources. There were marginal improvements in the coverage in a few of these towns during 1981–2001 but the overall situation remained deficient and alarming.

The figure for coverage of electricity in Delhi was 75 per cent in 1981. The figure has gone up, but by a smaller margin than that of the national level. As a consequence, as many as 40 per cent of the towns in the UA report figures less than the national average in recent years. Most of these towns are located in the periphery of Delhi.[13] The availability of electricity across towns or across core and periphery did, by and large, remain the same during the 1980s and 1990s.

The percentage of households having toilet facilities was 68 per cent in the 1981 census, which was about 10 percentage points higher than the national figure. What is surprising, however, is that while the national figure went up during 1981–1991, the share of households with toilet facilities in Delhi went down. The data from NSS confirm this pattern, reporting the figure coming down sharply by 10 percentage points during 1988–1993. The trend is further confirmed by the decreasing coverage of households with flush toilets.

Disparity in the availability of toilet facilities is very high within the Delhi UA. Most of the peripheral towns[14] have figures less than half of the NCT (63 per cent). This may be attributed to the higher incidence of slum populations in the peripheral areas. It may also be partly attributed to their maintaining rural

Table 7.3 *Households with access to basic amenities across towns in Delhi in 1991 (%)*

Town	Status	Access to electricity	Access to safe drinking water	Access to tap water	Access to toilet
Alipur	CT	85.52	98.43	78.43	48.84
Asola	CT	88.03	84.98	77.46	40.85
Babarpur	CT	45.37	98.08	34.15	81.79
Bawana	CT	74.22	89.15	68.87	27.16
Bhalswa Jahangirpur	CT	79.52	96.64	94.36	8.32
Bindapur	CT	78.17	86.22	37.82	88.81
Cantonment	Cantonment Board	79.62	87.65	78.86	67.23
Deoli	CT	15.66	96.76	16.71	52.15
DMC	Corporation	83.09	96.56	81.28	68.75
Ghitorni	CT	94.26	90.70	79.76	20.60
Gokalpur	CT	37.98	97.60	46.70	40.54
Jaffarabad	CT	97.47	96.14	66.09	98.97
Kanjhwala	CT	89.37	89.86	88.65	17.39
Molar Band	CT	80.22	96.00	62.25	64.59
Mundka	CT	57.87	93.88	21.98	23.89
Nanagal Dewat	CT	77.36	76.38	1.08	7.48
Nangloi Jat	CT	80.80	95.46	58.27	48.67
Nasirpur	CT	77.33	92.29	5.40	80.06
NDMC	Committee	79.85	97.13	85.13	78.61
Palam	CT	60.54	95.60	26.35	68.39
Patparganj	CT	61.73	76.21	39.20	51.50
Pehlad Pur Banger	CT	93.22	90.77	88.32	15.30
Pooth Khurd	CT	81.88	90.51	76.98	15.01
Pul Pehlad	CT	15.68	93.16	12.13	29.15
Rajokri	CT	52.04	60.34	37.39	15.13
Rangpuri	CT	95.20	91.06	84.93	26.32
Roshanpura	CT	56.97	97.70	26.22	56.90
Sultanpur	CT	90.10	86.72	69.95	51.69
Sultanpur Majra	CT	84.90	94.13	86.62	16.51
Tajpur	CT	22.52	96.58	62.19	8.97
Tigeri	CT	45.25	99.74	29.15	25.38
Yahya Nagar	CT	74.15	96.92	92.62	18.46

Table 7.3 *continued*

Town	Status	Access to electricity	Access to safe drinking water	Access to tap water	Access to toilet
	STDEV	22.94	8.12	28.63	26.67
	MEAN	69.87	91.54	56.73	42.61
	CV	32.83	8.87	50.47	62.59
Delhi	Total	79.48	95.78	75.73	63.38
	Rural	59.85	91.01	48.38	29.60
	Urban	81.38	96.24	78.37	66.64
All India	Urban	75.8	81.4	—	63.9

Note: STDEV = standard deviation; CV = coefficient of variation; CT = census town.

Source: Town Directory of Delhi, Census of India 1991. Unfortunately town level data on amenities are not yet available from the 2001 census

norms and habits, such as going to an open field for defecation, despite being included in the urban fold. Rural Delhi, nonetheless, has still lower coverage than the average of these towns.

The above analysis clearly brings out the disparity in the distribution of basic amenities between the city core and the peripheral towns. The degeneration of the periphery in the context of the micro-environment could largely be attributed to the policy of deliberate neglect or that of 'dumping the garbage in the backyard' (Kundu et al, 2002). The core has been receiving special attention, and consequently, the level of services has improved, while that in the periphery has either remained stagnant or declined over the years. Interestingly, the revised Master Plan recognizes the threat to health due to unplanned growth of polluting and non-conforming industries in the peripheral regions, but does not propose major investments to improve the basic amenities here. Consequently, the pressure on the existing amenities in these towns has increased tremendously, contributing to the process of degenerated peripheralization. This would be the major factor in explaining the outbreak of epidemics and high incidence of skin diseases in the peripheries, particularly in the low-income and slum areas. Further, denial of basic amenities in these slum colonies creates an atmosphere of individual and group violence, which, when a flashpoint is provided by socio-political exigency, leads to major riots, as the metropolis witnessed in 1984. Unfortunately, the costs of such disasters have not been brought within the framework of urban planning, and therefore remain a threat to the long-term sustainability of cities.

Disparity in housing and basic amenities between slum colonies of the core and periphery

Is the location of slums an important factor in determining the investments, economic well-being and quality of the micro-environment prevailing there? The question was investigated by collecting field data from 12 slum and resettlement colonies. These are located both in the core as well as the periphery of the city, making it possible to incorporate the locational factors within the explanatory framework. Selection of colonies was done in a manner so as to capture the impact of policies and programmes of different public agencies on the level of basic amenities (Kundu, 2003).

The data gathered through survey indicates that a large majority of slum dwellers in Delhi[15] have invested substantially for improving physical infrastructure and access to basic amenities, despite their not having formal land tenure. Many among them have established links with influential political leaders through their local leadership. They thus identify themselves with political parties and their leaders, and have become a part of the vote banks, which gives them a sense of security. This perceived security can also be attributed to no major dislocation of slums taking place during the 1980s and early 1990s.

The data further reveal that the above is true much more in the slums in the inner city than those in the periphery. The former have reported a much higher degree of community mobilization than the latter, possibly due to the long stay of the households and their proximity to the high- and middle-income neighbourhood. The local governments, too, have been less reluctant to extend formal or informal networks of civic supply in the case of inner slums, as that is less expensive than covering distant areas. Furthermore, a few *in situ* slum upgradation projects involving the community-based organizations and non-governmental organizations (NGOs) have also been launched here.

Unfortunately, such participatory projects are a rarity in peripheral areas. Investments in basic amenities made by the households living there, including those in resettlement colonies that have leasehold titles, have been very low. Furthermore, the basic amenities provided by public agencies are highly deficient. As a result, disparity between the centrally located and peripheral slums, including the resettlement colonies, in terms of basic amenities, works out as high and widening over the years.

One would thus argue that the porous character of the city with regard to in-migration and emergence of slums during the 1960s, 1970s and 1980s had, in a way, enabled the poor to get a foothold in certain pockets in central areas. This evidently improved the socio-economic conditions of these migrants, which, in turn, has helped in reducing the disparity between the core and periphery. Such opportunities are certainly on the decline in recent years. What is more important are the constant efforts being made to push these slums out of the city.

Conclusions and policy perspectives

Analysis of demographic data from the past five decades suggests that Delhi, along with its surrounding towns and villages, has attracted large-scale migration. This is primarily due to its strong economic base, which has increased in strength over the years due to the rapid growth of modern industries and commercial activities. The central city, however, has partially succeeded in diverting prospective migrants, pollutant industries and so on to marginal lands and peripheral areas. Using instruments such as Master Plans, the Industrial Policy Declaration and environmental legislation, the city has, off and on, relocated even the existing industries to its periphery. Policies and programmes of the state have thus facilitated the shifting of the environmental cost from high-income people in the core to those residing in marginal lands in the city or outlying towns and villages. Further, there has been widening of inequality in employment and income opportunities within the metropolis. Deteriorating economic conditions of the poor have acted as a disincentive to inmigration in recent years, helping select localities within the cities maintain a low population density. The disparity in terms of access to basic amenities has also been accentuated over recent years. All these suggest a process of socio-economic segmentation and unequal sharing of environmental costs at micro-level.

Despite this macro-scenario and a hostile policy regime, a section of the poor has devised ingenious ways of finding a foothold in the slums in the central city, resulting in reasonably high population growth in certain localities. To some extent, this has reduced the disparity between the core and the periphery. These people are able to take advantage of the politics of competitive populism at different layers of governance in the city, obtain informal assurances against eviction by building connections with political leaders, get covered under public utilities and plan programmes and, thereby, manage to stay within the city. This has enabled a small section of slum dwellers to partake in the benefits that accrue to regular city dwellers. They have benefited from the new income opportunities in the city and also gained access to basic services through extension of the existing system due to their proximity to formal colonies. Many among them have made investments in improving their housing and micro-environment due to their perceived sense of tenure security.

The scene has, however, changed dramatically in the early 2000s. Court orders favouring the landowning agencies and large-scale evictions of industries and slum colonies have shattered this security of tenure. There is a growing realization that social and political connections, informal assurances, a host of semi-legal documents and so on are not of much use in the event of a court order. In fact, many of the industrial units have closed down or moved to neighbouring towns or villages, despite having approval from the different departments of the local government and making hefty payments to them, formally or informally. Understandably, this has a serious environmental impact on the peripheral areas where the dislocated industries and slums are being relocated. This has led to an accentuation of core-periphery differential in terms of economic and social well-being and quality of life.

Importantly, this trend of declining inmigration of the poor, and urban segmentation can be observed at the national level as well, with a sharp fall in the rate of urbanization and poverty, particularly in metropolitan cities. Further, an analysis of the indicators of economic well-being, health, education and so on around urban centres at the national level shows that their values do not decline in smooth gradients, as distance from the city or town increases. On the contrary, these fall very sharply in the immediate periphery within a distance of 15 to 20km, which confirms the thesis of degenerated peripheralization in India. It suggests that the urban centres are generally shifting their environmental problems, making serious negative impacts on the periphery (Kundu et al, 2002).

Recent evictions in Delhi and in several other cities raise important policy questions. Do the poor have a place in the fast globalizing cities or would they have to be pushed into degenerating peripheries for considerations of macro-economic efficiency? Is the shifting of environmental costs to the slums and maintaining low-density and high-quality micro-environments in select localities sustainable in the long run? Should the planners not be made aware of the health and law and order problems associated with increasing concentration of the poor and polluting industries in marginal lands and city periphery?

The story of Delhi through five decades since independence has the moral that the shifting of environmental burdens onto the poor who can find a foothold only in the city periphery can be extremely costly in the long run. It is important that the challenges implicit in the questions posed above are analysed seriously at national level and immediate steps taken to deal with them. Research studies suggest solutions in the direction of providing formal or informal tenure to the poor and not dislocating them from their economic activities. Furthermore, the city authorities may be persuaded to launch, through financial and institutional support, measures for generating employment and income opportunities for the poor while ensuring compliance of environmental controls in location of industries in the periphery. Regional authorities should provide reasonable levels of basic amenities in the towns and villages (say within the NCR) so that they can attract a large part of the activities, and the process of degenerated peripheralization can be stalled. This approach should be effective in a strategy for tackling the problems of socio-economic segmentation in Delhi, and also in other cities in the developing world.

Acknowledgements

The author would like to gratefully acknowledge the assistance of Mr Bal Paritosh Dash in data collection, analysis and draft writing of this chapter.

Notes

1. The NCT of Delhi has 93 per cent of its population living in urban areas. The urban agglomeration of Delhi, which accounts for 99.5 per cent of the total urban population in the NCT, has 59 cities and towns. Delhi Municipal Corporation, New Delhi Municipal Corporation and Cantonment are its major constituents.
2. Employment rate or workforce participation rate may be defined as the percentage of workers to total population.
3. It may be more appropriate to compare the rate of employment for the 15+ age group (workers as percentage of population in the age group). Delhi has a smaller percentage of population below 15 years, which gives a slightly inflated value for the employment rate for the total population, in relation to the national average. The figure for *urban males (main+ marginal)* in Delhi was 79.6 per cent in 1993–1994 – higher than the national average of 76.8 per cent. However, this had gone down to 74.5 per cent in 1999–2000, slightly below the national figure of 75.2 per cent (Table 7.2).
4. MPD-2001 was approved by the government of India on 1 August 1990.
5. These units could use power not exceeding 5KW, occupy an area up to 30m^2 but must not create pollution or congestion. Subsequently, the Ministry of Urban Development and Poverty Alleviation further permitted F category industries in light food industries zones and removed the restriction of maximum power load.
6. The then union minister for urban development, who was also formerly responsible for large-scale eviction of slum dwellers during the period of the National Emergency, had given a firm policy directive for shifting the industries in non-conforming zones to industrial estates and to towns outside the NCT.
7. For accommodating these industrial units, the government of Delhi acquired 526ha acres of land at Bawana and Holambi Kalan and set up new industrial estates. Under the relocation scheme, 378 flatted factories were also constructed at Jhilmil Industrial Area.
8. A study conducted by the Indian Institute of Public Administration in 1980 revealed that the average time spent per person on the journey to and from the place of work was 50 minutes per day in case of the old colonies, and the monthly expenditure on travel was Rs. 15.46. The corresponding figures for the new colonies were 128 minutes and Rs. 32.31. The study further reveals that the re-settlers were only spending Rs. 9.22 on transport before shifting to the new colonies.
9. The Municipal Corporation of Delhi has been reluctant to service these areas due to deficiencies and lower standards of development, compared to the norms of the Master Plan, and also because these generate little revenue for the local government.
10. Only a few landowners could, however, start a new business or build an additional apartment/room for renting purposes with the money received as compensation for their land.
11. Town level data for Delhi have not yet been published for the 2001 Census.
12. Shockingly, the towns, such as Babarpur, Jaffarabad and Nasirpur, had percentage figures as low as 1.3, 1.5 and 2.4, respectively.
13. What is alarming is that the towns such as Pul Pehlad and Deoli report only 15 per cent of households having electricity.
14. The names of towns such as Alipur, Bawana, Bhalswa Jahangirpur, Kotla, Mahipalpur, Molar Band, Pehladpur Banger, Pooth Khurd, Roshanpura, Sutanpur Majra and Tigri, may be mentioned as illustrative cases.

15 The exceptions are those living in areas hazardous for human habitation due to proximity to railway tracks, industrial units, and so on.

References

Expert Group on the Commercialisation of Infrastructure Projects (1996) *The India Infrastructure Report: Policy Imperatives for Growth and Welfare*, Ministry of Finance, Government of India, New Delhi

Government of Delhi (1962) *Master Plan for Delhi (1960–88)*, Government of Delhi, New Delhi

Government of NCT of Delhi (1962) *Master Plan for Delhi, 2001*, MPD, 2001, Government of NCT of Delhi, New Delhi

Government of NCT of Delhi (2006) *Delhi Human Development Report 2006*, Oxford University Press, New Delhi

Kundu, A (2003) 'Provision of tenurial security for the urban poor in Delhi: Recent trends and future perspectives', *Habitat International*, vol 28, no 2, pp259–274

Kundu, A. (2006) 'Globalization and the emerging urban structure in India: An analysis with reference to regional inequality and population mobility', *India: Social Development Report*, Oxford University Press, New Delhi

Kundu, A., Bagchi, S. and Kundu, D. (1999) 'Regional distribution of infrastructure and basic amenities in urban India', *Economic and Political Weekly*, vol 34, no 28, pp1893–1906

Kundu, A. and Kundu, D. (2006) 'Urban land management, tenurial security and poor: An overview of policies and tools of intervention with special reference to Delhi', Paper presented at a seminar on *Secure Land Tenure: New Legal Frameworks and Tools in Asia & Pacific*, United Nations Economic and Social Commission for Asia and the Pacific (UNESCAP), 2006

Kundu, A., Pradhan, B. K. and Subramanium, A. (2002) 'Dichotomy or continuum: Analysis of impact of urban centres on their periphery', *Economic and Political Weekly*, vol. 37, no 14, pp5039–5046

National Sample Survey (NSS) Organisation (2001) *Employment Unemployment Situation in India 1999–2000*, Government of India, New Delhi

Planning Commission (1983) *Task Force on Housing and Urban Development – Shelter for Urban Poor and Slum Improvement*, Government of India, New Delhi

Planning Commission (2002) *Tenth Five Year Plan (2002–07)*, Government of India, New Delhi

Premi, M. K. (1985) *City Characteristics, Migration and Urban Development Policies in India*, Paper no 92, East West Centre, Honolulus

8
Motorization in Rapidly Developing Cities

Yok-shiu F. Lee

Introduction

The association between national income and environmental problems has been the subject of intellectual inquiries and policy debates in recent decades. One of the most popular postulations – commonly referred to as the Environmental Kuznets Curve (EKC) – suggests that as national incomes rise, environmental problems will first get worse and then get better. In contrast, the urban environmental transition (UET) theory argues that while many environmental problems are associated with the development process, not all of them follow the EKC trajectory (Marcotullio and Lee, 2003). Proponents of the UET framework claim that as cities develop, the priority environmental challenges would, in the first instance, shift from those relating to the 'brown agenda' – including water supply and sewage – to those pertaining largely to the 'grey agenda' – such as industrial pollution and vehicular emissions. And as cities enter the post-industrial phase, they are confronted with 'green' agenda challenges such as greenhouse gas emissions, ozone-depletion substances, non-point-source pollution and increasing volumes of municipal waste (McGranahan et al, 2001). In other words, as cities become wealthier, the nature of their environmental problems changes from being primarily localized to being globalized; their impacts transform from being mostly immediate to being delayed by years or even decades; and their consequences shift from being predominantly health threatening to being ecosystem damaging.[1]

A 'staged-type environmental evolution' (STEE) model, developed by Bai and Imura (2000) several years after the UET framework was formalized and based upon evidence gathered from the changing urban environmental conditions in East Asia, maintains that cities normally experience three distinctive sets of environmental problems sequentially: (i) poverty-related concerns; (ii) rapid growth-related crises; and (iii) wealthy lifestyle-related complications. However, the form of the sequence is not necessarily linear. Some cities could jump from the poverty-related problem stage directly to the consumption lifestyle-related problem stage, skipping the rapid growth-related problem stage altogether.

Combining elements from the UET model and the STEE theory, as well as empirical observations gleaned from rapidly growing cities in the Pacific-Asia region, a 'compressed and telescoped transition' (CTT) framework was more recently proposed by Marcotullio and Lee (2003) to deepen our understanding of the complexities of the fast-changing urban scenes in Asia. Drawing upon and comparing historical data on the development of the urban transport systems in Asia and the West, the CTT hypothesis claims that the environmental conditions in Asia's cities have been transformed at a much more rapid rate than their Western counterparts. First, they have gone through the three-stage urban environmental transition within a much shorter, compressed time-frame than cities in the West (Marcotullio et al, 2003). Second, environmental issues associated with a later stage have rapidly emerged, even though some or most of the problems of the earlier phase(s) have remained unresolved, resulting in a telescoping of the transition process where a city might be afflicted simultaneously by a set of overlapping problems associated with all three UET stages (see also Chapter 2).

One of the most glaring examples of such overlapping problems pertains to the impacts emanating from rapid motorization rates and rampant growth of the private automobile in middle-income cities found in the developing world.[2] In fact, the rise in personal mobility – facilitated primarily by the availability of cheap oil and affordable motor vehicles – has been one of the most dramatic changes recorded in the 20th century. The number of vehicles, at the global level, increased from about 40 million in the late 1940s to some 680 million by the end of the last century (UNEP, 1999). With new motor vehicle registrations increasing at an average rate of about 2 per cent to 3 per cent per annum, the global motor vehicle fleet will reach a level of between 885 million and 1100 million by the year 2010. Moreover, much of the growth in the number of vehicles and vehicular emissions will occur in developing countries. Even with major engineering breakthroughs in motor vehicle and fuel-use technology that can drastically reduce vehicular emissions, the long fleet turnover in developing countries (at 15–20 years) would ensure that an increasing share of global automotive emissions will be generated in growing cities in developing countries (Faiz, 1993).

Given that nearly 80 per cent of all transport-related energy is consumed by motor vehicles, and that most of the world's automobiles are found in cities, the transport sector has by now become, generally speaking, a major contributor to urban air pollution (Gan, 2003). Empirical evidence has shown that, since the mid-1980s, air pollution worsened substantially in most of the mega-cities as a result of rapid population growth, industrialization, increased urbanization and motorization, greatly extending its adverse health impacts from some highly exposed sub-groups to the general population (WHO

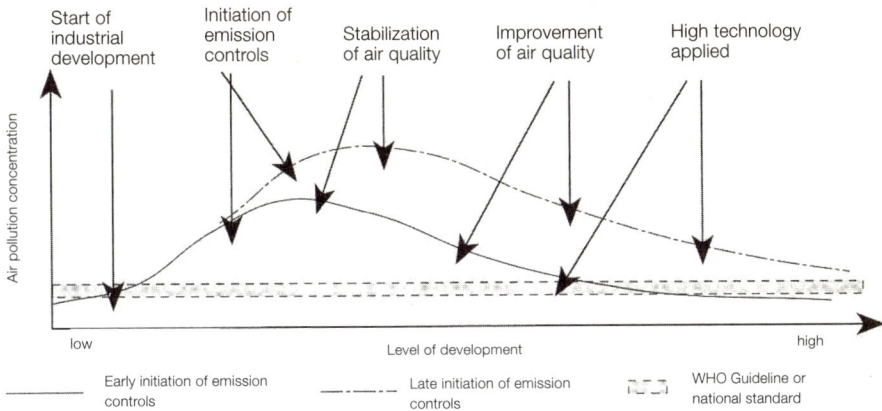

Source: WHO and UNEP (1992)

Figure 8.1 *Urban air pollution problems and level of economic development*

and UNEP, 1992). In fact, some cities have come close to experiencing the same record level of pollution reported in London in the 1950s. Figure 8.1 illustrates the general relationship between level and speed of development and environmental pollution.

Through a delineation of the global trends and patterns of motorization and its environmental impacts as well as an in-depth discussion of the causes and consequences of the unprecedented rapid rate of motorization reported in contemporary China, this chapter argues that technical solutions to reducing automobile dependence and its associated problem of urban air pollution aside, the fundamental key to reversing these negative trends is to persuade policy-makers in cities and countries of all income categories to re-think the fundamental objective of transportation development. Rapidly motorizing cities and countries, in particular, could only embark onto a sustainable path if accessibility, not mobility, is recognized and accepted as the ultimate goal of transport planning.

Motorization: A global view

Trends and patterns

Across the entire globe, the number of motor vehicles has been increasing by about 5 per cent per year, which is much faster than the world's population growth rate of 2 per cent per annum (Romieu, 1999). In 1950, there were about 53 million cars in the world. By 1994, the total number of the world's private automobiles had multiplied by a factor of 9 to 460 million. Adding to this annual increase of 9.5 million automobiles was an annual growth of 3.6

million trucks and buses for the same period. If two-wheeled vehicles were included in the calculation, the world's total number of motorized vehicles in 1994 was about 715 million (Romieu, 1999).[3]

Although the rate of increase of motor vehicles has slowed down considerably in the industrialized countries in recent years, sustained population growth, increasing rates of urbanization and rapid industrialization have hastened the growth of automobiles, trucks and buses elsewhere, particularly in developing countries. For instance, as correctly predicted by the World Bank in 1996, the number of motor vehicles in China, according to the *2005 Statistical Yearbook of China*, has tripled in the 1990–2000 decade (World Bank, 1996). Moreover, an analysis of the trends of global motor vehicle registrations and registrations per capita has shown that the world's total number of motor vehicles could reach a level of between 885 million and 1100 million by the year 2010 (Faiz, 1993).[4] These numbers seem to be in line with some other estimates that have been put forth, which suggested that if the current rates of expansion were to continue, there will be more than 1000 million vehicles on the road by 2025 (UNEP, 1999).

Furthermore, research has shown that the number of car-kilometres travelled has been forecasted to grow at an increasingly faster rate than the increase in the number of motor vehicles, which in turn would expand at higher rate than the growth of the world's urban population (Figure 8.2). Much of this growth in vehicle usage and demand for motor vehicles, however, is not evenly distributed across the globe. For instance, a study conducted in the early 1990s suggested that the annual demand for motor vehicles between 1988 and 2000 would increase at a rate of 220 per cent in developing countries, compared to only 10 per cent in the Organisation for Economic Co-operation and Development (OECD) countries (Faiz, 1993, p173).

There are enormous variations in the rate of motorization among different regions in the world, ranging from less than 14 private cars (for Asia excluding Japan) to more than 550 private cars (for North America) per 1000 persons. In certain parts of Asia and Africa, the level of motorization is even less than one per 1000 persons (Romieu, 1999). The latest available figures show that in 1996, a large proportion of the world's motor vehicles are found in Europe and Central Asia (37.9 per cent), North America (33.0 per cent), and the Asia-Pacific region (18.8 per cent). Motor vehicles found in Latin America and the Caribbean region accounted for only 6.5 per cent of the world's total fleet and Africa accounted for a mere 2.8 per cent (Table 8.1). Although the Latin American, Caribbean, and African regions respectively accounted for quite a small share of the world's total motor vehicle fleet, these regions have recorded substantially higher rates of growth than their European and North American counterparts. For instance, between 1980 and 1996, when the number of motor vehicles in North America grew by 20.8 per cent, the corresponding figure for Africa was a staggering 250.9 per cent (Table 8.1).

A more complete and interesting picture regarding the relationship between motorization and development level emerges when country-level statistics are categorized by income groups. For instance, in 1990, while high-income

Motorization in Rapidly Developing Cities 183

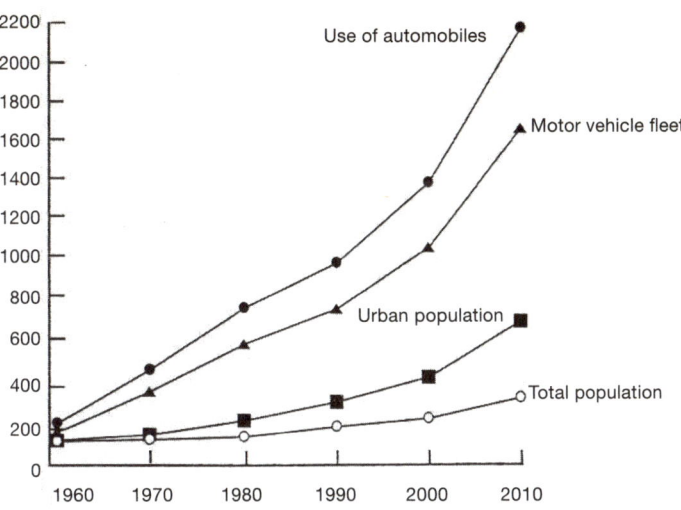

Source: World Bank (1996)

Figure 8.2 *Estimated growth in ownership and use of motor vehicles since 1950*

Table 8.1 *Geographic distribution of motor vehicles, 1980–1996*

	1980 (million)	(%)	1990 (million)	(%)	1996 (million)	(%)	△% 1980–1996
Europe and Central Asia	129.1	33.0	191.0	35.3	256.5	37.9	+ 98.7
North America	184.7	47.2	208.6	38.5	223.2	33.0	+ 20.8
Asia and the Pacific	52.3	13.4	93.2	17.2	127.3	18.8	+ 143.4
Latin America & the Caribbean	17.4	4.4	32.3	6.0	44.2	6.5	+ 154.0
Africa	5.3	1.4	11.1	2.0	18.6	2.8	+ 250.9
West Asia	2.3	*	5.5	1.0	6.4	1.0	+ 178.3
Total	391.1	100.0	541.7	100.0	676.2	100.0	+ 72.9

Note: * Less than 0.1 per cent.

Source: Compiled by author, using figures taken from UNEP (1999)

countries contained 22.7 per cent of the world's population, they accounted for 37.0 per cent of the world's urban population and an astounding 85.6 per cent of the world's motor vehicles. Middle-income countries, including both lower-middle and upper-middle income groups, contained 20.4 per cent of the world's population, 26.3 per cent of the world's urban population and 12.7 per cent of the global vehicle fleet. In contrast, low-income countries had to support 56.9 per cent of the world's population, with an urbanization rate of 36.7 per cent and operated only 1.7 per cent of the world's motorized vehicles (Table 8.2).[5]

Moreover, the high-income countries, while containing only 37 per cent of the world's urban population, accounted for a disproportionately high share of the world's daily vehicular trips in the urban areas. Over half of the daily 3400 million vehicular trips taken in the world's cities were recorded in high-income countries and 71.7 per cent of these were undertaken in private cars. Consisting of 26.3 per cent of the global urban population, residents in the lower-middle and upper-middle income countries were, altogether, responsible for 21.1 per cent of all vehicular trips and 14.1 per cent of all automobile trips. In contrast, the low-income countries, which contained 36.7 per cent of the world's urban population, accounted for only 27.2 per cent of the world's vehicular trips and just 14.2 per cent of its private car trips in cities (Table 8.3).

One comprehensive and extensive global study on automobile dependence conducted by Kenworthy et al (1999) reported that residents in US cities had the highest rate of car ownership (604 per 1000 persons), followed by the Australians (491 per 1000 persons), Canadians (524 per 1000 persons), Europeans (392 per 1000 persons), citizens in wealthy Asian cities (123 per 1000 persons) and, finally, people in developing Asian cities (102 per 1000 persons) (Table 8.4).

Table 8.2 *Population distribution and number of cars in different economic groups, 1990*

Economic group	Population estimate (10^3)	Proportion of world population (%)	Urban population (10^3)	Proportion of world urban population (%)	No. of cars (10^3)	Proportion of world car fleet (%)
Low income	3,019,510	56.9	879,379	36.7	6717	1.7
Lower-middle income	860,208	16.2	484,721	20.2	33,602	8.3
Upper-middle income	221,806	4.2	146,899	6.1	17,919	4.4
High income	1,201,748	22.7	887,929	37.0	346,801	85.6
World total	5,303,272	100.0	2,398,928	100.0	405,039	100.0

Source: Flachsbart (1999)

Table 8.3 *Estimates of daily vehicular trips made by different modes of travel within different economic groups, 1990*

Economic group	No. of urban vehicular trips per day (10^3)	Proportion of total urban vehicular trips per day (%)	No. of trips per day by car (10^3)	No. of trips per day by public or public-type transport (10^3)	Proportion of world total vehicular trips by car (%)	Proportion of world total vehicular trips by public and public-type transport (%)
Low income	927,517	27.2	176,802	750,716	14.2	34.6
Lower-middle income	538,645	15.8	127,009	411,637	10.2	19.0
Upper-middle income	181,498	5.3	48,708	132,791	3.9	6.1
High income	1,765,290	51.7	892,265	873,025	71.7	40.3
World total	3,412,950	100.00	1,244,784	2,168,169	100.00	100.00

Source: Flachsbart (1999)

Table 8.4 *Patterns of private transportation in cities of different levels of development, 1990*

Group of cities	Total vehicles per 1000 people	Cars per 1000 people	Total private vehicle km per capita	Private pass. vehicle km per capita	% of workers using transit	% of workers using private transport	% of workers using foot or bicycle
US cities average	751	604	12,336	11,155	9.0	86.3	4.6
Australian cities	595	491	8034	6571	14.5	80.4	5.1
Canadian cities	598	524	7761	6551	19.7	74.1	6.2
European cities	452	392	5026	4519	38.8	42.8	18.4
Wealthy Asian cities	217	123	2950	1487	59.6	20.1	20.3
Developing Asian cities	227	102	2337	1848	37.8	43.9	18.4

Source: Kenworthy et al (1999)

There is, as a consequence, a big contrast in terms of the volume of private vehicular traffic between the most well-off countries and the least developed ones – with Americans logging annual per capita passenger vehicle kilometres of 11,155, residents in developing Asian cities were generating a corresponding figure at only 1848. While it is not surprising that 86.3 per cent of the workers in US cities drove to and from work in their private cars, one would expect to see a much lower percentage of workers in the developing Asian cities *not* using private cars on a daily basis. It is therefore somewhat puzzling to find that, on average, 43.9 per cent of workers in developing Asian cities were relying on private automobiles in their daily commutes, which was twice the level recorded for wealthy Asian cities at 20.1 per cent (Table 8.5).

A closer look at the breakdown of the city-specific data helps reveal the fact that there were some wide variations in the statistics among the developing Asian cities included in Kenworthy et al study (1999), and these variations were masked by the average figures reported for this category. For instance, whereas only 20.6 per cent of Seoul's workers drove to and from work in their private cars, up to 60.0 per cent of Bangkok's residents were getting around their city in some form of private transport (Table 8.5). It is also worth noting that, on average, a much higher per centage of workers in wealthy Asian cities (59.6 per cent) were using mass transit on their daily trips, whereas only 37.8 per

Table 8.5 *Patterns of private transportation in wealthy and developing Asian cities*

Group of cities	Total vehicles per 1000 people	Cars per 1000 people	Total private vehicle km per capita	Private pass. vehicle km per capita	% of workers using transit	% of workers using private transport	% of workers using foot or bicycle
Wealthy Asian cities							
Hong Kong	78	43	1459	493	74.0	9.1	16.9
Singapore	200	101	3597	1864	56.0	21.8	22.2
Tokyo	374	225	3795	2103	48.9	29.4	21.7
Average	217	123	2950	1487	59.6	20.1	20.3
Developing Asian Cities							
Bangkok	348	199	3198	2664	30.0	60.0	10.0
Jakarta	201	75	1597	1112	36.3	41.4	22.3
Kuala Lumpur	403	170	4944	4032	25.5	57.6	16.9
Manila	86	66	901	732	54.2	28.0	17.8
Seoul	119	66	1899	1483	59.6	20.6	19.8
Surabaya	209	40	1483	1064	21.0	55.7	23.5
Average	227	102	2337	1848	37.8	43.9	18.4

Source: Kenworthy et al (1999)

cent of the residents in developing Asian cities were taking advantage of such a means of public transport. What the figures in Tables 8.4 and 8.5 have revealed are (a) generally speaking, as broad categories, and with a few exceptions, automobile ownership is positively correlated with a city's level of economic development; but (b) there are some wide variations in terms of automobile usage *between* different groups of cities categorized by level of development as well as *among* cities of a seemingly similar level of development; and (c) automobile usage is *not* necessarily positively correlated with a city's level of economic development.

These observations are further reinforced by the findings of another global study on motorization and automobile usage in selected metropolitan areas in different parts of the world. As shown in Table 8.6, where the metropolitan areas are arranged in an ascending order in terms of their respective per capita gross national product (GNP), the rate of automobile ownership, viewed from a global perspective, tends to increase as per capita GNP goes up. There are several prominent exceptions, however, such as Osaka and Tokyo, where the rates of car ownership for the 1983-1989 period were less than half of that recorded in US cities, despite the fact that their respective per capita GNP figures were quite comparable.

Using the current automobile growth rates as the basis to project the future size of the world's motor vehicle fleet, the United Nations Centre for Human Settlements (UNCHS) (2001) has shown in its *Global Report on Human Settlement 2001* that the share of motorized vehicles in low- and middle-income countries will gradually rise from 25 per cent in 1995 to 48 per cent in 2050 (Table 8.7). Indeed, leading researchers on the issue of motorization have reached a consensus that the world's motor vehicle population, with reasonable certainty, will continue to increase rapidly in the coming decades – for instance, doubling within a 30 year time-frame between 1990 and 2020 – and that much of this growth will take place in the developing countries (Faiz, 1993; Flachsbart, 1999; Romieu, 1999; Schwela, 1999).

Environmental impacts

Before we proceed to examine the environmental impacts of vehicular emissions per se, it is instructive to note that anthropogenic air pollutants originate from a variety of sources, including households, vehicles, industries, agriculture, and forest burning. Besides power generation plants and factories, domestic consumption of fossil fuels – such as heating oil, biomass and brown coal – is a significant source of ambient air pollutants, such as suspended particulate matter and sulphur dioxide, particularly in parts of China and Eastern Europe, which have to endure long, cold winter months (Kojima and Lovei, 2001). Motor vehicles, therefore, are not the only source, nor even frequently the major source, of urban air pollution, but they are a primary source of some categories of pollutants, which could result in some major impacts at the local, regional and global levels (Table 8.8). Given that both the number of automobile dependent cities and the rate of automobile dependence

Table 8.6 *Modal split, motorization and per capita GNP for selected metropolitan areas*

Economic group	Metropolitan area	Percentage of urban trips by car	Cars per 1000 population (1983–1989)	Per capita GNP (1988 US$)
Low income	Bombay, India	8	3	340
	Karachi, Pakistan	3	5	350
	Nairobi, Kenya	45	5	370
	Jakarta, Indonesia	27	6	440
Lower-middle income	Manila, Philippines	16	13	630
	Cairo, Egypt	15	15	660
	Abidjan, Ivory Coast	33	15	770
	Bangkok, Thailand	25	15	1000
	Bogota, Colombia	14	26	1180
	Medellin, Colombia	6	26	1180
	Tunis, Tunisia	24	40	1230
	Ankara, Turkey	23	25	1280
	Amman, Jordan	44	54	1500
	Mexico City, Mexico	19	62	1760
	Kuala Lumpur, Malaysia	37	72	1940
	Rio de Janeiro, Brazil	24	91	2160
	São Paulo, Brazi	32	91	2160
Upper-middle income	Budapest, Hungary	15	171	2460
	Seoul, South Korea	9	22	3600
High income	Barcelona, Spain	53	270	7740
	Madrid, Spain	37	270	7740
	Singapore	47	100	9070
	Wellington, New Zealand	56	412	10,000
	Liverpool, England	58	322	12,810
	London, England	61	322	12,810
	Milan, Italy	33	421	13,330
	Naples, Italy	58	421	13,330
	Brussels, Belgium	43	374	14,490
	Amsterdam, Netherlands	80	350	14,520
	Vienna, Austria	56	355	15,470
	Paris, France	63	409	16,090
	Copenhagen, Denmark	56	292	18,450
	Hamburg, Germany	50	364	18,480
	Munich, Germany	61	364	18,480
	Stuttgart, Germany	44	364	18,480
	Stockholm, Sweden	48	393	19,300
	USA	93	567	19,840
	Oslo, Norway	57	380	19,990
	Osaka, Japan	31	248	21,020
	Tokyo, Japan	32	248	21,020

Source: Schwela and Zali (1999)

Table 8.7 *Projected growth of global motor vehicle fleet by national income level, 1995–2050*

Year	Low- and middle-income countries		High-income countries		Total motor vehicle fleet (millions)
	(millions)	(%)	(millions)	(%)	
1995	164	25	487	75	651
2000	209	27	565	73	774
2010	340	31	759	69	1099
2020	555	35	1020	65	1575
2030	905	40	1370	60	2275
2040	1470	44	1840	56	3310
2050	2400	48	2475	52	4975

Source: UNCHS (2001)

in these cities – which are mostly middle-income cities – are rapidly growing, the world's consumption of fossil fuels and total emissions of greenhouse gases from vehicle use will also increase dramatically in the coming decades.

In fact, should the middle-income cities reach the degree of motorization found in US and European cities, the problems of urban air pollution in these cities will definitely be greatly exacerbated.[6] Moreover, such problems would be compounded by the fact that even if technological breakthroughs in vehicle and fuel technology were able to greatly reduce emissions from motor vehicles, the long turnover period for automobiles in developing countries – estimated at between 15 and 20 years – would in all likelihood result in these vehicle fleets accounting for an increasingly larger share of the world's total automotive emissions (Faiz, 1993). Furthermore, in these cities, gasoline often has a high sulphur content and vehicle engines are not always properly maintained. Coupled with the fact that a significant proportion of their trucks are powered by diesel and motorbikes powered by two-stroke engines, the overall air quality in the middle-income cities would thus continue to deteriorate and – particularly for those cities subject to continental high pressure systems for most parts of the year, such as Beijing, Delhi, and Mexico – smog would continue to be a serious and worsening region-wide problem (Romieu, 1999).

For a variety of reasons, comprehensive data on air pollution emissions from the transport and other sectors are not available for all countries and cities (WHO and UNEP, 1992). Nevertheless, the World Bank has estimated that, at the global level, vehicle emissions accounted for about one-fifth of the incremental increase in carbon dioxide, one-third of the chlorofluorocarbons, (CFCs) and half of the nitrogen oxides (World Bank, 1996). More detailed data collected on the OECD countries, as well as on several other European countries, have shown that, in the early 1990s, the transport sector accounted for about 4 per cent of sulphur dioxide emissions recorded in their cities, 11 to 14 per cent of the suspended particulate matters, more than 50 per cent

Table 8.8 *Emissions from transport: Local, regional, and global effects*

Pollutant	Type of impact						Source of emission	Health effects of pollutant
	Local	Regional		Global				
	High concentrations	Acidification	Photochemical oxidants	Indirect greenhouse effect	Direct greenhouse effect	Stratospheric ozone depletion		
Suspended particulate matter	X		X				Products of incomplete combustion of fuels; also from wear of brakes and tyres	Irritates mucous membranes; respiratory/pulmonary effects; carcinogenic
Lead	X			X			Added to gasoline to enhance engine performance	Affects circulatory, reproductive, and nervous systems
Carbon monoxide	X		X	X			Incomplete combustion product of carbon-based fuels	Reduced oxygen-carrying capacity of red blood cells
Nitrogen oxides (NO$_x$)	X	X	X	X		X	Formed during fuel combustion at high temperatures	Irritates lungs; increases susceptibility to viruses
Volatile organic compounds (VOCs)	X		X	X			Combustion of petroleum products; also evaporation of unburned fuel	Irritates eyes, causes intoxication; carcinogenic
Tropospheric ozone		X	X	X			Not an exhaust gas; product of photochemical reaction of NO$_x$ and VOCs in sunlight	Irritates mucous membranes of respiratory system; impairs immunities
Methane (CH$_4$)					X		Leakage during production, transport, filling and use of natural gas	
Carbon dioxide					X		Combustion product of carbon-based fuels	
Nitrous oxide				X	X	X	Combustion product of fuel and biomass; also formed in catalytic converters	
Chlorofluorocarbons					X	X	Leakage of coolant from air conditioning systems	

Source: OECD (1997)

of the nitrogen oxides, 50–60 per cent of the hydrocarbons and, in some city centres, over 90 per cent of the carbon monoxide (Romieu, 1999). The extent of exceedances of air concentrations of some key transport-related pollutants at the global level is summarized in Table 8.9, and the level of contribution of vehicular traffic to conventional pollutant emissions in some selected cities in Asia, Latin America and Eastern Europe is reported in Table 8.10.

Moreover, in cities where leaded gasoline is still used, the transport sector has been found to be a major contributor to fine particulate emissions – which are partly generated by lead in gasoline – and often produces as much as 80–90 per cent of the atmospheric lead (Kojima and Lovei, 2001). Research conducted by the World Bank has also shown that in some large city centres vehicular emissions may account for as much as 90–95 per cent of lead and carbon monoxide, 60–70 per cent of nitrogen oxides and hydrocarbons, and a major share of suspended particulate matters (World Bank, 1996). These emissions damage human health, especially in urban areas where human exposure to high concentrations of motor vehicle air pollutants is much more prevalent than in rural areas because the former has more roadways, parking garages and street canyons.[7]

The proportions of roadside population who are prone to a higher degree of exposure to urban air pollutants than those residing or working at some distances away from major roads in countries of different levels of development income are presented in Table 8.11. As shown in this table, an estimated 62 to 103 million people in cities in low-income countries were spending a considerable amount of their working hours in roadside locations. Their roadside population was about 1.7 times as large as that found in cities in middle-income countries and 2.3 to 2.7 times as large as those found in high-income countries. Put differently, whereas only about 7 per cent of the population in upper-middle income countries and around 22 per cent of the residents in lower-middle income countries were found to be living near highways, up to half of the population in the low-income countries were living next to major traffic thoroughfares and presumably exposed to a higher degree of concentrations of vehicular pollutants than their rural counterparts.[8] One major reason why the proportion of urban residents exposed to motor vehicle air pollutants in roadside settings was higher in developing countries than that in developed countries may be due to the fact that the former contained a much larger percentage of informal-sector labour force living and working in the streets than the latter (Flachsbart, 1999).

Although there are many risk factors other than air pollution, respiratory diseases are the major cause of death and illness in developing countries (Smith and Lee, 1993). In cities in the developing world, urban air pollutants have been blamed for causing an estimated 500,000 to 1 million people to die prematurely each year and for causing millions of cases of respiratory illness. In Mexico City, an estimated 12,500 people die prematurely each year due to exposure to high levels of particulate matter. Several comparative risk assessment and health studies conducted in a number of cities such as Bangkok, Cairo, Mexico City, Quito and Santiago de Chile also confirmed that fine particulate matter

Table 8.9 *Extent of exceedance of air concentrations of transport-related pollutants*

Pollutant	Extent of exceedance
Suspended particulate matter	World Health Organization (WHO) guidelines are exceeded by up to or more than a factor of 2 in 17 of 21 cities considered in one survey; in another, the guidelines were exceeded in 20 of 37 cities, with only 5 cities having concentrations within both annual and daily guidelines; the US Environmental Protection Agency (EPA) has designated 82 in 1994 areas as non-attainment areas.
Lead	People in about one-third of the world's cities are exposed to levels above WHO guidelines.
Carbon monoxide	Short-term WHO guideline values are often exceeded in many urban areas in Europe and in southern California; in the US, the EPA designated 36 regions as non-attainment areas for CO in 1994, with Los Angeles being classified as serious.
Nitrogen oxides	Major cities and metropolitan areas in Europe, the US, and Japan continue to experience high episodic values exceeding applicable standards; concentrations exceeding WHO guidelines by a factor of 2–4 have been measured in some non-OECD megacities.
Volatile organic compounds (VOCs)	Emissions and exceedances vary according to the compound. Acceptable emission levels for carcinogens may be zero, as in the case of two of the most important VOCs, 1, 3-butadiene and benzene, which respectively account for 32 and 5 per cent of US cancer cases related to air pollution and of which transport is responsible for 94 and 85 per cent of all emissions.
Tropospheric ozone	WHO guidelines for short- and long-term exposure are frequently exceeded in large areas of OECD Europe, North America, and Japan; the US EPA designated 77 areas as non-attainment areas in 1994.

Source: OECD (1997)

Table 8.10 Contribution of motor vehicles to conventional pollutant emission in selected cities

City	Year	Population (million)	Total human-made emissions from all sources		Per cent attributable to motor vehicles					
			('000 tons/year)	kg/capita	CO	HC	NO_x	SO_x	PM	Total
Asia										
Beijing	1989	9.7	n.a.	n.a.	39	75	46	n.a.	n.a.	n.a.
Bombay	1981	8.1	546	67.83	86	20	44	n.a.	3	31
Calcutta	1978	7.9	537	68.15	87	15	25	n.a.	n.a.	n.a.
Delhi	1981	5.6	360	64.29	84	81	67	9	24	52
	1987	7.0	428	61.58	90	85	59	13	37	57
Hong Kong	1984	5.2	550	106.59	93	n.a.	47	2	18	n.a.
Kuala Lumpur	1987	1.7	435	260.95	97	95	46	1	46	79
Manila	1987	7.1	496	69.96	93	82	73	12	60	71
Seoul	1983	10.1	n.a.	n.a.	15	40	60	7	35	35
Latin America										
Mexico City	1987	16.7	5027	301.92	99	89	64	2	9	80
	1989	19.4	4806	248.10	96	54	64	8	3	90
Santiago	1988	4.7	214	45.53	81	48	90	13	3	63
São Paulo	1977	10.1	2629	261.59	94	72	73	9	7	74
	1981	12.5	3150	252.00	96	83	89	26	24	86
	1987	15.5	2110	135.78	94	76	89	59	22	86
	1988	15.5	2080	133.85	94	77	82	73	31	n.a.
Eastern Europe										
Budapest	1987	2.1	n.a.	n.a.	81	75	57	12	n.a.	n.a.
Sarajevo	1983	0.3	n.a.	n.a.	82	35	68	5	23	n.a.

Note: n.a. = not available; CO = carbon monoxide; HC = hydrocarbon; NO_x = nitrogen oxide; SO_x = sulphur oxide; PM = particulate matter.

Source: Faiz (1993)

Table 8.11 *Estimates of roadside populations by economic group, 1990*

Economic group	Urban labour force (10^3)	Roadside population (10^3)		Proportion of total roadside population (%)	
		Low estimate	High estimate	Low estimate	High estimate
Low income	413,379	62,007	103,345	51.3	49.6
Lower-middle income	183,307	27,496	45,827	22.8	22.0
Upper-middle income	55,751	8,363	13,937	6.9	6.7
High income	427,554	23,024	45,226	19.0	21.7
World total	1,079,991	120,890	208,335	100.0	100.0

Source: Flachsbart (1999)

was causing the greatest damage to human health (Kojima and Lovei, 2001).[9] And in Cairo, as a result of high concentrations of lead in the outdoor air in some parts of the city, the lead content in the blood of children was found to be three to five times as high as that of their counterparts in rural Egypt (World Bank, 1996). Translated into economic terms, the monetary losses linked to urban air pollution in Asian cities were estimated to reach from US$1 billion to US$4 billion each year and up to US$6 billion in the newly independent states. In some of Asia's capital cities, such as Bangkok, Kuala Lumpur, and Jakarta, up to 10 per cent of the urban income were reportedly lost as a result of air pollution problems (Kojima and Lovei, 2001).

Despite these enormous costs in economic and human health terms associated with growing vehicular traffic, researchers have concluded that the total number of the world's motor vehicles, with a high degree of certainty, would double its 1990 level by the year 2020 (Faiz, 1993). More recent estimates focusing on China even suggested that its overall motorization rate would multiply five-fold between 1997 and 2020. Such a rapid increase in motor vehicles in that country is expected to lead to an annual growth rate of 6.9 per cent in transport-related gasoline consumption, vastly outpacing the world's average rate of 2.3 per cent (Riley, 2002). Given that the rapid increase in motor vehicles in China would in all likelihood result in some significant impacts on air quality at the city, regional and global scale, it is instructive for us to examine in greater detail the contours, causes and consequences of rapid motorization in this country.

Rapid motorization in China's cities

Motorization: Patterns and drivers

China's current transportation revolution – a shift to motor vehicles, and private automobiles in particular, as a dominant means of transport – has substantial implications not only for the state of its environment but for the global climate as well. Although China's motor vehicle fleet is still relatively small, it is already a significant source of urban pollution (World Bank, 1997). While it is difficult to distinguish vehicular emissions from other types of emissions in their contribution to air pollution, recent estimates of increased emissions in China suggested that motor vehicles had become a significant source of air pollutants (Riley, 2002). At the moment, for instance, non-point sources account for 45–60 per cent of nitrogen oxide and 85 per cent of carbon monoxide emissions in China's major cities (Pek, 2002). Ample empirical evidence has been gathered to show that air pollution from vehicles has become the major source of China's critical urban public health crisis, even though this claim may appear to be counterintuitive because common knowledge informs us that coal is the predominant source of energy in China (Lin et al, 1998; Weisbrod, 1999).[10]

The transport revolution in China is all the more significant because motorization is proceeding rapidly and the process is essentially irreversible. This rapid growth of private cars deserves closer scrutiny and systematic examination because such a drastic increase will certainly lead to substantial impacts on the environment, not only in China but also throughout the region and the globe. Even though the number of motor vehicles in China has tripled from 2.4 million in 1984 to 13.2 million in 1998, China is still one of the least motorized major world economies. In 1996, for example, 11.5 million motor vehicles were formally registered, yielding a motorization rate of 9.3 vehicles per 1000 persons (Riley, 2002).[11] Nevertheless, with the motor vehicle population expected to expand by a factor of 13–22 between 2000 and 2020, China is experiencing one of the highest annual motorization growth rates in the world (Gan, 2003). For instance, a recent World Bank study reported that China's passenger vehicles recorded an average annual growth rate of 19 per cent between 1990 and 1999. Some other recent estimates suggested that the country's overall motorization rate would increase by 500 per cent between 1997 and 2020, largely as a consequence of rapid growth of passenger vehicles (Riley, 2002). Moreover, if China should reach the private car ownership rate of 24 per 1000 persons – the average for low- and middle-income economies – by 2030, passenger vehicles could expand to as many as 33.6 million units, or the equivalent of 25 per cent of the US level in the year 2000 (Gan, 2003).[12]

There is abundant evidence that an increasing number of Chinese households have recently attained the magic income threshold – estimated at about US$4000 a year – that allows them to purchase their own private cars (Liu, 2002). In the late 1990s, private automobile ownership was growing by almost 50 per cent per year (World Bank, 1997). In fact, passenger vehicles have been growing nearly twice as rapidly as trucks since the mid-1980s.

While up until the recent past most of these cars were owned by government agencies or private firms (80 per cent in 1995), privately owned passenger vehicles are now the most rapidly growing segment of the automobile market (World Bank, 1997). The falling prices of automobiles, precipitated by China's entry into the World Trade Organization (WTO) in December 2001, have helped an expanding army of middle-income urban households to join the rank of car owners. In the early 1990s, less than 10 per cent of the automobiles were bought by private individuals. By the year 2000, the corresponding figure had jumped to 50 per cent (Liu, 2002). All these trends helped push up the rate of private ownership of motor vehicles, which rose from 6.7 per cent in 1984 to 25.0 per cent in 1995 (Riley, 2002).

In the case of China, there is a consensus among researchers that this rapid growth in the number of passenger vehicles, and private cars in particular, can be attributed more to a set of government policies than to the country's overall rapid economic growth (Stares and Liu, 1996; Pek, 2002; Riley, 2002; Gan, 2003). Even before the promulgation of the 1994 New Automobile Industrial Policy – which prescribed an overall strategy for the development of the automobile industry up to the year 2000 and beyond – the central government has been intimately involved in the expansion of the motor vehicle sector. During the 1995–2000 period the state invested a total of 58.8 billion yuan (equivalent to approximately US$7.5 billion) in 13 major state-owned enterprises which altogether controlled 90 per cent of the automobile market (Gan, 2003).

In the Tenth Five-Year Plan (2001–2005), the central government assigned a high priority to the development of the auto industry. The annual vehicle production output target for 2005 was set at 3.2 million units, with passenger cars accounting for about one-third of this figure. Considering the fact that passenger car production was enumerated at around 600,000 units per year in 2001, the emphasis placed on its rapid growth is remarkable (Gan, 2003). In fact, the Automobile Industrial Policy has called for the annual production of motor vehicles to increase by over 300 per cent between 1994 and 2010, and that of private cars to expand by no less than 1500 per cent for the same period (Spencer, 1999). The policy also envisioned that, by 2010, 3.5 million cars would be produced each year and 90 per cent of this output would be sold in the domestic market (Stares and Liu, 1996).

The priority given to the development of the automobile industry sector was motivated by the belief, on the part of central and local government officials alike, that the car industry could serve as a pillar of the economy. In other words, the increase in the use of the private car is largely the result of the government's strategy to promote economic development through the growth of the automotive sector and road-related infrastructure projects. Thus, between 1980 and 1998, the output value of the automobile industry increased 32.8 times, its fixed assets increased 25.5 times, and the workforce more than doubled (Gan, 2003). Moreover, by 2020, the Chinese government planned to invest in and complete a 54,700-km national highway network to connect all the country's major cities (Pek, 2002).

With regard to private cars, a number of policies have been deliberately designed and introduced to stimulate the middle-income households to purchase, own and use the automobiles: availability of car purchase loans; reduced fees for vehicle use; the development of roads and other road-related infrastructure to support motor vehicles; and lengthening the useful life of passenger cars from 10 to 15 years through better design and maintenance programs (Stares and Liu, 1996; Gan, 2003). For example, in 1993, after the Guangzhou municipal government decided to allow 5 per cent of the local households to own private cars by 2005, plans were put forth on how to eliminate two-thirds of the bicycles (reducing their number from 3 million to 1 million) on the city's streets by 2010 (Riley, 2002).

Motorization and air pollution

China's rapidly growing fleet of motor vehicles has had a substantial damaging impact on the country's environment, particularly in major urban centres where traffic jams are now becoming a daily feature. Although insufficient information does not allow us to assess the impact of motorization on the emissions factor, research has shown that automobiles have by now become the primary source of overall air pollution in China's major cities, accounting for 50–60 per cent of total emissions (World Bank, 2001; Riley, 2002). In 1999, only one-third of China's 338 cities – where air quality was monitored – were able to meet the country's residential ambient air quality standards (Pek, 2002). Findings from the national environmental monitoring network showed that 34 out of 94 cities monitored had exceeded the country's standard for nitrogen oxide emissions, which already allowed 10 times more emissions than those in Europe (Riley, 2002).

With air pollution figures in major industrial cities such as Taiyuan and Jinan regularly exceeding the WHO's safety limits by a factor of 10, a host of China's cities has continuously joined the ranks of the world's top 10 filthiest settlements (Pek, 2002). The WHO has recently stated that stricter enforcement of air quality standards in China's major cities could help prevent approximately 178,000 deaths and 346,000 hospital admissions each year. And the World Bank estimated that in the late 1990s the health impacts emanating from air pollution have shaved 3.5–7.7 per cent off China's overall economic output (Pek, 2002).

The shift to using the private car as the major means of transportation and its corresponding impact on urban air pollution is most apparent in China's major metropolitan regions. For instance, in Beijing, the number of private cars increased from 260,000 in 1986 to 1.65 million in 2000. Although this represented less than one-tenth of the number of private cars in Los Angeles, the emissions from Beijing's motor fleet were substantially greater (Gan, 2003). In Guangzhou, motor vehicles have become the principal source of several major pollutants: 67 per cent of nitrogen oxides and 87 per cent of carbon monoxides (Spencer, 1999). In Beijing, the corresponding figures reached 46 per cent and 30 per cent in the mid-1990s; and they have been on the rise as motorization advanced in that city and replaced industrial and domestic pollution as the most significant source of urban air pollution (Stares and Liu, 1996). In 2000, according to the World Resources Institute, up to 70 per cent of carbon monoxides emission in Beijing and Shanghai were attributed to motor vehicles. Thus, even though China's vehicle fleet, in terms of total absolute number, is still small compared to that of industrialized countries, its major metropolitan regions are already blanketed with heavy smog (Gan, 2003).

There are three inter-related aspects of the motorization trend that have helped contribute to higher levels of vehicle emissions: increasing transportation intensity, exacerbating traffic congestion, and high fuel intensity of China's transport sector. First, in most urban centres, the traffic volume of motor vehicles has expanded at a much higher rate than the growth of vehicles. In Beijing, for instance, the daily traffic volume has grown by 20 per cent annually in recent years, even though the number of motor vehicles has increased by only 14 per cent each year (Lin et al, 1998). That is, total vehicle emissions are on an upwards trend not simply because there are more cars but that each car is driven more.

Secondly, in recent years, the number of motor vehicles has grown nearly twice as fast as the expansion in highway networks, leading to increased road congestion in major urban centres and much lower average driving speeds. Coupled with the fact that the expansion of the road networks could not keep pace with the rapid rise in traffic volume, poor road quality and the mix of different speed vehicles have helped lower the average speed of urban traffic to 15–20km per hour and down to 10–15km per hour in city cores (Lin et al, 1998). Lower driving speeds increase emissions per kilometre because motors are less efficient at lower speeds and because they take longer to travel the same distance. Automobiles have thus become the predominant source of emissions in major metropolitan regions in China (World Bank, 1997).

Thirdly, motor vehicles have become the principal source of major air pollutants in urban China partly because the designs of the current fleet of motor vehicles are outdated and partly because the emission standards for new vehicles are inadequate.[13] By the mid-1990s, most of the vehicles produced were still following technologies from the 1970s and the 1980s, which had been transferred from industrialized countries (Walsh, 1996; Gan, 2003). For example, in the late 1990s, truck engines were still based on designs that were more than 20 years old, and most car engine designs were at least 10 years old. Chinese emission standards for cars allow 40 times more carbon monoxide, 6 times as many hydrocarbons, and 8 times as many nitrogen oxides than US standards. As a consequence, emissions from in-use vehicles were often 10–50 times those of vehicles in the US and Japan (World Bank, 1997).[14]

Air pollution is not the only environmental problem resulting from increased motorization in China's cities. A host of other problems pertaining to increased motor vehicle use includes traffic noise, especially from high-speed traffic on urban expressways, decreased local amenities associated with excessive vehicular flows through local streets, visual intrusion from poorly designed highway structures, and reduced traffic safety conditions due to the mixture of intrinsically incompatible vehicles, bicycles and pedestrians (Stares and Liu, 1996).

The public health problem emanating from automobile emissions has also led to the emergence of an unforeseen social equity issue. On the one hand, the increasing use of private automobiles has mostly benefited the well-off households in terms of advancing their social status and improving their mobility. On the other hand, a disproportional share of the health impact of pollution resulting from increased road transport tends to fall upon marginalized social groups such as the older generation, children and the working class. These groups do not enjoy any of the benefits resulting from policies that promote car ownership and subsidize infrastructure projects. In fact, they have been victimized by such policies, which tend to restrict or reduce their access to public transportation services.[15] Moreover, these groups have to pay for their own hospitalization and health care costs resulting from exposure to automobile emissions and traffic accidents. However, this inequity issue stemming from the government's public transportation policy has not yet been fully recognized and addressed by China's policymakers (Chang, 1999/2000; Gan, 2003).

Future prospect

The Chinese government has only recently come to acknowledge the importance of pollution control in the urban transport sector (He and Cheng, 1999; Walsh, 1999). For instance, automobile-related environmental protection issues were formally included in China's latest Five-Year Plan and leaded gasoline is being phased out throughout the country (Pek, 2002). A repertoire of policy options – albeit largely technically oriented – has been identified by the research community to help reduce vehicular emissions: instituting vehicle-emission standards; imposing fuel economy standards; encouraging fuel-efficiency improvements; and improving accessibility to reduce transport growth (Lin et al, 1998; Gan, 2003). Techniques in road management and transport demand management have also been offered as partial solutions to the traffic congestion problem (Stares and Liu, 1996). To be sure, China would have to overcome major barriers in introducing and implementing such technical fixes to the urban vehicular emissions problem. Such barriers – such as weak research and development (R&D) capacity to manufacture fuel-efficient models and a low rate of technology dissemination – are, however, not insurmountable, particularly given the fact that China could 'leap frog' the technology ladder by adopting from abroad advanced clean-vehicle technologies and products (Gan, 2003).

From a larger political economy perspective, it is institutional constraint, not technology gaps per se, that stand in the way of building a sustainable transport system in China's cities. Attempts to reduce emissions from automobile use in major metropolitan areas are hampered by enormous contradictions between leading government policies. At the national level, powerful ministries responsible for the economic development portfolio have identified the automobile manufacturing industry as one of the major pillars of the national economy. Apparently convinced by its capacity to stimulate economic growth through its many backwards and forwards linkages, the motorization strategy has received full blessing from the central government. Thus, despite concerns pertaining to constrained oil production and road transport capacities as well as adverse environmental impacts associated with the use of private cars, the national government still gives enormous support to building up the automobile industry. Moreover, regulations and action programmes designed to reduce automobile dependency, emissions and their impacts on the environment, mostly promulgated by the national-level environmental protection authorities, are rendered largely ineffective because they are not enforced by provincial- and city-level governments keen on hosting and gaining from the presence of automobile manufacturing plants within their jurisdictions (Shen, 1997; Menz, 2002).

Judged by the momentum of the unprecedented, rapid growth of automobile manufacturing and sales activities throughout China in recent years, it is evident that total vehicular emissions will continue to increase even if tougher emissions standards are implemented and traffic management improved. Given that the growth in automobile use is inevitable under China's government-sanctioned mass private motoring strategy, perhaps the need for developing innovative urban transport policies as well as sustainable urban land-use planning practices is all the more pressing.

Summary and conclusion

Given that virtually all the carbon monoxide and more than 800 million tonnes of carbon emitted in cities each year could now be traced to all types of motor vehicles, their environmental and health-related damages at both the local and global scales are no longer debatable (Walsh, 1996; Walsh, 2003). Researchers are, however, intrigued by the fact that the urban transport sector in both rich and poor countries has remained highly resistant to policy reform measures, despite an increasing body of evidence linking transport activities to urban air pollution, which in turn is associated with costly human health impacts. For instance, globally, the number of motor vehicles has recently been growing at a rate of 5 per cent per year, which is much faster than the human population growth rate of 2 per cent per annum (Romieu, 1999). And most of this increase in automobiles is taking place in middle-income cities and countries. Hence, while the absolute number of private automobile ownership in developing countries is on average still relatively low when compared with those registered in North America, middle-income cities – particularly those located in China's coastal and most prosperous provinces – have nevertheless reported the fastest rate of growth in the past 10 years or so.

The complexity of the causes and sources of urban air pollution requires policy-makers to adopt a multi-pronged approach, necessarily envisioned and implemented at the local level, which embodies a broad spectrum of instruments and actors and could thus gain from their synergetic and complementary impacts. For instance, an effectual reduction of vehicular emissions would have a much better chance of being fully accomplished through the implementation of a broad and yet coherent policy programme that comprises improvements in land-use planning and transport planning practices; tightening of the standards of fuel quality, vehicle technology and maintenance regime; and enhancement of traffic management and infrastructure investment (Murray and McGranahan, 2003). For the purposes of formulating and implementing either short-term or longer-term approaches, there is then an urgent need to improve the monitoring and emissions inventory capabilities of cities – which are glaringly lacking in middle-income cities with the fastest growth rates of automobiles – to help them establish the scientific basis of air pollution management strategies. Thus, while it is true that effective policies need to be locally driven, it is also critical that they be internationally informed in the sense that successful experiences in urban air pollution management approaches adopted by middle-income cities elsewhere be referred to in the policy formulation process.

Research has shown that effective strategies and measures have indeed been devised in some middle-income cities and countries to tackle successfully the problem of vehicular pollution (Walsh, 2003). Based upon the analysis of extensive empirical data gathered at the global level, for instance, a study that focused on the issue of automobile dependence has identified three key priority public policy directions that are crucial in reversing such a worldwide trend: increase investment in public transit infrastructure to help shape the city as well as ease traffic congestion; shift the orientation in transport and land-use planning towards facilitating pedestrian flow and bicycle traffic; and build high-density urban villages to encourage walking and the use of public mass transit systems (Newman, 1999).

In addition to these automobile dependence reduction strategies and air pollution reduction policy options, such as the institution of stringent vehicle-emission and fuel-efficiency standards, both of which are primarily technically-oriented, a fundamental key to reducing the adverse environmental and health impacts of rapid motorization in middle income cities and countries lies in persuading policy-makers to re-think and re-define the ultimate objective of transportation development. Accessibility, not mobility, needs to be recognized as the basic goal of transport planning (Lin et al, 1998). Specifically, this means that the motorization strategy – which emphasizes the expansion of the transport system to facilitate increased level of mobility – would need to be replaced by an end-user approach that emphasizes accessibility, efficiency and sustainability. However, given that most middle-income cities and countries are committed to an official policy to promote or facilitate economic development through automobile sector growth, convincing them to do otherwise would require sustained efforts, at the local, national and global levels, to support scientific and policy research as well as develop innovative political manoeuvres that are aimed at changing the mindset of all the relevant stakeholders.

Notes

1 The UET framework has been applied to examining and identifying some broad patterns of contemporary urban conditions in a cross-section of Southeast Asian cities: While poor cities such as Hanoi and Phnom Penh were afflicted with largely 'brown' issues such as lack of potable water and sanitation facilities, air pollution and congestion problems were identified as the dominant concerns for middle-income cities such as Bangkok and Kuala Lumpur (Webster, 1995).
2 The compressed and telescoped process of development of the transport systems prominently evident in many middle-income cities could be traced to three major causes. Firstly, the rise of foreign direct investment (FDI) and transfer of technology between developed and developing countries have helped fuel the internationalization of the motor vehicle industry in the late 1980s and the 1990s. Secondly, coupled with the availability of new consumer items such as the automobiles, the enormous size of the rapidly expanding middle and upper classes in the Asia-Pacific region – 400 million Chinese were classified as high-income (fifth quintile) earners in 1997, for instance – has helped to exacerbate the telescoping effect of the UET. Thirdly, a number of developing countries in Asia, such as South Korea, Malaysia and China, have rushed into developing the motor industry as a backbone of their national economies. The congregation of motor vehicles in cities that were not designed to accommodate modern transport thus contributes to the aggravation of the overlapping impacts of different types of urban environmental problems (Marcotullio and Lee, 2003).
3 A 1996 World Bank forecast suggested that the world's motor vehicles would grow 34 per cent between 1989 and 2000, from 557 million units to 745 million units (World Bank, 1996, p24).
4 This analysis has shown that, since 1970, 16 million motor vehicles have been added to the world's motor vehicle fleet each year and that worldwide registrations have grown by about two vehicles per 1000 persons (Faiz, 1993, p176). If this trend were to continue until 2010, there would be 215 motor vehicles per 1000 persons (excluding motorcycles) compared with 178 in 1994 (Romieu, 1999, p6).
5 This statistic highlights the fact that the most common forms of travel in low-income countries are by foot and mass transit such as pedal and motorized rickshaws, small minibuses, shared taxis, motorcycles and mopeds, and converted vans, pickups and jeeps (Flachsbart, 1999, p112).
6 Motor vehicles are the major source of a number of pollutants, such as carbon monoxide, nitrogen oxides, unburned hydrocarbons and lead and, in smaller proportions, suspended particulate matter, sulphur dioxide and volatile organic compounds.
7 It is worth noting that although a country's level of urbanization could be used as an indicator of the number of people exposed to motor vehicle air pollution, studies have shown that stationary monitoring sites have always underestimated human exposure to some types of vehicular emissions (Flachsbart, 1999, p111).
8 This assertion is based on these two premises: (a) that the number of people exposed to air pollutants from motor vehicles depends upon the extent to which a country is motorized; and (b) that countries with higher levels of motorization can potentially

lead to a greater number of people exposed to vehicular pollutants (Flachsbart, 1999, p111).

9   It should be noted that emission trends in the two best studied pollutants, sulphur dioxide and particulates, have been mixed; some developing nations showed slow downward trends (China), whereas others showed significant (India) or spectacular (Thailand) increases. Even leaving aside the question of uncertainty of such estimates, Smith and Lee (1993) have cautioned that the shift in temporal and spatial patterns of these emissions makes it difficult to draw conclusions about their potential health impacts.
10 A recent nationwide survey conducted by the State Environmental Protection Administration has revealed that, in two out of every three cities, the residential ambient air quality failed to meet national standards (Peng et al, 2002). Given that a substantial proportion of the population in major metropolitan regions is exposed to excessive health risks such as chronic bronchitis, pulmonary heart disease and lung cancer, it is not surprising to find that a recent national study of mortality has concluded that respiratory diseases are the second leading cause of premature deaths in China (Xu, 1998).
11 In comparison, 206 million vehicles were registered in the US in the same year, generating a motorization rate of 770 vehicles per 1000 persons (Riley, 2002). It should be noted that, after industrial and commercial vehicles are excluded in the calculation, the figures of passenger car ownership rate for both China and the US are much lower. In 1999, they were 3 per 1000 persons and 478 correspondingly (Gan, 2003).
12 Some other analysts have predicted a much higher level of motorization rate: 54 vehicles per 1000 persons by 2020. This would translate into an average annual increase of 7.3 per cent between 2000 and 2020 (Riley, 2002).
13 Although China follows the Euro I standards for urban automobile emissions, actual emissions often exceed the standards because of a lack of monitoring stations, advanced monitoring equipment, and effective law enforcement measures (Gan, 2003).
14 According to the World Resources Institute, Chinese-made vehicles emit 2.5–7.5 times more hydrocarbons, 2–7 times more nitrous oxides, and 6–12 times more carbon monoxide than foreign vehicles (Gan, 2003).
15 According to a Woodrow Wilson Center report, despite government policy support for public transport backed by investment in vehicles and services, the effectiveness of increased capacity in public transport was eroded by a decrease in public transport efficiency and quality of services (Chang, 1999/2000).

References

Bai, X.-M. and Imura, H. (2000) 'A comparative study of urban environment in East Asia: Stage model of urban environmental evolution', *International Review for Environmental Strategies*, vol 1, pp135–58

Chang, D. T. (1999/2000) 'A new era for public transport development in China', *China Environment Series*, Woodrow Wilson Center, Environmental Change and Security Project, issue 3, pp22–27

Faiz, A. (1993) 'Automotive emissions in developing countries – Relative implications for global warming, acidification and urban air quality', *Transportation Research*, vol 27A, no 3, pp167–186

Flachsbart, P. G. (1999) 'Exposure to exhaust and evaporative emissions from motor vehicles', in D. Schwela and O. Zali (eds), *Urban Traffic Pollution*, E. & F. N. Spon, London and New York, pp110–119

Gan, L. (2003) 'Globalization of the automobile industry in China: Dynamics and barriers in greening of the road transportation', *Energy Policy*, vol 31, pp537–551

He, K. and Chang, C. (1999) 'Present and future pollution from urban transport in China', *China Environment Series*, Woodrow Wilson Center, Environmental Change and Security Project, issue 3, pp38–50

Kenworthy, J. R., Laube, F. B., Raad, T., Poboon, C. and Guia, B. (1999) *An International Sourcebook of Automobile Dependence in Cities, 1960–1990*, University Press of Colorado, Boulder, CO

Kojima, M. and Lovei, M. (2001) *Urban Air Quality Management: Coordinating Transport, Environment, and Energy Policies in Developing Countries*, World Bank Technical Paper No. 508, Pollution Management Series, World Bank, Washington, DC

Lin, X., Polenske, J. and Polenske, K. R. (1998) 'Energy use and air-pollution impacts of China's transportation growth', in M. B. McElroy, C. P. Nielsen and P. Lydon (eds), *Energizing China: Reconciling Environmental Protection and Economic Growth*, Harvard University Press, Cambridge, MA, pp201–238

Liu, M. (2002) 'Road warriors, middle-class Chinese are going car crazy, buying autos and hitting the road as never before,' *Newsweek*, 29 April, p26

Marcotullio, P. J. and Lee, Y.-S. F. (2003) 'Urban environmental transitions and urban transportation systems; a comparison of the North American and Asian experiences', *International Development Planning Review*, vol 25, no 4, pp325–354

Marcotullio, P. J., Rothenberg, S. and Nakahara, M. (2003) 'Globalization and urban environmental transitions: Comparison of New York's and Tokyo's experiences', *Annals of Regional Science*, vol 37, pp369–390

McGranahan, G., Jacobi, P., Songsore, J., Surjadi, C. and Kjellen, M. (2001) *The Citizens at Risk: From Urban Sanitation to Sustainable Cities*, Stockholm Environment Institute, Stockholm, and Earthscan, London

Menz, F. C. (2002) 'The US experience with controlling motor vehicle pollution: Lessons for China', *International Journal of Environment and Pollution*, vol 18, no 1, pp1–21

Murray, F. and McGranahan, G. (2003) 'Air pollution and health in developing countries – The context', in G. McGranahan and F. Murray (eds), *Air Pollution and Health in Rapidly Developing Countries*, Earthscan, London, pp1–20

Newman, P. (1999) 'Transport: Reducing automobile dependence', in D. Satterthwaite (ed), *The Earthscan Reader in Sustainable Cities*, Earthscan, London, pp173–198

OECD (1997) 'Conference highlights and overview of issues,' in *Towards Sustainable Transportation*, Organisation for Economic Co-operation and Development Proceedings, conference organized by the OECD hosted by the government of Canada, Vancouver, British Columbia, 24–27 March 1996

Pek, J. E. (2002) 'China's tailpipe tally: The world's biggest nation "modernizes" with more cars', *The Environmental Magazine*, vol 13, issue 6, p16

Peng, C., Wu, X., Liu, G., Johnson, T., Shah, J. and Guttikunda, S. (2002) 'Urban air quality and health in China', *Urban Studies*, vol 39, no 21, pp2283–2299

Riley, K. (2002) 'Motor vehicles in China: The impact of demographic and economic changes', *Population and Environment*, vol 23, no 5, pp479–494

Romieu, I. (1999) 'Epidemiological studies of health effects arising from motor vehicle air pollution', in D. Schwela and O. Zali (eds), *Urban Traffic Pollution*, E. & F. N. Spon, London and New York, pp1–7

Schwela, D. (1999) 'Conclusions and recommendations', in D. Schwela, and O. Zali (eds), *Urban Traffic Pollution*, E. & F. N. Spon, London and New York, pp237–240

Schwela, D. and Zali, O. (eds) (1999) *Urban Traffic Pollution*, E. & F. N. Spon, London and New York

Shen, Q. (1997) 'Urban transportation in Shanghai, China: Problems and planning implications,' *International Journal of Urban and Regional Research*, vol 21, issue 4, pp589–606

Smith, K. and Lee, Y.-S. F. (1993) 'Urbanization and the environmental risk transition', in J. D. Kasarda and A. M. Parnell (eds), *Urbanization, Migration, and Development*, Sage Publications, Newbury Park, CA, pp161–179

Spencer, A. H. (1999) 'Challenges to sustainable transport in China's cities', in G. P. Chapman, A. K. Dutt and R. W. Bradnock (eds), *Urban Growth and Development in Asia, Vol I: Making the Cities*, Ashgate, Aldershot, Brookfield, Singapore, Sydney, pp426–443

Stares, S. and Liu, Z. (1996) 'Theme Paper 1: Motorization in Chinese cities: Issues and actions,' in S. Stares and L. Zhi (eds), *China's Urban Transport Development Strategy*, East Asia and Pacific Region Series, World Bank Discussion Paper No. 352, World Bank, Washington, DC, pp43–104

UNCHS (United Nations Centre for Human Settlements) (2001) *Cities in a Globalizing World: Global Report on Human Settlements 2001*, Earthscan, London

UNEP (United Nations Environment Programme (1999) *Global Environment Outlook 2000*, Earthscan, London

Walsh, M. P. (1996) 'Theme Paper 2: Motor vehicle pollution control in China: An urban challenge', in S. Stares and L. Zhi (eds), *China's Urban Transport Development Strategy*, East Asia and Pacific Region Series, World Bank Discussion Paper No. 352, World Bank, Washington, DC, pp105–151

Walsh, M. P. (1999) 'Transportation and the environment in China', in *China Environment Series*, Woodrow Wilson Center, Environmental Change and Security Project, issue 3, pp28–37

Walsh, M. P. (2003) 'Vehicle emissions and health in developing countries', in G. McGranahan and F. Murray (eds), *Air Pollution and Health in Rapidly Developing Countries*, Earthscan, London, pp146–175

Webster, D. (1995) 'The urban environment in Southeast Asia: Challenges and opportunities', *Southeast Asian Affairs*, Institute of Southeast Asian Studies, Singapore, pp89–107

Weisbrod, R. E. (1999) 'Solving China's urban crisis: China's transportation energy future', *Journal of Urban Technology*, vol 6, no 1, pp89–100

WHO (World Health Organisation) and UNEP (1992) *Urban Air Pollution in Megacities of the World*, Blackwell, London

World Bank (1996) *Sustainable Transport: Priorities for Policy Reform*, World Bank, Washington, DC

World Bank (1997) *Clear Water, Blue Skies, China 2020 Series*, World Bank, Washington, DC

World Bank (2001) *China: Air, Land and Water*, World Bank, Washington, DC

Xu, X. (1998) 'Air pollution and its health effects in urban China,' in M. B. McElroy, C. P. Nielsen and P. Lydon (eds), *Energizing China: Reconciling Environmental Protection and Economic Growth*, Harvard University Press, Cambridge, MA, pp267–285

9
A Comparative Perspective on Urban Transport and Emerging Environmental Problems in Middle-income Cities

Jeff Kenworthy and Craig Townsend

Introduction

It is easy to find anecdotal horror stories about the traffic chaos and danger, unattractive public transport systems, vehicular pollution and generally degraded public environments that exist in many cities in developing countries. There is general agreement that urban transport systems and environmental conditions in very large and rapidly developing cities, including Bangkok, Beijing, Jakarta, Manila, Mumbai, and São Paulo, are in great need of improvement. Often, the sheer presence and impact of traffic in such cities lead to a perception that they must be highly dependent on private motorized transport. In like fashion, the high-quality public transport systems, good conditions for walking and cycling, and attractive environments of many European cities are often compared favourably to the auto-dependent cities and less attractive public environments of some American cities. Whatever the comparisons being made, people have firm opinions and perceptions about the strengths and weaknesses of urban transport systems, the differing levels of environmental quality in cities around the world, and the root causes of transport problems and how to fix them. However, it is rather more difficult to find systematic and reliable comparative transport research and standardized data on a wide range of cities with which to back up or refute assertions and prescriptions for urban transport and environmental improvements.

This chapter utilizes standardized data on cities from 50 nations in order to compare transport, land-use, economic, and environmental characteristics in 1995. Particular attention is paid to the passenger transport characteristics and related environmental implications in middle-income cities. Given the importance of reliable and comparable data to such an exercise, the chapter commences with a brief overview of methodological issues behind the research data that underpin the chapter. The sections that follow analyse differences between cities across six dimensions:

1 private motor vehicle ownership and use;
2 evolution of urban transport systems and urban form;
3 private transport infrastructure;
4 public transport service and usage;
5 environmental impacts of urban transport;
6 economic aspects of urban transport.

Findings are evaluated in the context of the Urban Environmental Transition (UET) thesis, which argues that the type and intensity of environmental impacts in cities change reasonably systematically with growing affluence (McGranahan et al, 1999, 2001).

Methodology

This chapter draws on the Millennium Cities Database for Sustainable Transport, the only large, standardized source of data on urban transport and related characteristics of cities (Kenworthy and Laube, 2001). The database offers the potential for quantitative comparisons of cities (together with their suburbs) on a global scale. At this macro-scale, some relationships and patterns that challenge findings of single-city or micro-scale studies emerge.[1] In order to compare cities on the basis of wealth, this chapter groups cities into clusters.

Income classification

The city data is arranged first by gross regional product (GRP) per capita, and then by national boundaries and continents according to clusters of characteristics (Table 9.1). These groupings are based on a hierarchical cluster analysis and other exploratory methods described in Barter et al (2002). This chapter utilizes income classifications established by Barter et al (2002), and these classifications differ from the World Bank Atlas country income categories, which are used as the basis for income classifications in the other chapters of this book. Barter et al (2002) divide cities into high income ($US16,000 or more), middle income ($US3000–$US15,999), and low income ($US2999 or less). According to the World Bank Atlas method, the groups are high income ($US9206 or more), upper middle income ($US2976–$US9205), lower middle income ($US746–$US2975), and low income ($US745 or less).

Table 9.1 Urban areas in the Millennium Cities Database for Sustainable Transport grouped according to region and income groups, and populations (millions)

Western Europe (WEU) (69.41)			United States (US) (57.36)	Canada (CAN) (11.49)
Munich (1.32)	Copenhagen (1.74)	Berlin (3.47)	San Francisco (3.84)	Vancouver (1.90)
Frankfurt (0.65)	Stockholm (1.73)	London (7.01)	Washington (3.74)	Calgary (0.77)
Zurich (0.79)	Ruhr (7.36)	Barcelona (2.78)	New York (19.23)	Toronto (4.63)
Geneva (0.40)	Nantes (0.53)	Madrid (5.18)	Denver (1.98)	Ottawa (0.97)
Düsseldorf (0.57)	Graz (0.24)	Glasgow (2.18)	Chicago (7.52)	Montreal (3.22)
Bern (0.30)	Marseilles (0.80)	Manchester (2.58)	Atlanta (2.90)	
Lyon (1.15)	Helsinki (0.89)	Newcastle (1.13)	Houston (3.92)	**Aust/NZ (ANZ) (9.98)**
Paris (11.00)	Amsterdam (0.83)	Athens (3.46)	Los Angeles (9.08)	Sydney (3.74)
Stuttgart (0.59)	Brussels (0.95)		Phoenix (2.53)	Perth (1.24)
Vienna (1.59)	Bologna (0.45)		San Diego (2.63)	Melbourne (3.14)
Oslo (0.92)	Rome (2.65)			Wellington (0.37)
Hamburg (1.70)	Milan (2.46)			Brisbane (1.49)
High Income Asia (HIA) (60.23)	**Middle Income Asia (MIA) (36.00)**	**Low Income Asia (LIA) (68.16)**	**Assorted Middle Income (AMI) (33.78)**	**Assorted Low Income (ALI) (30.76)**
Tokyo (32.34)	Taipei (5.96)	Guangzhou (3.85)	Tel Aviv (2.46)	Bogotá (5.57)
Osaka (16.83)	Seoul (20.58)	Shanghai (9.57)	Prague (1.21)	Teheran (6.80)
Sapporo (1.76)	Kuala Lumpur (3.77)	Manila (9.45)	Curitiba (2.43)	Tunis (1.87)
Hong Kong (6.31)	Bangkok (6.68)	Jakarta (9.16)	Riyadh (3.12)	Cairo (13.14)
Singapore (2.99)		Beijing (8.16)	Budapest (1.91)	Dakar (1.94)
		Ho Chi Minh City (4.81)	Saõ Paulo (15.56)	Harare (1.43)
		Mumbai (17.07)	Johannesburg (2.25)	
		Chennai (6.08)	Cape Town (2.9)	
			Krakow (0.74)	

Notes: For consistency with the results of a cluster analysis designed to group the cities, the cut-off points in terms of gross regional product (GRP) per capita (1995 prices) between high-income and middle-income cities and between middle-income and low-income cities, have been chosen to be US16,000 and US$3000, respectively. US, Canadian, Western European and Australian/New Zealand cities are all considered to be high-income cities.
The following cities are also included in the database but could not be included in the analysis here due to incomplete data sets: Lille, Turin, Lisbon, New Delhi, Buenos Aires, Rio de Janeiro, Brasilia, Salvador, Santiago de Chile, Mexico City, Caracas, Abidjan, Casablanca, Warsaw, Moscow, Istanbul.

Source: Barter et al (2002)

The high-income city clusters created by Barter et al (2002) and used in this chapter all correspond with cities that would have been high income according to the World Bank Atlas method. These clusters are Western Europe (WEU), the United States (US), Canada (CAN), Australia/New Zealand (ANZ) and High Income Asia (HIA), which comprises Hong Kong, three cities in Japan and Singapore (see Table 9.1). Had the World Bank Atlas method been used, Seoul, Taipei and Tel Aviv, which are classified as middle income under the Barter et al (2002) classification, would be high income (and presumably added to the High Income Asia cluster).

The Middle Income Asia (MIA) and Assorted Middle Income (AMI) clusters comprise mainly cities that would have been classified as upper middle income by the World Bank Atlas method. In similar fashion, the Low Income Asia (LIA) and Assorted Low Income (ALI) clusters comprise cities that would have been classified as lower middle income by the World Bank Atlas method. The cities included in the two 'Assorted' clusters come from a range of continents and nations and are classified here as residual clusters. The Indian city of Chennai, included in the Low Income Asia cluster, is the only city in the database that would have been classified as low income using the World Bank Atlas method. The lack of low-income cities in the Millennium Cities Database, according to World Bank income rankings, raises city selection issues which are addressed below. It thus needs to be borne in mind that where reference is made to the low-income clusters of cities in this chapter, the cities would virtually all be lower-middle income cities, according to World Bank classifications.

These discrepancies in income classification raise important issues about the types of cities that were selected for the Millennium Cities Database. One is that most of the cities from developing countries selected for the database are those undergoing industrialization and economic growth. Largely absent are large low-income cities from Asia, the Middle East, Central America and sub-Saharan Africa. The cities chosen for the database are mainly from wealthy nations (60 per cent), and within the cities of wealthy nations, are skewed towards Western Europe (35 out of 60).

Another issue that is raised by these income classification methods is the difference between national and urban statistics. The Millennium Cities Database provides standardized figures for physical urban areas (that is, it controls for the wide variations in the forms of urban administration and accompanying statistics that can otherwise be non-compatible), rather than for national averages. In more developed and socially equitable nations, the figures for nations and cities do not differ a great deal. However, in many developing nations, large cities are often far wealthier than more peripheral and smaller cities and rural areas. For example, the large capital cities of Thailand (Bangkok) and India (Delhi) are one income class higher than their whole nation.

City selection and data issues

In addition to issues surrounding income classification, there are also difficulties posed by city size. The Western Europe cluster includes cities ranging in size from 0.25 million to 11 million inhabitants, while US and Japanese cities of under 1.5 million are absent from the database (see Table 9.1). The range of cities from nations of the developing world is in most cases limited to the large capital cities, but it is unclear whether those cities are representative of the wider urban conditions found in smaller cities in those developing nations. While to some extent this reflects the location and mission of the organization that funded the Millennium Cities Database,[2] it also reflects differing demographic and settlement patterns between nations and continents.

Apart from size, the choice of cities is also influenced by data availability. Cities with very limited or inaccurate data are excluded from the database, and 16 of the 100 cities in the database are excluded from the cluster analysis in this chapter as a result of incomplete data sets. Despite these exclusions, there remain inaccuracies such as the over-counting of motor vehicles and under-counting of urban populations in many cities of developing nations.[3] These problems are systematic and are largely beyond the scope of a cross-national urban transport and land-use database. Such problems are inevitable in any study of this geographic scope and magnitude of data collection.

Notwithstanding these overall caveats on the data, many of the observed patterns convey such stark differences that the error bands are judged acceptable for this chapter's purpose. The main aim is to convey some general differences between clusters of cities classified as middle income compared to other clusters, and to explore reasons for those differences.

Urban transport in middle-income cities

The following analysis of urban transport in middle-income cities is based on the data set out in Table 9.2 and is organized into six categories. Based on these comparisons and reference to other research, some suggestions are also made about how cities may compare in the future. In particular, an attempt is made to assess whether environmental problems associated with urban transport change in ways corresponding with the UET thesis. The thesis suggests that cities go through similar stages and 'As affluence increases, there tends to be a spatial shifting of environmental problems from the local to the regional and global' (McGranahan et al, 1999, p109).

Private motor vehicle ownership and use

The use of privately owned motor vehicles in cities and the attendant environmental, economic and social problems, have been recognized as critical issues since the early 20th century. The eminent American urbanist Lewis Mumford recognized it as early as the 1930s and vividly articulated his concerns in *The City in History* (Mumford, 1961). Also in 1961, Jane Jacobs passionately criticized the actions being taken by city planners and others to turn city streets over to fast moving cars. She spoke of the need to make conditions less convenient for cars in order to restore critical elements of city life, and called this needed process the 'attrition of automobiles by cities' (Jacobs, 1961). Just before the first oil crisis in 1973, which was to dent the modernist assumptions and unquestioning optimism of many Americans toward the automobile as a part of daily life, Owen (1972) identified a 'worldwide conflict between cities and cars'. To a large extent, the middle-income cities in 1995 faced similar problems but, for a variety of reasons that are addressed in this chapter, the problems were and continue to be even more intense than those faced earlier in the US and Europe.

Vehicle ownership
With only a fraction of the GRP per capita of the five high-income city clusters, Middle Income Asia and Assorted Middle Income clusters have distinctly high levels of car ownership. In addition, Middle Income Asia has a very high level of motorcycle ownership, which appeared only in the late 20th century, and which is not a prominent feature of the Assorted Middle Income and Assorted Low Income clusters (outside Asia). High Income Asia has 3.6 times higher GRP per capita than Middle Income Asia ($34,797 compared to $9776). However, the combined car and motorcycle ownership of High Income Asia is significantly less than (283 per 1000 people compared to 352 per 1000 people). The contrast is even more stark when the Assorted Middle Income cluster comprising cities outside Asia is compared with High Income Asia: GRP per capita is five times higher in High Income Asia ($34,797 vs $6625) but combined car and motorcycle ownership per capita is virtually identical (283 vs 280).

However, it is not only the middle-income cities that have high vehicle ownership. Combined car and motorcycle ownership ranges from 134 to 86 per 1000 people in the Low Income Asia and Assorted Low Income clusters respectively in 1995. Considering the very low per capita GRPs of $1689 and $1949, these vehicle ownership levels are particularly noteworthy.

Vehicle use
Both Middle Income Asia and the Assorted Middle Income clusters are significantly higher in combined car and motorcycle use per capita than High Income Asia. Middle Income Asia and Assorted Middle Income clusters record 4682 and 4211 passenger km per capita respectively, compared to 3824 per capita in High Income Asia. Both Low Income Asia and Assorted Low Income have markedly less private motorized travel (1201 and 1262 passenger km per capita) than Western Europe (6321 passenger km) and the US (18,200 passenger km). Private passenger travel in US cities in 1995 is some four times higher than in middle-income cities and fourteen times higher than in the low-income cities.

If we consider all trips by all modes of transport, including foot and bicycle, the automobile dependent, hyper-mobile cities of the US, Canada and Australia/New Zealand have only 10–20 per cent of all trips by public transport and non–motorized modes. On the other hand, High Income Asia has on average 61 per cent of all daily trips by public transport and non-motorized modes. Indeed, wealthy cities have 50 per cent of daily trips by these 'green modes'. By contrast, 45 per cent of daily trips in Middle Income Asia are by public transport and non-motorized modes and 54 per cent are

Table 9.2 Land-use and transport system characteristics by groupings of cities, 1995

		US	ANZ	CAN	WEU	HIA	MIA	LIA	AMI	ALI
Land use and wealth										
Urban density	persons/ha	14.9	15.0	26.2	54.9	134.4	164.3	205.6	53.7	122.1
Proportion of jobs in central business district (CBD)	%	9.2	15.1	15.7	18.7	20.1	13.1	31.8	16.8	21.2
Metropolitan gross domestic product (GDP) per capita	US$	$31,386	$19,775	$20,825	$32,077	$34,797	$9776	$1689	$6625	$1949
Transport investment cost										
Per cent of metro. GDP spent on public transport investment	%	0.18	0.30	0.18	0.41	0.47	1.22	0.53	0.39	0.62
Per cent of metro. GDP spent on road investment	%	0.86	0.72	0.87	0.70	0.96	1.34	1.82	0.70	0.75
Private transport infrastructure indicators										
Length of expressway per person	m/person	0.156	0.129	0.122	0.082	0.022	0.027	0.004	0.043	0.009
Parking spaces per 1000 CBD jobs		555	505	390	261	121	164	55	374	134
Public transport supply and service										
Public transport seat kilometres of service per capita	seat km/person	1557	3628	2290	4213	5535	2734	2057	3283	3322
Rail seat kilometres per capita (Tram, LRT, Metro, Sub. rail)	seat km/person	747	2470	677	2609	2720	362	250	1684	120
Per cent of public transport seat kilometres on rail	%	34.2	65.2	27.8	55.5	57.2	13.1	12.9	33.6	10.1
Overall average speed of public transport	km/h	27.4	32.7	25.1	25.7	33.2	16.4	16.6	24.8	21.1
Ratio of public versus private transport speeds		0.58	0.75	0.57	0.79	1.08	0.78	0.80	0.70	0.71
Private transport supply (cars and motorcycles)										
Passenger cars per 1000 persons		587	575	530	414	217	198	38	265	71
Motorcycles per 1000 persons		13	13	9	32	66	154	96	15	15
Mode split of all trips										
non-motorized modes	%	8.1	15.8	10.4	31.3	29.1	19.8	50.1	27.9	36.3
motorized public modes	%	3.4	5.1	9.1	19.0	32.3	25.6	28.3	26.6	32.8
motorized private modes	%	88.5	79.1	80.5	49.7	38.6	54.6	21.6	45.5	30.9
Private mobility indicators										
Passenger car passenger kilometres per capita	p.km/person	18,155	11,387	8645	6202	3724	3517	785	4133	1172
Motorcycle passenger kilometres per capita	p.km/person	45	81	21	119	100	1165	416	78	90

Table 9.2 *Continued*

		US	ANZ	CAN	WEU	HIA	MIA	LIA	AMI	ALI
Public transport mobility indicators										
Total public transport boardings per capita	bd./person	59.2	83.8	140.2	297.1	464.1	274.2	267.3	340.5	234.4
Rail boardings per capita (Tram, LRT, Metro, Sub. rail)	bd./person	21.7	42.5	44.5	162.2	284.8	38.9	30.0	159.0	15.6
Proportion of public transport boardings on rail	%	25.7	48.8	28.9	50.0	62.0	12.8	11.0	33.1	7.6
Proportion of total motorized pass. km on public transport	%	2.9	7.5	9.8	19.0	50.3	26.9	51.1	36.6	54.2
Public transport productivity										
Public transport operating cost recovery	%	35.5	52.7	54.4	59.2	138.5	98.8	138.6	82.9	107.9
Overall transport cost										
Total passenger transport cost as % of metropolitan GDP	%	11.79	13.47	13.72	8.30	5.41	13.60	13.63	15.45	17.66
Total private pass. transport cost as % of metro. GDP	%	11.24	12.39	12.87	6.75	3.81	11.52	11.19	13.11	13.50
Total public pass. transport cost as % of metro. GDP	%	0.55	1.08	0.85	1.55	1.60	2.08	2.44	2.34	4.16
Traffic intensity indicators										
Private passenger vehicles per km of road	units/km	98.7	73.1	105.8	181.9	118.1	290.4	169.3	137.5	139.7
Passenger vehicles per km of road	units/km	98.9	73.3	106.1	183.1	121.7	300.4	184.4	138.9	154.5
Average road network speed	km/h	49.3	44.2	44.5	32.9	31.3	20.9	20.5	35.9	30.4
Transport energy indicators										
Private passenger transport energy use per capita	MJ/person	60,034	29,610	32,519	15,675	9556	10,555	2376	10,569	4052
Public transport energy use per capita	MJ/person	809	795	1044	1118	1500	1583	607	1012	1696
Energy use per private passenger kilometres	MJ/p.km	3.25	2.56	3.79	2.49	2.42	2.03	1.63	2.39	2.10
Energy use per public transport passenger kilometres	MJ/p.km	2.13	0.92	1.14	0.83	0.44	0.74	0.46	0.53	0.69
Air pollution indicators										
Total emissions per capita (CO, SO_2, VHC, NO_x)	kg/person	264.6	188.9	178.9	98.3	31.3	97.2	69.1	157.5	81.8
Total emissions per urban hectare	kg/ha	3563	2749	4588	5304	3894	12,952	13,357	7236	9211
Emissions per kilometre of motorized vehicle travel	kg/p.km	0.020	0.025	0.027	0.021	0.012	0.026	0.069	0.052	0.071
Transport fatalities indicators										
Total transport deaths per 100,000 people		12.7	8.6	6.5	7.1	5.9	20.7	10.4	18.3	13.2
Total transport deaths per billion passenger kilometres		7.0	6.8	7.1	9.6	7.4	29.2	37.4	29.3	34.0

Source: Kenworthy and Laube (2001)

by these modes in the other middle-income cities. From the point of view of transport economists, who generally argue that consumer preferences tend towards motorized private transport as wealth increases, and that this is an 'irresistible force' (Lave, 1992), we thus have a somewhat counterintuitive situation in some cities. A number of high-income cities have higher usage of the non-motorized and public modes than middle-income cities.

One likely reason for this discrepancy is the public response to negative impacts of large numbers of private motor vehicles in urban areas. External costs of motorized urban transport are high in cities, and where local environmental quality and quality of life for a majority of citizens are high priorities, motorized transport is restrained or managed by governments and communities using a range of means. This is despite the fact that people can afford private motorized mobility and that it may be desirable from the point of view of individual consumers. The patterns also relate to the availability and quality of public transport services and conditions for pedestrians and cyclists, which either encourage or mitigate the usage of green modes. In many middle- and low-income cities, conditions for pedestrians and cyclists can be very poor and dangerous and public transport can be unreliable, crowded and slow, so that fewer people use these modes than might otherwise be the case.

In addition, the sheer size and density of many large and rapidly growing cities of the developing world, particularly in Asia, place some spatial constraints on how much private car use can be accommodated. Barter (1999) suggests that there are physical limits to motorization in large, high-density cities, which will not be able to change enough in time to accommodate the level of space and decentralized activities demanded by private motorized transport. The data on urban transport infrastructure, discussed later, would appear to support this link: the level and type of transport infrastructure provision, much of it provided by governments, is linked to patterns of usage of different modes.

Evolution of urban transport systems and urban form

While the database that forms the basis of this chapter is for one year, 1995, it is useful to consider how urban transport systems and urban forms change over time. Based on the same 1995 data used in this chapter, along with panel data for other years, Barter (2000) presents a simple generic model of transport and land-use evolution in cities. The model is based on empirical analysis of urban transport data on the industrializing nations of East and Southeast Asia between 1960 and 1995 and helps to explain the transport patterns described above (Figure 9.1). It suggests that the level of motorization in Asia's middle- and low-income cities in the 1990s had not been reached by the now wealthy cities when they were at comparable income levels. Similarly, Hayashi et al (1994) find that based on car and motorcycle ownership, motorization measured in terms of passenger car units was approximately 30 per cent higher in Bangkok in 1988 compared with Tokyo and Nagoya when they were at similar levels of wealth. These differences are even more surprising given that the ratio of car prices to average annual household income were about 10 times higher in

1989 Bangkok than in 1972 Tokyo and Nagoya (Hayashi et al, 1994). All of this evidence suggests that motor vehicle ownership and usage in middle-income cities in the late 20th century were not found in high-income cities decades earlier. This would suggest that middle- and low-income cities in the 1990s were on fundamentally distinct transport trajectories.

According to Barter (2000), some middle-income cities in Asia are moving rapidly from transport systems in which walking, non-motorized vehicles and rudimentary, low-cost, bus-based public transport systems cater for the majority of transport needs, to a situation where cars and motorcycles dominate. The result, in cities such as Bangkok and Kuala Lumpur, is what Barter calls 'traffic saturated bus cities and motor-cycle cities' or 'traffic disasters'. This outcome differs from the wealthy cities in industrialized nations, which Barter (2000) classifies dichotomously as either 'automobile cities' or 'modern transit cities'. The outcomes are influenced not only by the decision of individual consumers and private firms, but also by government policies towards motor vehicle ownership and use, road building, urbanization and suburbanization, traffic restraint and relative levels of investment in roads, public transport and non-motorized modes.

Recent evidence suggests that middle-income cities at lower levels of wealth are becoming like Bangkok and Kuala Lumpur rather than low-density automobile cities or high-density modern transit cities. In Delhi, a rapidly expanding mega-city, ownership of motorized vehicles and particularly motorcycles is still low by international comparison, 'but remarkably high considering the population's relatively low income' (Bose and Sperling, 2001). In addition to India's rapid motorization and emerging 'transport crisis' (Pucher et al, 2005), cities in China in particular have become a focus of global attention from those concerned with environmental implications of motor vehicles in cities (Hook and Ernst, 1999; Kenworthy, 2001).[4]

Whether they are now 'automobile dependent' cities (e.g. Los Angeles) or 'transit metropolises' (e.g. Tokyo), 1995's high-income cities all went through periods in which extensive rail and mass transit systems were in place. While many advocates of sustainable transport (including the authors of this chapter) are concerned with the rise of automobile dependent cities, it is acknowledged that earlier in the 20th century there was a certain order to the process of motorization. Institutions and processes were developed to mediate the negative impacts of motorization and problems such as the conflicts between people and cars on streets, air pollution and displacement of homes and businesses for roads and parking.

In contrast, many of the low- and middle-income cities of the developing world are now comparatively large and lack effective and representative public institutions. So, unlike the historical experience of today's cities in industrialized nations, low- and middle-income cities are not experiencing an ordered and relatively slow evolution from walking to public transport to automobile cities, described by Newman and Kenworthy (1999), and conceptualized further in Barter (1999). As suggested in Figure 9.1, middle-income cities are not experiencing a period involving the development of extensive and competitive public transport systems, especially fixed-track systems. Unlike in industrialized nations before them, they have little physical infrastructure which provides an alternative to road-based motorized transport.

Another major difference is that an 'interim' form of private mobility is now offered in many middle-income cities by a proliferation of cheap motorcycles. When this is combined with poor vehicular emissions controls, lax vehicle standards and the fact that some of these cities are either at or approaching mega-city status (>10 million people), transport is imposing environmental burdens on huge numbers of people (Badami, 2001). For example, traffic intensity data in Table 9.2 show that the middle- and low-income cities, especially those in Asia, have the highest average number of passenger vehicles per kilometre of road of all cities (139 up to 300). By contrast, the dense cities of High Income Asia have remarkably low totals of passenger vehicles per kilometre of road (122), which greatly ease the overall intensity of some traffic-based environmental problems, which are nonetheless highly localized.

Despite the growing decentralization of development to the fringes of many of these low- and middle-income cities, they still have dense settlement forms, averaging 136 persons per hectare (ha), compared to an average of only 28 per ha across the US, Australia and New Zealand, Canada and Western Europe clusters. There is therefore a serious 'mismatch' between their urban form and their limited road and parking infrastructure and the huge demands of the new 'eating' private vehicle technologies (Dimitriou, 1992; Poboon, 1997; Barter, 1999).

Private transport infrastructure

The use of private motor vehicles in cities is linked to the provision of urban transport infrastructure to accommodate those vehicles. Freeway and central business district (CBD) parking availability are clearly highest in the automobile-oriented cities of the US and Australia/New Zealand, followed by cities in Western Europe. What is also very clear is the relatively low supply of these facilities in the rail-oriented cities of Hong Kong, Japan, and Singapore comprising the High Income Asia cluster. In stark comparison to High Income Asia, less affluent Middle Income Asia is distinguished in 1995 by 23 per cent higher freeway provision and 35 per cent higher CBD parking.

In the Assorted Middle Income cluster, freeway provision is twice as high and CBD parking three times as high as in High Income Asia. Although the low-income cities generally have fewer freeways and less CBD parking, freeway construction, which is often justified and rationalized as a solution to congestion problems, continues apace. For example, in Bangkok (Middle Income Asia), and more recently in Mumbai (Low Income Asia), city governments enthusiastically build elevated 'flyovers' designed to move private motor vehicles more quickly over intersections. While virtually useless from the point of view of addressing wider transport problems, they are nonetheless highly popular among politicians and influential constituencies of car owners. The flyovers reduce the relative and absolute provision of public transport infrastructure and facilities for non-motorized mode users (Hook and Replogle, 1996), and exacerbate the mismatch between car-based travel and dense urban form. On a wider scale, this is evident from a comparison of two of Asia's most densely populated islands:

> *The island of Java ... has a population density more than double that of rail-intensive Japan, but unlike Japan, its transportation system is dominated entirely by road transport. [W]hile rising per capita incomes have played a role in increasing the rate of motorization, Government of Indonesia policy has strongly encouraged motor vehicle ownership and use through massive subsidies to road users, underinvestment in public transport and rail systems, and a general public sector hostility to non-motorized modes of transportation.* (Hook and Replogle, 1996, p80)

Even without consideration of their high density and mixed use character, which makes them well suited to mass transit, a systematic bias in infrastructure towards roads over public transport stands out. In Low Income Asia, road investment as a percentage of gross domestic product (GDP) (averaged over five years) exceeds public transport investment by a factor of 3.4. In China the factor is 3.7, despite the fact that only 16 per cent of daily trips in 1995 were by private transport (Hook and Ernst, 1999; Kenworthy and Laube, 2001). Only the US and Canada clusters exceed this orientation to road investment, with 4.8 times more investment in roads than in the development of public transport. But unlike in China, private motor vehicle trips in these cities account for 80–90 per cent of all daily trips. So, despite the generally high densities of urban development in all of the cities of the middle- and low-income clusters (Table 9.2), the fact that this makes them ideally suited to public and non-motorized transport, and the fact that they are not automobile dependent, there is a bias towards transport infrastructure for private motor vehicles. There is a strong negative equity or regressive dimension to the emphasis on building infrastructure which first and foremost serves the small, wealthy segment of cities with low- and middle-income levels of GRP. That many of these infrastructure investments are financed by multilateral development banks has become a contentious issue raised by the increasingly vocal critics of institutions such as the World Bank.

Public transport service and usage patterns

Some deeper insights into reasons for motorization patterns can be gained by considering features of the public transport systems of different groups of cities. Table 9.2 shows that automobile dependent US, Canada, and Australia/ New Zealand clusters have the lowest use of public transport, ranging from 59 to 140 trips per person per annum, with only 3–10 per cent of their motorized passenger travel on public transport. By contrast, the Western Europe cluster has nearly 300 trips per annum and almost 20 per cent of motorized travel by public transport. High Income Asia, despite high GRP, achieves by far the

218 Scaling Urban Environmental Challenges: From Local to Global and Back

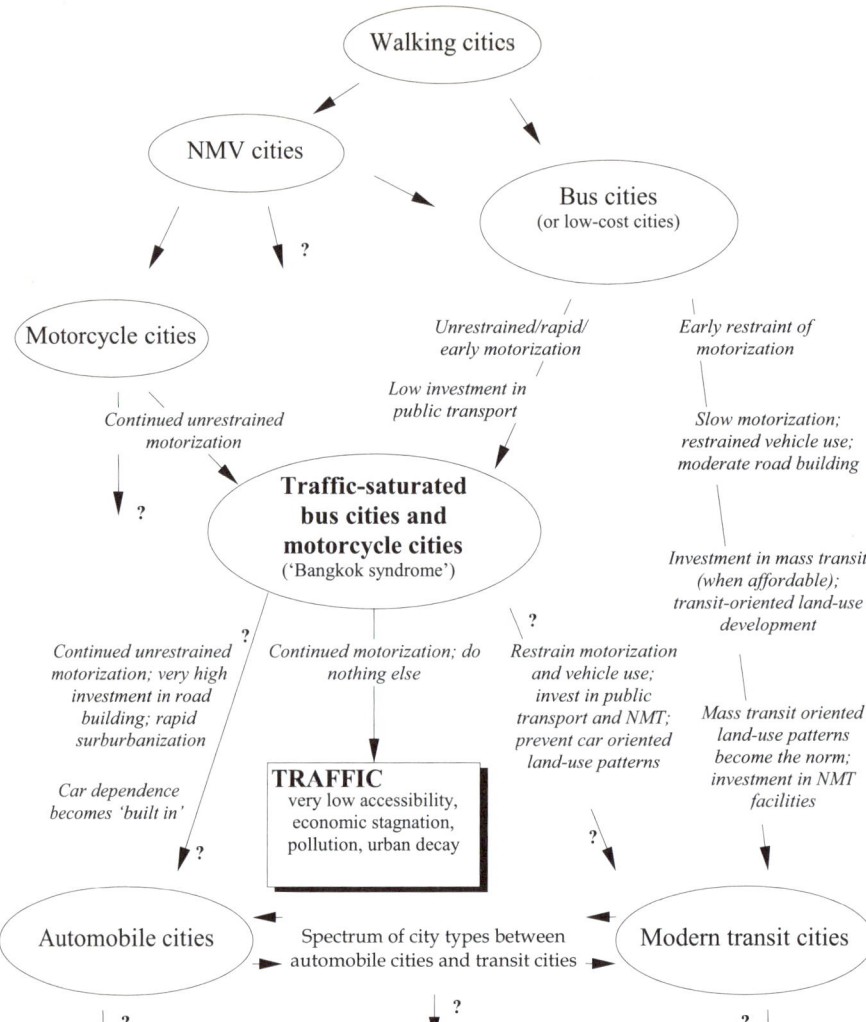

Note: This scheme is intended to describe the paths taken or potentially to be taken by cities that are in the so-called developing world or which were in the 'developing world' until the 1960s or so.

NMV = non-motorized vehicle; NMT = non-motorized transport.

Source: Barter (2000)

Figure 9.1 *Simple generic model of urban transport and land-use evolution in developing cities*

highest use of public transport in the international sample (464.1 trips per capita and 50 per cent of motorized travel).

It might be expected that through a combination of much lower GRP per capita and a larger captive rider population, that middle-income cities would achieve very high levels of public transport use. However, in Middle Income Asia, on average only 27 per cent of motorized travel (passenger km) is by public transport and in the Assorted Middle Income cluster, 37 per cent. Low Income Asia and Assorted Low Income clusters have 51 per cent and 54 per cent, or about the same as High Income Asia. This may be partly related to the relatively high use of non-motorized modes. In Low Income Asia, 50 per cent of daily trips are made by non-motorized modes, compared with 36 per cent in the Assorted Low Income cluster, and 29 per cent in High Income Asia. Of course, a partial reason for high foot and bicycle use can be the poor service delivery and speed of public transport systems. Where buses are stuck in traffic, it can be quicker and more direct to just walk or ride a bike.

On this point it is clear from the public transport supply and service data in Table 9.2 that there are significant differences in the extent and quality of public transport services in different groups of cities. US cities, which at 14.9 persons per ha are the lowest density, also have the lowest seat km of service per capita of all cities (1557 per year).[5] In the US, a self-reinforcing cycle of declining densities and declining public transport has operated for many decades, in combination with transport policies strongly geared to the automobile (Schneider, 1979). It is only since 1995 up to the present time, that there have been steady gains in US transit ridership (Pucher, 2002). Densities are also stabilizing or increasing slightly in some US cities (Newman and Kenworthy, 1999; Kenworthy and Laube, 2001). However, it is clear that Low Income Asia, comprising the highest density cities in the sample (205.6 persons per ha), is the next lowest in public transport service (2057 seat km per person per year), due partly to reasons already discussed. This is even less public transport service than in largely automobile dependent Canadian cities (2290 seat km per person per year), which have an average density of only 26.2 persons per ha.

But it is not just Low Income Asia that has comparatively low public transport service. The relatively dense low-income cities in other regions and middle-income cities around the world, with average densities ranging from around 54 to 164 persons per ha, all have annual per capita seat km below even the levels found in cities in Australia and New Zealand. These latter cities have low densities of only 15 persons per ha. Finally, the middle- and low-income cities have public transport service levels that are very significantly below levels in Western Europe and High Income Asia, where densities are comparable with the poorer cities. This supports the previous suggestion that middle- and low-income cities have yet to see significant development of publicly owned, or in the case of some systems in wealthy cities, privately owned and operated, but publicly regulated, mass transport systems.

The other issue here is quality of service, which is partly reflected in the proportion of public transport service that is by rail modes (tram, Light Rail

Transit (LRT), metro and suburban rail). All metro and suburban rail services, and some tram and LRT services, operate on physically segregated rights of way, which means that they are free of traffic congestion and achieve higher operating speeds that are more competitive with cars. Low Income Asia and Assorted Low Income clusters have only 10.1–12.9 per cent of their public transport seat kilometres by rail, which requires mobilization of large sums of capital to construct. The Middle Income Asia cities again have only 13.1 per cent, while the Assorted Middle Income cities do somewhat better with 33.6 per cent of service on rail. These figures, however, are in stark contrast to the other dense cities in the study (Western Europe and High Income Asia), where rail accounts for over 55 per cent of public transport seat kilometres.

In High Income Asia, high-quality rail systems, in some cases privately owned and operated, make public transport competitive with the private motor vehicle. In these densely populated, rail-oriented cities the overall speed of the public transport system is highest (33.2km/h) and this is the only group of cities where public transport speed actually exceeds road traffic system speed (8 per cent higher). In the low- and middle-income Asian cities where congestion is very high, the public transport systems average only 16 to 17km/h.

Railway infrastructure is clearly a factor in explaining the very different levels of public transport use between these groups of cities. Indeed, the data in this paper highlight the importance of urban rail systems as components of public transport systems that provide a successful alternative to movement by private motor vehicles. There are no regions where the average speed of bus systems exceeds 26km/h, and the overall average across the nine clusters is only 19km/h. In Chinese cities buses operate at an average of 12km/h, or about the same speed as cycling. In São Paulo and Curitiba, authorities find it difficult to operate the extensive busways at more than an average speed of 25km/h (Vasconcellos, 2002). On the other hand, metro systems operate between 30 and 37km/h (average 34km/h), while suburban rail systems average 43km/h across the regions. When these speeds are compared to general road traffic speed, which averages 34km/h across all regions, it can be seen that only rail systems can compete (Kenworthy and Laube, 2001).

While there clearly are cities in the world (e.g. in Latin America, Africa and China), where overall levels of public transport use are high and there is little or no rail, these examples rely heavily on a large population base of poor, 'captive' users.[6] As incomes rise and motorcycle and car ownership levels grow, the public transport systems of these low- and middle-income, bus-based cities tend to get hit hardest for market share because they cannot compete in speed or comfort with private transport. Because of their manoeuvrability, motorcycles in particular tend to compete heavily with bus systems that are engulfed in traffic (Barter, 1999).

One conclusion is that despite their dense, mixed-use urban forms, which are often strongly corridor-based and ideally suited to high-frequency railway or busway mass transit services, low- and middle-income cities suffer from inferior public transport systems. They thus have much lower levels of public transport utilization than they could support, based purely on their urban form, but institutional capacity would most likely need to be expanded for them to significantly upgrade public transport services. The fact that public transport in these cities is generally not offering any serious competition to the motorcycle or car, is an important factor in understanding their relatively high motorization levels and large environmental burdens. This suggests that while urban form is strongly correlated with certain types of transport (Kenworthy et al, 1999), there are other factors that influence transport patterns. Where politics and institutions do not provide the conditions and support necessary for development of high-quality public transport, then usage will fall short of its potential.

Environmental impacts of urban transport

The significant level of motorization already apparent in middle-income cities and rapidly growing in low-income cities has some important global and local environmental implications. In considering these implications, it is important to keep in mind the distinction between vehicle ownership and vehicle use. High vehicle ownership does not necessarily equate to commensurately high vehicle use. Nor does vehicle ownership alone impose environmental burdens, which in urban areas result from the high land requirements for roads and parking (neighbourhood severance, visual intrusion, urban blight), vehicular air and noise emissions, and traffic accidents. We see this to some extent in the high-income Western Europe cities, which have only 30 per cent lower car ownership than US cities, but 66 per cent lower car use. This helps to explain significant differences in energy use, emissions and transport deaths evident in Table 9.2 between these two groups of cities, as well as differences in environmental quality of the public realm (Beatley, 2000).

Energy

Energy use per capita in private passenger transport in both middle-income clusters already exceeds that of High Income Asia by 10 per cent, in spite of lower per capita GRPs. Continued increases in this factor in highly populated middle- and low-income cities potentially has large implications for global greenhouse emissions, though clearly it is the North American and Australian cities that are the largest energy consumers. For example, a US city of less than 400,000 people consumes annually as much energy in private passenger transport as a low-income Asian mega-city of 10 million people.

Despite the huge range in per capita energy use for private passenger transport (2376 MJ per capita in Low Income Asia compared to 60,034 in the US, or a 25-fold difference), there is a comparatively small range in public transport energy use per capita. This ranges from 607MJ per capita in Low Income Asia up to 1696 in the Assorted Low Income cluster, or only a three-fold difference. This small difference in energy use for public transport systems is despite a huge range in the role played by public transport, which varies from only 2.9 per cent of motorized travel in US cities to 54.2 per cent in the non-Asian low-income cities, or a 19-fold difference. This helps to conceptualize the extraordinary capacity of public transport systems to carry people with very low energy requirements, and to some extent the huge underutilized capacity of existing public transport services to save energy through higher passenger loadings.

In terms of the energy consumed to move one person 1km, public transport systems as a whole range from 1.5 times more efficient than cars in the US, up to 5.5 times more efficient in High Income Asia. In the US, buses for the most part chase thinly spread passengers across transit-hostile suburban landscapes, whereas in High Income Asia, public transport, especially rail, services dense, high-demand nodes and corridors. In most cities, public transport is somewhere between three to four times less energy consumptive per passenger kilometre than cars, though at present the Middle Income Asian cities are the poorest outside the US (2.7 times less energy consumptive public transport). This is some reflection of the relatively poor state of public transport in these cities and the considerable positive influence of motorcycles on private transport energy efficiency.

Emissions

One key impact of urban transport is air pollution, which is growing steadily, especially in low- and middle-income cities due to rapidly rising vehicle fleets, traffic volumes and urbanization per se (WHO/UNEP, 1992; Faiz, 1993; Kirby, 1995; Anderson et al, 1996). The data in Table 9.2 show that in 1995 the auto cities in North America and Australia clearly have the highest per capita generation rates for transport emissions, ranging from 178.9kg to 264.6kg per capita (shown here as combined kilograms of CO, SO_2, VHC and NO_x).[7] High Income Asia has the lowest level of per capita transport emissions which are at 31.3kg per capita. The middle- and low-income cities have high rates of per capita transport emissions relative to their private transport usage. For example, the Assorted Middle Income cluster, composed of cities outside Asia, has 60 per cent of the per capita emissions of an average US city, whereas the combined car and motorcycle use is only 23 per cent that of a US city. Middle Income Asia generates over three times more per capita emissions from transport than High Income Asia, yet the car/motorcycle use is 1.2 times higher.

Clearly, these middle- and low-income cities are generating much greater emissions per kilometre of travel than those in wealthier cities due to inferior emissions control standards on vehicles (both private and public transport), poorer enforcement of standards and high levels of emissions from motorcycles. High Income Asia has the lowest rate of emissions per passenger kilometre, with the middle- and low-income cities having between two and six times higher rates.

Middle Income Asia cities have a very high level of motorcycle use (25 per cent of combined car and motorcycle passenger kilometres). This imposes peculiar and severe environmental impacts through high levels of emissions from 2-stroke motor cycles, very high noise disturbance and greatly increased traffic danger and threat of death (Badami, 2001). This pattern is being followed in Low Income Asia, where 35 per cent of private passenger kilometres is accounted for by motorcycles. However, middle- and low-income cities outside Asia have not yet begun to manifest anywhere close to the same degree of motorcycle orientation (only 2–7 per cent of private passenger kilometres by motorcycles).

Perhaps of greater concern is the picture of emissions on a spatial basis. This is measured here in total emissions from passenger transport per urbanized hectare. It reflects to some extent differences in intensity of exposure to transport emissions experienced by the population. The low-density auto cities, despite having higher emissions per capita, are lowest in emissions per hectare, averaging around 3600kg/ha. The High Income Asia and Western Europe cities are a little higher at around 4600kg/ha. However, the Middle and Low Income Asia cities average close to 13,200kg/ha (12,952 and 13,357kg/ha respectively), which emphasizes how even low levels of motorization can lead to highly concentrated and severe environmental impacts in dense cities.

The other middle- and low-income cities are comparatively high in transport emissions as well, with 7236kg/ha and 9211kg/ha respectively. Private transport plays the major role in this problem, and in poorer cities higher-income people are the main users of private transport. The poor majority, who are mainly public transport and non-motorized mode users, are exposed to this regional-scale air pollution problem, in addition to other serious emissions hazards that occur at a household and neighbourhood level (McGranahan, 1993; McGranahan et al, 1999, 2001). Merely negotiating the public environments of such cities on foot or bicycle, in paratransit-type vehicles such as motorized three-wheel vehicles, on motorcycles, or in buses, or working along the street in the informal economy, can be a significant health issue.

The combination of large cities with high urban densities, relatively high rates of car and motorcycle use, and high rates of emission per kilometre of travel, has led to well-documented air pollution problems in a number of low- and middle-income cities (WHO/UNEP, 1992).[8] Due to high densities, rising numbers of motor vehicles and higher private motorized mobility, technical approaches such as fuel improvements and fuel substitution, and inspection and maintenance programmes, cannot fundamentally address the air pollution problem in middle- and low-income cities. These efforts must be combined with programmes to address rising private vehicle use, otherwise the gains made in reducing per kilometre emissions rates are simply eaten up in higher vehicle kilometres (Badami, 2001).

One response to the problem of high exposure to concentrated emissions in dense cities is to disperse the problem through spatial decentralization, in order to spread environmental loads. This approach, however, has to be balanced with the extra travel by private transport that is associated with lower densities, which in turn is reflected in higher per capita emissions from transport (Newman and Kenworthy, 1988, 1989).

Transport deaths
Deaths that are attributable to transport in cities vary widely around the world.[9] Among the clusters of wealthy cities, transport deaths are highest in the US (12.7 deaths per 100,000 people) and lowest in High Income Asia (5.9 deaths

per 100,000 people). Deaths per billion passenger kilometres of travel are, on the other hand, uniformly low in the wealthy cities (around 7.5 per billion passenger kilometres). The picture is quite dramatically different in the four low- and middle-income city clusters. The minimum value of 10.4 deaths per 100,000 people in the Low Income Asia cluster is due primarily to low rates in the Chinese cities, where motorization is lowest and facilities for cyclists and pedestrians are still comparatively good, though deteriorating rapidly (Hook and Ernst, 1999; de Boom et al, 2001).

The Assorted Low Income cities have higher transport deaths at 13.2 deaths per 100,000 people, while both middle-income clusters have even more serious loss of life in transport at 18.3 and 20.7 deaths per 100,000 people. Across all the low- and middle-income cities there is also a comparatively serious problem of deaths per unit of distance travelled. The average is 32.5 deaths per billion passenger kilometres, which means that it is over four times as dangerous moving around in these poorer cities than it is in wealthy cities. However, this risk is distributed unequally. In many middle- and low-income cities, car owners walk in environments where pedestrians are protected from motor vehicles (such as gated communities and air-conditioned shopping centres). At the same time, they impose great risks of injury and fatality on the non-driving public utilizing sidewalks and public streets.

We thus see in the low- and middle-income cities a disproportionately high rate of deaths in transport relative to their motorization level. This is attributable to a host of factors. These include inadequate vehicle safety standards and driver education, insufficient traffic regulation, lack of laws or law enforcement, low quality of road networks, insufficient space for non-motorized transport, and the rapid increase in motorized vehicles, whose drivers by and large have little respect for pedestrians and cyclists belonging to lower social classes.

Economic aspects of urban transport

Table 9.2 summarizes in a few simple indicators, a very large amount of economic data in the database on urban transport systems. It shows the overall expenditure on urban passenger transport (private, public and total) calculated as a proportion of a city's GDP. The overall expenditure includes all fixed and variable costs and investment and operating costs for all public and private modes. Table 9.2 also includes road and public transport investment expressed as proportions of metropolitan GDP, and the public transport operating cost recovery.

As would be expected, some very high expenditures occur in the wealthy automobile dependent cities where between 11.79 per cent and 13.72 per cent of their GDP is consumed by urban transport. In these cities, only a tiny fraction of this overall expenditure is accounted for by public bus and rail transport (about 0.5 to 1 per cent of GDP). In contrast, public transport-oriented Western Europe and High Income Asia collectively spend amounts that are equivalent to only 5–8 per cent of their GDPs on passenger transport. In the case of High Income Asia, where governments have actively discouraged car ownership and usage to a greater extent than in other wealthy nations, only

1.6 per cent of GDP is spent on public transport, but this caters for 50 per cent of motorized passenger travel. On the other hand, 5.41 per cent of GDP is spent to move the other half, using private transport.

Additionally, in cities where overall incomes are relatively low, urban transport accounts for a relatively large proportion of the cities' GDP. In the four middle- and lower-income clusters, urban passenger transport as a whole accounts for between 13.6 per cent and 17.6 per cent of GDP. Corresponding with the relatively high level of private motor vehicle ownership and use described earlier in the chapter, the data indicate that between 11 per cent and 13.5 per cent of GDP is spent on private transport. This is equal to or in excess of the share in the high cost automobile cities in the US, Australia/New Zealand, and Canada. While private transport expenditure is high, public transport carries much larger volumes of passengers at much lower cost. In the middle-income cities, private transport expenditure accounts for about three times as much as public transport expenditure, but approximately equal numbers of passengers are moved. In the two low-income clusters, private transport is between four and five times more expensive in achieving a similar amount of mobility.

While in industrialized cities public and private urban passenger transport is subsidized as an input into production, in the middle- and low-income cities the revenues generated from public transport fares hover around the 'breakeven' point. In Middle Income Asia, 98.8 per cent of operating costs are recovered by fares, and in the Assorted Middle Income cluster the figure is 82.9 per cent. In the two low-income clusters it is evident that there are profits from public transport because farebox collections exceed the operating costs (107.9 per cent in the Assorted Low Income cluster and 138.6 per cent in the Low Income Asia cluster). However, while these figures may indicate financial strength of transport operations, the perennial indebtedness of governments in these nations would suggest that these profits are not invested into public facilities. The results in the middle- and low-income cities do not necessarily indicate that the services are high quality, but simply that these cities have large numbers of captive riders who ensure that a good proportion of the services that are provided are highly utilized. In addition, because many of the services are provided by less-well regulated operators, standards and thus costs are kept relatively low. The negative side of this is, however, that vehicles are often highly polluting, old and in many cases also unsafe. As a result of this and other factors, such as lack of dedicated rights of way for buses and chronic overcrowding, people are more than willing to abandon public transport as soon as they can afford a motorcycle or a car.

Some operators of public transport services in High Income Asia actually make profits in environments where public transport is still 3.4 times cheaper than private transport in moving an identical proportion of the city's passenger kilometres. However, in these cities, the standard of public transport service is much higher and there are many 'choice riders'. In the other wealthy cities public transport recovers 50–60 per cent of its operating costs from fares, except in US cities where the figure is only 35.5 per cent.

Urban environmental transitions and urban transport

The UET thesis suggests that as the aggregate level of affluence in a city rises, urban environmental problems become more physically dispersed, less intensive and less direct. McGranahan et al (1999) summarize the argument as follows:

- The urban environmental hazards causing most ill-health are those found in poor homes, neighbourhoods and workplaces, principally located in the South.
- The most extreme examples of city-level environmental distress are found in and around middle-income mega-cities and the industrial cities of the formerly planned economies.
- The largest contributors to global environmental problems are the affluent, living predominantly in the urban areas of the North.

This examination of the urban transport characteristics of cities, which have been clustered on the basis of wealth, gives some empirical support to the points above. The low-income and middle-income clusters (which comprise cities that, based on the World Bank Atlas method, would have been mostly upper-middle income and lower-middle income cities) have relatively high levels of private motor vehicle use and associated environmental impacts, which are magnified in these large and dense metropolitan areas. One way of illustrating some of the key data in terms of urban environmental transitions is by normalizing it according to GRP per capita. Table 9.3 contains a selection of the data normalized per $1000 of GRP, or the economic output of the cities. The following points can be made:

- Vehicle ownership per unit of economic output is much higher in all the low-income and middle-income clusters than in the high-income clusters. However, low-income clusters (lower-middle income cities using the World Bank Atlas method) are particularly high.
- The provision of freeways relative to economic output is most consistently high in the high income clusters of the US, Canada and Australia/New Zealand and lowest in the Western Europe and the High Income Asia clusters. However, the Assorted Middle Income cities exceed US levels and equal the highest cities in Australia and New Zealand. The Middle Income Asia cluster exceeds Western Europe levels, while the Assorted Low Income cluster almost equals levels of the US cluster.
- In terms of the level of motorized private mobility undertaken relative to the economic output achieved by the cities, the US and Australia/New Zealand clearly lead the affluent clusters, while Western Europe and High Income Asia are by far the lowest. However, the two low-income clusters and the Assorted Middle Income cluster all exceed the US/Australia/New

Zealand usage, while the Middle Income Asia cluster is approaching US levels.
- The total amount of energy used for passenger transport in cities (private public transport) follows a similar pattern. The middle- and low-income clusters are either similar to or in excess of the transport energy used in affluent cities to generate an equal amount of economic output. The Assorted Low Income cities are extreme in this factor relative to all the other groups.
- Transport emissions per unit of economic output in the middle- and low-income clusters are either equal to the worst of the affluent cities (Australia/New Zealand), or way in excess of them. The two low-income clusters (or lower middle-income cities according to the World Bank classification) show up in this factor as being significantly above the middle-income clusters.
- Finally, the transport deaths that occur in realizing the GRP of cities are massively higher in the middle- and low-income clusters compared to affluent cities. Again, however, it is the two Low Income clusters that distinguish themselves in having the most pervasive problem, two to three times higher than the middle-income clusters.

In general, we see a pattern of pervasive disadvantage in both middle- and low-income clusters of cities with respect to some key transport factors that most directly relate to the environmental integrity of cities and some direct measures of environmental problems. The earlier discussion has suggested that, in line with the UET argument, the direct per capita data on factors such as vehicle ownership, vehicle use, freeway provision, energy use, emissions and transport deaths, are more problematic in middle- than in low-income cities. This will tend towards more intense city-level environmental stress in middle-income cities. However, the data in Table 9.3 moderate this conclusion a little by suggesting that relative to what the cities are achieving in terms of economic output (one measure of what the city is delivering in terms of well-being), the low-income cities (or lower-middle income cities, according to the World Bank), experience higher levels of stress.

Given the spatial dimension of the UET argument, there is another argument that suggests that the lower-middle income cities may have at least as bad, or in some cases worse, transport-based city-level air pollution than upper-middle income cities. The main basis of support for this comes from the total air pollution emissions per hectare of urban land. These data show that the emissions intensity is highest (13,357kg/ha) in the cluster of Low Income Asia cities, while the figures in the affluent cities are radically lower (2749–5304kg/ha). However, the next highest level (12,952kg/ha) is in the Middle Income Asia cluster and although this is lower than in the Low Income Asia cluster, it is considerably higher than in the Assorted Low Income and Assorted Middle Income clusters. The discrepancies can be partially explained by the effect of population density.

Table 9.3 *Selected transport factors with a key influence on environmental quality, normalized according to affluence*

Transport factors	US	ANZ	CAN	WEU	HIA	MIA	LIA	AMI	ALI
Cars and motorcycles per $1000 of GRP	19	30	26	14	8	36	79	42	44
Freeway length (10^{-4} metres) per $1000 of GRP	50	65	59	26	6	28	24	65	46
Private motorized mobility per $1000 of GRP	580	580	416	197	110	479	711	636	648
Total pass. trans. energy use per $1000 of GRP	1939	1538	1612	524	318	1242	1766	1748	2949
Transport emissions per $1000 of GRP	8	10	9	3	1	10	41	24	42
Transport deaths per $ billion of GRP	4	4	3	2	2	21	62	28	68

Source: Calculated from Kenworthy and Laube (2001)

This raises the question of whether densities in middle- and low-income cities, and the concentration of pollutants, will decline to the extent that urban air pollution will decrease substantially along with economic growth. Evidence would suggest that this is not occurring to the extent that would be necessary. It seems almost impossible in any reasonable time-frame for metropolitan densities, which are mostly in excess of 100 per ha up to 350 per ha, to decline to automobile city levels of 10 to 20 per ha. Furthermore, motorization is increasing rapidly and emission rates per kilometre remain much higher than in affluent cities where air pollution regulations are much tougher.[10] In addition, from a sustainability perspective, a strategy of decreasing densities, even if it could be achieved in already very dense cities, would lead to other problems facing low-density North American and Australian and New Zealand cities. These problems would also be on a much larger scale due to the mega-city status of many less affluent cities. Nevertheless, successful examples of dealing with the high density issue in relation to urban transport and environmental impacts do exist in cities such as Singapore, Hong Kong and Tokyo (e.g. Cervero, 1998).

In addition to density, the issue of equity could play a role in explaining the discrepancies in the urban transport data. A modification to the UET thesis could be that affluence *with equity* results in decreased concentration of urban environmental problems. This could explain the relatively high levels, with few signs of abatement, of transport-related urban environmental problems in the low- and middle-income city clusters in this chapter (lower- and upper-middle-income cities, based on the World Bank classification). In these cities, the environmental externalities of car use by wealthy élites and middle classes are imposed on the general public of cities in which the majority of the inhabitants are poor.

It is the poor who are disproportionately the victims of urban pollution and who are 'less buffered than the non-poor' from these impacts (Brandon, 1996, p202). Evidence from Asia suggests that due to low incomes and high concentration of land in the hands of wealthy landowning classes (Communist China and Vietnam are exceptions), the poor are excluded from secure housing tenure and are forced to cluster in centralized squatter or slum settlements. These settlements, which are located close to (or in many cases next to) their places of work are relatively exposed to air and noise pollution. This pollution is generated to a high degree by cars transporting people who mainly live farther out of the central city or in walled inner city compounds or condominiums. The provision of health care in most low- and middle-income nations is on a user pays basis which excludes the poor from seeking medical assistance for problems such as upper respiratory tract infections exacerbated by air pollution. Also, as mentioned earlier in the chapter, it is the poor users of non-motorized transport and motorcycles who are most exposed to traffic accidents.

Conclusions

The analysis in this chapter suggests that while large cities experience similar problems, the intensity of problems differs substantially among cities. The evidence from the Millennium Cities Database, and from some other sources, suggests that large middle-income cities in developing countries have distinctly higher levels of motor vehicle use than wealthy cities in years past (though still much lower than these wealthy cities today). Furthermore, many of the cities in the developing world have reached comparatively high population and density levels, which exacerbate many environmental problems. On the other hand, cities in the developed world have extraordinarily high overall levels of private motorized mobility, high energy use and high per capita emissions from transport. However, these cities are dispersed over much larger areas at lower densities so that emissions and general traffic impacts on a spatial basis are lower and population exposure less intense and less direct than in many middle-income cities. Thus while needing to address such issues from a local environment perspective, their urban environmental quality generally exceeds that of middle-income cities. Environmental concerns over high resource use and waste output of wealthier cities, while still retaining a strong local dimension, have a global resonance, through, for example, the greenhouse gas and global climate change issue.

In simple per capita terms, most of the critical indicators suggest that cities in the middle-income city clusters in this chapter are in a more advanced state of motorization with more serious and pervasive city-wide levels of environmental stress from transport than cities in the low-income city clusters. On the one hand, this and the findings in relation to the higher-income cities support the UET thesis. On the other hand, the level of motorization and the impacts from motorization are similar or higher in low-income city clusters compared to middle-income city clusters, on the basis of comparable economic output or wealth. In addition, the spatial intensity of transport emissions is greater in the Low Income Asia city cluster than in all groups of cities.

While higher densities in the cities of developing countries are therefore associated with some problems, high density has benefits in terms of urban transport. However, it is clear from this analysis that cities classified by the World Bank as lower- and upper-middle income cities, are not exploiting their urban form advantage by prioritizing the modes most suited to them (i.e. public transport, foot, bicycle and other non-motorized modes). Rather, they are emphasizing motorized private transport, which requires more space than these cities can realistically provide. This results in high levels of environmental impact, even at relatively low levels of motorization. Given the perennially poor state of public institutions in developing countries, it is perhaps not surprising that public transport services are not commensurate with levels that could be supported by prevailing urban densities.

Those middle- and high-income cities that have developed relatively good overall public transport systems have taken measures to improve the quality and quantity of public transport (usually by creating segregated rail or busway systems) as household incomes rise. Where public institutions have been unable to meet this challenge, or where vested interests and others have prevented this from occurring, the result has been the syndrome of high congestion and high pollution endemic to cities like Bangkok or São Paulo.

Should decision makers have the inclination and autonomy to introduce changes benefiting the wider public, including the poor, significant opportunities still exist to slow the motorization process in middle-income cities and particularly the lower-middle income cities (or low-income city clusters in this study), in which the penetration of the automobile has been limited in depth. There is little doubt that this would have widespread environmental benefits, especially to lower-income people. However, with every passing year without action, such opportunities are receding.

Acknowledgements

The authors wish to acknowledge the financial support (1998–2001) of the International Union (Association) of Public Transport (UITP) in Brussels, in developing the data used in this paper. Mr Jean Vivier of UITP is especially thanked for his collaboration with the authors in carefully checking and verifying data in each city. We also wish to acknowledge the major data collection effort of Dr Felix Laube in producing a high proportion of the data contained in this chapter. We also acknowledge the efforts of Paul Barter, Michelle Zeibots, Gang Hu, Momoko Kumagai, Chamlong Poboon, Benedicto Guia (Jr) and

Antonio Balaguer in helping to collect data in specific cities. Finally, but not least, we wish to thank the literally hundreds of people worldwide who cooperated over a long period of time by providing data and assisting with innumerable requests for clarification and follow-up, to ensure all data were of the best available quality.

Notes

1. There are up to 175 data items per city depending on the number of modes involved in each public transport system. From these primary data a standardized data set was created, consisting of 229 indicators of land use and transport in cities. A partial list of these standardized variables forms the basis of Table 9.2 and the discussion in this chapter. Data collection commenced in 1998 and lasted 3.5 years, indicative of the long periods required for collection, release, acquisition and collation before analysis of reliable data can even begin.
2. The database was funded by the International Association of Public Transport (UITP), which is based in Brussels.
3. For example, in both Bangkok and Shanghai (and this problem likely faces most cities in the developing world), numbers of unregistered migrants living in the city comprised an estimated 22 per cent of the official population figures (Shen, 1997; MVA, Asia et al, 1998). These discrepancies have significance for urban transport research as they generally understate the use of non-motorized modes used by the migrants, who are generally poor, and overstate the level of car use.
4. With the support of institutions including the World Bank and Ford Motor Company, China is now aggressively pursuing the development of what could become some of the biggest 'automobile cities' in the world. While the Communist party leadership has embraced automobile production as a pillar of an industrial economy and bicycles in cities as something to be literally and figuratively driven aside, there is evidence that they have yet to ponder the negative environmental and social consequences of these changes (Hook and Ernst, 1999). Or perhaps, as Jacobs identified in the US over 40 years ago, decision makers shaping urban transport in China have a 'sheer disrespect for other city needs, uses, and functions' (1961, p353).
5. Seat kilometres of service is calculated for *all* modes of public transport in the city by multiplying the vehicle kilometres of service for each mode and model of vehicle by the number of seats in the vehicle. This is a better measure of service than vehicle kilometres alone because it reflects the size of the vehicle (mini bus, standard bus, tram, LRT, train, ferry) and hence is a better measure of the capacity of the service being provided.
6. Captive public transport users are people who have no choice but to use public transport, either because they cannot afford other means and their destinations are beyond walking (or cycling distance), or they have some other limitation related to age or disability. Choice public transport riders are those who have a private transport option available, but choose to travel by public transport because for one reason or another, it provides a competitive service (this can involve speed, convenience and comfort factors, non-availability or expense of parking, or cost savings). Captive and choice public transport users occur in different proportions in different cities depending upon socio-economic and demographic factors, the quality of the public transport system offered and other 'sticks and carrots' that may exist (e.g. Webster and Bly, 1980).

7   CO = carbon monoxide, SO_2 = sulphur dioxide, VHC = volatile hydrocarbons, NO_X = nitrogen oxides.
8   Bangkok, for example, has developed a reputation for air pollution problems. In the period leading up to 1995, reports indicate very high levels of suspended particulate matter (SPM), lead, CO, hydrocarbons (HC) and NO_X, considerably above World Health Organization (WHO) standards (Boontherawara et al, 1994). The severe health impacts of air pollution (respiratory problems, lung cancer, high blood lead levels) and severe congestion (nervous disorders and anxieties caused by traffic) are also widely reported (USAID, 1990; Aekplakorn et al, 1991; Magistad, 1991; Mallet, 1992; Sayeg, 1992).
9   Data here are taken from the WHO International Classification of Diseases (ICD codes 810–820), which is a much more reliable source of data than police records, which generally record only deaths at the scene of an accident (Kenworthy and Laube, 2001).
10 Metropolitan density data from Kenworthy et al (1999) and Kenworthy and Laube (2001) show that between 1990 and 1995 Kuala Lumpur declined from 59 to 58 per ha, Bangkok from 149 to 139 per ha and Seoul from 245 to 230. However, Manila rose from 198 to 206, Jakarta from 171 to 173, Singapore from 87 to 93, Hong Kong from 301 to 320. Densities in already dense cities are by no means on inevitable downward trajectories, but can and do increase. The point is that it is almost impossible for large established urban regions to negate existing vast areas of high density development by outward low density expansion or by inward rebuilding. Barter (1999) develops this argument in some detail.

References

Aekplakorn, W., Metadilogkul, O., Sawanpanyalert, P. and Rugronnayuth, K. (1991) *Comparison of Respiratory Health Between Traffic and Non-traffic Policemen*, National Epidemiological Board, Thailand

Anderson, W. P., Kanaroglou, P. S. and Miller, E. J. (1996) 'Urban form, energy and the environment: A review of issues, evidence and policy', *Urban Studies*, vol 33, pp7–36

Badami, M. G. (2001) 'A multiple objectives approach to address motorized two-wheeled vehicle emissions in Delhi, India', PhD Dissertation, Faculty of Graduate Studies, School of Community and Regional Planning, The University of British Columbia, Vancouver

Barter, P. A. (1999) 'An international comparative perspective on urban transport and urban form in pacific Asia: Responses to the challenge of motorisation in dense cities', PhD Thesis, Murdoch University, Murdoch, Western Australia

Barter, P. A. (2000) 'Urban transport in Asia: Problems and prospects for high density cities', *Asia-Pacific Development Monitor*, vol 2, no 1, pp33–66

Barter, P., Kenworthy, J. and Laube, F. (2002) 'Lessons for Asia on Sustainable Transport', in N. P. Low and B. J. Gleeson (eds), *Making Urban Transport Sustainable*, Palgrave (Macmillan), Basingstoke, pp252–270

Beatley, T. (2000) *Green Urbanism: Learning from Cities in Europe*, Island Press, Washington DC

Boontherawara, N., Paisarnutpong, O., Panich, S., Phiu-Nual, K. and Wangwongwatana, S. (1994) 'Traffic crisis and air pollution in Bangkok', *TEI Quarterly Environment Journal*, vol 2 no 3, pp4–37

Bose, R. and Sperling, D. (2001) *Transportation in Developing Countries: Greenhouse Gas Scenarios for Delhi, India*, Prepared for the Pew Center on Global Climate Change, Arlington, VA

Brandon, C. (1996) 'Confronting the growing problem of pollution in Asia', *The Journal of Social, Political and Economic Studies*, vol 21 no 2, pp199–204

Cervero, R. (1998) *The Transit Metropolis: A Global Inquiry*, Island Press, Washington, DC

de Boom, A., Walker, R. and Goldup, R. (2001) 'Shanghai: The greatest cycling city in the world?' *World Transport Policy and Practice*, vol 7, no 3, pp53–59

Dimitriou, H.T. (1992) *Urban Transport Planning: A Developmental Approach*, Routledge, London

Faiz, A. (1993) 'Automotive emissions in developing countries – Relative implications for global warming, acidification and urban air quality', *Transportation Research*, vol 27A, pp167–186

Hayashi, Y., Rithika, S., Mackett, R., Doi, K., Tomita, Y., Nakazawa, N., Kato, H. and Anurak, K. (1994) 'Urbanization, motorisation and the environment nexus: An international comparative study of London, Tokyo, Nagoya and Bangkok', *Memoirs of the School of Engineering, Nagoya University*, vol 46, no 1, pp55–98

Hook, W. and Ernst, J. (1999) 'Bicycle use plunges: The struggle for sustainability in China's cities', *Sustainable Transport*, vol 10, no 6–7, pp18–19

Hook, W. and Replogle, M. (1996) 'Motorization and non-motorized transport in Asia: Transport system evolution in China, Japan, and Indonesia', *Land Use Policy*, vol 13, pp69–84

Jacobs, J. (1961) *The Death and Life of Great American Cities*, Pelican Books, Harmondsworth

Kenworthy, J. (2001) 'Public transport cities are successful cities: An international perspective on motorisation in Urban China', *Proceedings of the Third Sino-Swiss Symposium on Sustainable Urban Development and Public Transportation Planning, Kunming, China, 24–26 October, 2001*, Industrielle Betriebe, Zurich, pp40–65 (Chinese version pp20–39)

Kenworthy, J. and Laube, F. (2001) *The Millennium Cities Database for Sustainable Transport*, International Union (Association) of Public Transport, Brussels, and ISTP, Perth (CD-ROM publication)

Kenworthy, J., Laube, F., Newman, P., Barter, P., Raad, T., Poboon, C. and Guia, B. (Jr) (1999) *An International Sourcebook of Automobile Dependence in Cities 1960—1990*, University Press of Colorado, Boulder, CO

Kirby, C. (1995) 'Urban air pollution', *Geography*, vol 80, pp375–392

Lave, C. (1992) 'Cars and demographics', *Access*, vol 1, pp4–11

Magistad, M. K. (1991) 'Bangkok's progress marked by health hazards', *The Washington Post Health Magazine*, 7 May, p13

Mallet, V. (1992) 'Third world city, first world smog', *The Financial Times*, 25 March, p11

McGranahan, G. (1993) 'Household environmental problems in low-income cities. An overview of problems and prospects for improvement', *Habitat International*, vol 17, no 2, pp105–121

McGranahan, G., Jacobi, P., Songsore, J., Surjadi, C. and Kjellén, M. (2001) *The Citizens at Risk: From Urban Sanitation to Sustainable Cities*, Earthscan, London

McGranahan, G., Songsore, J. and Kjellén, M. (1999) 'Sustainability, poverty and urban environmental transitions', in D. Satterthwaite (ed), *The Earthscan Reader in Sustainable Cities*, Earthscan, London, pp107–130

Mumford, L. (1961) *The City in History: Its Origins, its Transformations, and its Prospects*, Penguin Books, London

MVA, Asia, Comsis Corporation, Padeco Co., Genie Engineering Consultant Co., Asian Engineering Consultants Corp. (1998) *Final Report Volume 1: Executive Summary: Urban Transport Database and Model Development Project*, Office of the Commission for the Management of Land Traffic, Bangkok

Newman, P. W. G. and Kenworthy, J. R. (1988) 'The transport energy tradeoff: Fuel-efficient traffic versus fuel-efficient cities', *Transportation Research*, vol 22A, no 3, pp163–174

Newman, P. W. G. and Kenworthy, J. R. (1989) *Cities and Automobile Dependence: An International Sourcebook*, Avebury Technical, Aldershot

Newman, P. W. G. and Kenworthy, J. R. (1999) *Sustainability and Cities: Overcoming Automobile Dependence*, Island Press, Washington, DC

Owen, W. (1972) *The Accessible City*, The Brookings Institution, Washington, DC

Poboon, C. (1997) 'Anatomy of a traffic disaster: Towards a sustainable solution to Bangkok's transport problems', PhD Dissertation, Murdoch University, Murdoch, Western Australia

Pucher, J. (2002) 'Renaissance of public transport in the United States', *Transportation Quarterly*, vol 56, no 1, pp33–49

Pucher, J., Korattyswaropam, N., Mittal, N. and Ittyerah, N. (2005) 'Urban transport crisis in India', *Transport Policy*, vol 12, pp185–198

Sayeg, P. (1992) *Assessment of Transportation Growth in Asia and its Effects on Energy Use, the Environment, and Traffic Congestion: Case Study of Bangkok, Thailand*, The International Institute for Energy Conservation, Washington, DC

Schneider, K. (1979) *On the Nature of Cities: Towards Creative and Enduring Human Environments*, Jossey-Bass, San Francisco

Shen, Q. (1997) 'Urban transportation in Shanghai, China: Problems and planning implications', *International Journal of Urban and Regional Research*, vol 21, no 4, pp589–606

USAID (1990) *Ranking Environmental Health Risks in Bangkok, Thailand*, vol 1, December, United States Agency for International Development, Bangkok

Vasconcellos, E. A. (2002) Personal communication on a tour of São Paulo's busways with the CEO of Brazil's Associacao Nacional de Transportes Publicos (ANTP)

Webster, F. V. and Bly, P. H. (1980) *The Demand for Public Transport: Report on the International Collaborative Study of Factors Affecting Public Transport Patronage*, Transport and Road Research Laboratory, Crowthorne

WHO/UNEP (1992) *Urban Air Pollution in Megacities of the World*, World Health Organization, United Nations Environment Programme, Blackwell, Oxford

10
Fixing Environmental Agendas in Mexico

Priscilla Connolly

Introduction

Like much of the literature on the subject, this book uses Mexico City to illustrate environmental problems generated by mega-cities in middle-income countries. With a population of slightly less than 20 million, Mexico's capital is unquestionably a very large city, while a per capita gross national income of US$5500 qualifies its economy as 'upper-middle income' according to World Bank criteria. But to what extent does a specific case, such as Mexico City, really illustrate the basic set of proposals set forward in earlier chapters? This chapter takes up this question, but not with a view to either verifying or refuting these models: an unfeasible task on the basis of a single case study. Instead, it will be shown that while the Urban Environmental Transition (UET) model and related propositions are useful for exploring urban environmental change, an understanding of the environmental burdens, agendas and policies of a particular city needs additional, place-specific, dimensions. Most importantly, the social construction of these burdens, agendas and policies has to be considered.

Before proceeding, it is useful to recall the basic models or propositions we are referring to. Firstly, the dominant theme is the spatial and temporal displacement of environmental burdens as wealth increases (environmental transition) and the challenge this poses to environmental justice and the possibility of sustainable development (see Chapter 1). In this model, lack of clean water and sanitation is typical of localized environmental burdens found in poor cities or those in early stages of industrialization, generating the need for a 'brown agenda'. Metropolitan- and regional-level pollution, particularly atmospheric pollution from automobiles, is symptomatic of large cities in middle-income industrialized countries, while greenhouse gas emissions are a prime example of how cities in rich countries export environmental burdens to the whole world. Both these scales of impact relate to the need for 'green agendas'. Secondly, the additional complication of overlapping agendas created by accelerated environmental transitions in developing countries (time–space telescoping) renders ineffective the dominant environmental management approach to policy solution (Chapter 3). The third proposal (Chapter 13), more prescriptive than analytical, emphasizes the role of the different levels of state and governance, favouring the 'bottom-up' contribution of local governments for deriving agendas of public interest out of both localized environmental burdens and global environmental risks (The Local Agenda for urban sustainability).

At a first-cut analysis, Mexico City does indeed conform to the first two proposals outlined above, while providing some positive examples of local policy solutions. The unhappy combination of developing and developed world environmental problems has become a catchphrase often used to characterize this city. I myself have made the point that Mexico City compounds environmental deprivation arising from poverty with ecological degradation derived from economic progress (Connolly, 1999). Lack of access to clean water and sanitation is still a major environmental burden for the poor. At the same time the whole city, albeit unequally, suffers the effects of atmospheric pollution caused by industry and transport. Other examples may easily be found to illustrate how this city must cope with an environmental agenda in which global, national, regional, metropolitan and localized concerns compete for limited economic and political resources. The ambitious reforestation programme funded by international agencies increases filtration of water to the subsoil, but does not directly benefit those without access to a tap. Eradicating settlements from surrounding hilltops prevents ecological degradation, but may create the ultimate environmental disaster of homelessness for the affected families (which is perhaps why in Mexico City this has only occurred in exceptional cases). Undoubtedly, the telescoped environmental transition model provides an invaluable tool for exploring these conflicting environmental agendas.

However, a closer look at the history and evolution of these environmental agendas reveals some limitations of the model for explaining or predicting the city's environmental problems in time. Neither is it immediately useful as a tool to understand, evaluate or inform decisions on environmental policies for achieving more sustainable urban development. Instead, what emerges from the analysis of Mexico City is a need to focus more on if, when and how 'environmental burdens' get onto 'environmental agendas' and if, when and how these 'environmental agendas' become 'environmental policies' at different levels of government. These questions are the main focus of this chapter.

The inverted commas are used here because, before proceeding further, it is necessary to define what is meant by environmental 'burdens', 'agendas' and 'policies'. 'Environmental burdens' could be defined as the costs and risks provoked by human interaction, in this case by urban development, with the existing environment. It is important here to stress the idea of outcomes of human interaction with environment, rather than strictly anthropogenic impacts on a 'natural' ecology. 'Nature' is, in fact, part of the burden, and Mexico City provides a good example of extreme high risk urbanization from natural causes. As a city built in the middle of a lake in a highly seismic zone, it is threatened by 'natural' phenomena such as floods, earthquakes and occasional volcanic eruptions. Geomorphic conditions, latitude and altitude are other natural factors that aggravate the city's propensity to thermal inversions of polluted air and photochemical smog. Following the same argument, the pathogens in dirty water are also part of nature. 'Environmental burdens' include, then, a wide array of impacts, which affects human beings in different times and places. Theoretically, they may be objectively identified, located and measured; if they cannot, then the UET model has no substance.

Just as environmental burdens existed prior to the relatively recent preoccupation with the future of the environment, there is a whole host of unidentified environmental burdens out there waiting to be discovered. In fact, most burdens are only recognized as such after they are lifted onto someone's agenda: that is after they have become an issue, the subject of political and scientific concern. This brings us to a major problem in the practical application of the 'UET model': the difficulties of objectively identifying environmental burdens at any moment in time. The environment, like beauty, is in the eye of the beholder. And what any particular eye sees is heavily dependent on its own point of view and coloured by the state of scientific knowledge, available information, cultural beliefs about what is harmful, emphasis given to certain problems by the media and in political discourse, funding priorities for academics, social conflict and so forth. Certainly general consensus exists on some issues, and there are official versions of the existence and scale of the city's environmental problems, informed both by expert opinion and political expediency. But these accepted wisdoms are constantly shifting and so, too, are the environmental agendas they inform. This is partly in response to changing perceptions of the problems, but also due to real modifications of environmental conditions over time, in some cases over a relatively short length of time. The fact that some of the parameters are millennial, while others are in a constant state of flux, questions further the explanatory capacity of both the UET model, and notion of 'time–space' telescoping of environmental problems, specifically concerning the significance of the x-axis. Does this axis refer to increments in age or dates, economic development phases, the city's own historical evolution or progressive global technology cycles?

The third proposal, 'environmental policy', may be defined as concerted government action, which, in principle, should respond to environmental agendas in order to alleviate environmental burdens. As I will try to demonstrate in the case of Mexico City, the distinction between the theoretically separate concepts of burdens, agendas and policies is rather blurred, and the relationships between them are complex and far from linear. Environmental policies do not necessarily respond to either environmental burdens or agendas, but often define them. If a problem is not on the agenda, it is possible that there is insufficient information about it to be considered a burden. Finally, it can be shown that much of the environmental policy implemented effectively in Mexico City is not determined locally, but imported from abroad in the form of international agreements, project finance and technology provided by transnational companies. This inverse sequence relating policy to problems, or how the burdens are identified and understood, would seem to work in exact opposition to the recommendations prescribed by the 'bottom-up' approach of the local agenda for urban sustainability.

To illustrate the above arguments, the following section looks at how Mexico City's environmental problems and solutions have been perceived over the past few decades, contrasting the successive visions of environmental burdens, agendas and policies between themselves and against subsequent events and factual information. The remaining sections take up in more detail the social construction of two of Mexico City's prominent environmental agendas: water and air pollution. The enormous difference in the timescale of these two problems is itself significant. Mexico City was founded on hydrological conditions that still define the ongoing water agenda seven centuries later. Concern about atmospheric pollution stretches back only a few decades, but during this short spate of time, the nature of the problem has been constantly transformed. Mexico City has a long history of transferring water-related environmental problems elsewhere, while this process in the case of air pollution, or at least cognizance of it, has only just been initiated. Neither has air pollution substituted for water in the environmental agenda. Both these problems continue to evolve simultaneously, but this has not resulted in compressed or competing agendas. The policies that address them are organized at different levels of government, implicate different types of resource, and operate at separate time and spatial scales.

Whose burdens and which agenda? Changing perceptions of Mexico City's environmental problems

1958: An early ecological agenda for restoring natural equilibrium

> This is the Mexico basin, the 'valley which is not a valley', seat of the proud and ancient Tenochtitlan, of the peaceful colonial Mexico City, named by the Spanish 'the very noble and very loyal city', and which today is one of the biggest metropolis of the Continent.
>
> The excessive growth of this Metropolis, our pride in more ways than one and unfailingly admired by visitors, by altering the ecological equilibrium has created a series of tremendous problems that threaten its very existence...

> *Mexico City and the surrounding valley – fortunately – still provide a desirable place to live. But their future depends on us: whether they become so degraded that they are no longer inhabitable, or to the contrary, whether they can be improved to become increasingly hospitable and pleasant.* (Beltrán, 1958, pp22, 253)

Enrique Beltrán, the distinguished biologist and protagonist of the incipient Mexican ecologist[1] movement was writing about a city whose estimated population of 4.5 million in 1957 was registering its highest ever annual growth rate of 5.7 per cent (Camposortega 1992, p8) and expanding out of the nation's capital, the Federal District, into the adjacent municipalities belonging to the State of Mexico. The country's economy that year was growing annually, in real terms, at the rate of over 7 per cent per capita (Nacional Financiera, 1981); total industrial growth was in the order of 13 per cent, almost half of which was concentrated in Mexico City (Garza, 1985, p141), along with almost all the country's further education facilities, political opportunities, as well as the best chances of both social mobility and economic survival. The wages of such economic success, according to Beltrán, were the life-threatening ecological imbalances provoked by the urbanization of the Mexico Valley. The alteration of the valley's hydrological cycle by the combined extraction of stormwater and sewage, and exacerbated by deforestation and erosion, had given rise to the city's well-documented problems of lack of drinking water, sinking subsoil and persistent seasonal floods. Although many pages describe how these problems had affected successive human settlements in this valley since the Palaeolithic period, the main culprit was the 'immoderate growth of large populations'. Logically, the agenda responding to this diagnosis was to decentralize industry and population, perhaps creating satellite towns, as well as urban parks. The then-recent exodus of industry out of central Mexico City towards the State of Mexico was seen as a step in the right direction, although Beltrán deplored the rapid slum growth there. In contrast, the schemes for importing water from distant sources were applauded (no concern about ecological footprints (EFs) here); these, together with other 'ecological remedies' such as sewage treatment plants and reservoirs, would ensure the city's water supply while protecting the aquifer. But this was not enough, admonished Beltrán, unless demographic growth was controlled. So he was less enthusiastic about the government's efforts to provide services in the new irregular peripheral settlements or 'proletarian colonies', which he called 'tragic manifestations of environmental misery':

> *The problem is too complex to be solved with mere palliatives. An appreciable proportion of the population lacks adequate means of subsistence – as a pathological result of the capital's excessive growth – they cannot contribute to the maintenance of these services and, consequently, weigh on the rest of the population. The total solution of the problem involves much more than just urbanising these zones of permanent poverty.* (Beltrán, 1958, pp193–194)

Beltrán did not elaborate on what 'much more' might imply, but went on to recommend the total eradication of the first 30,000 inhabitants of the dried-up saline Texcoco lake bed, to the east of Mexico City: the area now occupied by Ciudad Nezahualcóyotl, a metropolitan municipality constituted in 1967 that is now home to 1.2 million inhabitants from all walks of life, except the very rich.

Beltrán's agenda was finalized with an exhortation for collective action by Mexico City's inhabitants themselves:

> *Official action, by itself, is not enough if it does not count on the backing and collaboration of all. Whatever sacrifice will be compensated by the thought that we are working ... toward the collective benefit, in which the first beneficiaries will be us ourselves.* (Beltrán, 1958, pp254–255)

This recognition of the environment as a public good, whose restoration requires not only government measures but effort, sacrifice and civic action on the part of the population is, perhaps, Beltrán's most important, but mostly forgotten, contribution to Mexico City's environmental agenda.

Beltrán's other recommendations, notably his insistence on the need to control the demographic growth – of the city, but not in general – and especially the redistribution of urban development, would remain on the environmental and urban planning agendas, although in practical terms, nothing was done to curb Mexico City's expansion. In the absence of any alternative, the informal settlements were allowed to establish and consolidate, among other things because public investment favoured the water supply and drainage, or the 'brown agenda', over the 'restoration of hydrological balance agenda'. Effectively, piped water and drainage provision increased from 43 to almost 70 per cent of dwellings in the following three decades, benefiting mostly the new middle-class developments, but also the 'squatter settlements' and 'proletarian colonies' on the ever-expanding outskirts. An ample supply of cheap land for irregular occupation with the support of authorities, public investment in electricity, schools, paved roads and regularization programmes, not to mention the settlers' own investments in housing and small businesses, converted many of the 'tragic manifestations of environmental misery' into fairly good places to live. By 1970, more than half of Mexico City had been urbanized by unauthorized, but tolerated, irregular settlements. Then, a large proportion of these still lacked services, legal recognition of property and paved roads, often being located in high risk areas from flooding or landslides.

The 1980s environmental NGO agendas: From conservationism to neighbourhood defence, environmental democracy and the urban movement

> *The years from 1984 to 1988 could be called the golden age of the Mexican environmental movement. The environmental demands contained an integral vision, emphasizing the question of democratic participation, justice and social*

equality, regional equilibrium, respect for ethnic diversity, land rights and use, ecological sustainability: in theory and practice. (Barba, 1998, p705)

After Beltrán's pioneering efforts in the first half of the century, various environmentalist non-governmental organizations (NGOs) sprung up in Mexico. Before the mid-1980s, however, these organizations, networks and academic interest groups had been largely concerned with green or protectionist agendas, such as nuclear power[2] (Schoijet, 1985) and other nuclear risks (Calvillo and Velázquez, 1985), industrial pollution, the destruction of biotic resources due to economic development projects (Toledo, 1985) and, in general, education, research and conscious-raising about matters ecological. All these issues were either nationwide or not specifically urban. The 1980s did indeed see not only the proliferation of NGOs concerned with the environment (González Dueñas, 2000, p36), but also the emergence of agendas specifically related to the city, and particular neighbourhoods. In Mexico City, where a disproportionate number of these NGOs are located (González Dueñas, 2000), the ecological banner was mostly unfurled by organizations concerned with the territorial defence or conquest, resistance to public works, but also with grass-roots participation in government. For example, The Ecological Associations of Coyoacán and Tlalpan were motivated, initially at least, by concern for the environmental deterioration from land-use changes in these rather nice residential areas, including the conservation of woodland and green space. Other localized groups successfully opposed the construction of a public transit skyrail through their neighbourhoods, illustrating a direct conflict between the 'democratic participation' and green agendas.

At the other end of the social scale, the Popular Urban Movement incorporated ecological considerations into their demands with relative success. The El Molino housing project, promoted by the UCISV-Libertad (Cananea Housing Movement),[3] obtained authorization for their green-field housing project partly because of the incorporation of SIRDOs (Sistema Integral de Reciclamiento de Desechos Orgánicos, a dry system for recycling organic waste): an 'ecological' solution to the problem of sanitary drainage in an area where conventional drainage was thought to be impossible due to the low-lying terrain (CENVI/UCISV-Libertad, 1986). In another case, carefully documented by Pezzoli (1997), settlers on the Ajusco foothills resisted being evicted from their 'ecological protection site' with an 'ecological urban development' counter-proposal. This project, however, ceased to mobilize the population once the threat of eviction was removed (Pezzoli, 1997, pp292–297). Other 'ecological' measures implemented by popular urban organizations with varying degrees of success and permanence have been organic farming on empty lots, reforestation, water and garbage recycling. In recent years, environmental concerns, including both the conservation of local resources and the implementation of alternative services solutions, have become a prime issue around which some communities are organized and mobilized: a kind of approach that might be termed 'greening the brown agenda'.[4] These examples are, perhaps, important exceptions to the more general tendency for poor neighbourhood organizations to demand the strictly brown-agenda solutions offered by local and federal authorities informed by urban planning professionals.

1983: Urban planners' (mostly brown) environmental agenda for Mexico City

> *The magnitude of the concentration (of economic activities and population in Metropolitan Mexico City) limits the capacity to solve the serious urban problems that are generated. Some of these, like air pollution, affect the whole population, without distinguishing income or place of residence.*
>
> *The development strategy proposed in the Programme aims at breaking the inertia that has caused this anarchic form of growth.*
>
> - *Stimulating social participation.*
> - *Controlling and regulating the growth of the Metropolitan Zone.*
> - *Coordinating the actions of all government entities and dependencies involved in the development of the Metropolitan Zone.*
>
> (SPP, 1985, pp183–185)

The above citation is typical of the wave of urban development plans arising out of the institutionalization of human settlements planning in Mexico that followed the first United Nations (UN) Conference on Human Settlements held in Vancouver in 1976. In this case, the 'plan' refers to the whole metropolitan area and, as this covers more than one federal entity, had no effective legal mandate. However, it reflects the emerging metropolitan environmental agenda of that time. Twenty-five years after Beltrán's book (1958), Mexico City had almost tripled in size, but its problems were diagnosed much in the same way. The prime concern was still the 'brown agenda', or the inability to meet the demand for housing and urban services due not, we should note, to demographic growth per se, but to the excessive concentration of this growth, and the resulting 'anarchic' urban development. In hindsight, this emphasis is rather surprising, given the extraordinary advances in services provision, which was still better in Mexico City than in almost any other city in the country, and far superior to the smaller towns and rural areas. At that time, however, the population of Mexico City was grossly overestimated. The above-cited programme estimates it at 'about 17 million' and growing at a faster rate than the country as a whole. In fact, in 1983, the population of Metropolitan Mexico City was around 14 million, and the annual growth rate for the decade was 1.95 per cent, slightly less than the national average. The miscalculation was partly due to errors in the 1980 census, but also to a lack of understanding of women's response to the birth control policy initiated in the early 1970s; the fertility rate in the Federal District dropped from 5.3 in 1970 to 1.7 in 2000 (CONAPO, 1982, p59, 2001, p37). The incipient reversal of migration trends was also undetected. As it turned out, the financial crisis of 1982 and subsequent economic re-structuring, coupled with the effects of a devastating earthquake in 1985, would convert Metropolitan Mexico City into a net exporter of population by the end of the decade.

Even after adjusting the population estimates, the need for a 'brown agenda' to provide urban services for at least half a million homes was clearly justified. But meanwhile, new problems were being added to a greener agenda. Air pollution, believed to be affecting 'rich and poor alike', now topped the list of an emerging 'environmental control and ecological protection' line of action, prescribed in the Urban Development Programme (SPP, 1985, pp219–221). Here, emissions from transport were recognized as the prime culprit, followed by certain industrial uses, such as the oil refinery and two cement factories, as well as loss of green areas from urban development. The diagnosis and policies proposed to remedy the situation coincide with the ongoing policy to combat air pollution, initiated in 1978, which will be analysed later in this chapter.

The second item on the 'environmental control and ecological protection' agenda was solid waste collection, transport and disposal. The main problem here was thought to be some 4150 tonnes of uncollected garbage of an estimated total of 11,400: no doubt an exaggerated amount, as the calculation was based on the inflated population of 17 million, instead of 13 million, generating 700g daily. Proposed lines of action included updating the legal framework in the Federal District and surrounding municipalities in order to make the service more efficient; avoidance of collection of dangerous refuse, making it obligatory for this to be incinerated in situ; improvement of municipal tips; completion of one processing plant and further introduction of recycling technology and reduction of per capita generation of refuse. Most of these recommendations have been patchily adopted, except the last one. Average daily per capita refuse generation in Mexico City was recently estimated at 1.2kg (Castillo, 2000): almost double that of the 1983 estimation.

The problem of toxic and noxious liquid waste was the third item on the 1983 environmental control and ecological protection agenda. Apart from various measures to improve the general drainage system, the main line of action proposed was to legislate for controls of toxic emissions: an agenda that would shortly after be institutionalized in successive legal reforms.

Global burdens, international agendas and Mexican environmental legislation

> *Every person has the right to a healthy environment (1996 version: that guarantees his or her development, health and well-being). The authorities, according to this and other laws, shall apply the necessary measure to guarantee this right.* (Art. 15-XII General Law of Ecological Equilibrium and Environmental Protection, 1988, reformed 1996)

> *Between 1940 and 1993, Mexico signed 68 agreements on matters relating to the conservation and protection of the environment. Most of these agreements obliged Mexico to incorporate environmental protection legislation... The two outstanding agreements referring to general environmental policy commitments*

on which Mexico should legislate were the 1992 UN Summit in Río de Janeiro ... and the North American Trade Agreement and Parallel Accords (signed in 1994). (González and Montelongo, 1999, p41)

The first piece of Mexican legislation[5] concerned *ex profeso* with environmental protection was passed in 1971, in the context of the preparatory meetings leading up to the 1972 UN Conference on the Human Environment in Stockholm the following year. The agenda here was essentially the prevention or control of air, water and subsoil pollution by regulatory procedures and fiscal incentives (González and Montelongo, 1999, pp16–27). The administration of these measures was set squarely in the hands of federal government, specifically in the Health Secretariat, for which constitutional reform was required, also in 1971. Both the timing of these initiatives and the general philosophy of public regulation as a solution to environmental problems follow closely on the heels of major environmental legislation breakthroughs in the US: President Nixon's National Environmental Policy Act of 1969–1970 and subsequent regulatory policy (see Wallace, 1995, p112; Yeager, 1991).

Eleven years later, in 1983, a reform of the Mexican Constitution emphasized the State's guiding role in the national economy and included 'protection of productive resources' and 'care of the environment' as conditions where private enterprise should be subjected to the public good. State enterprises, which included some of the worst polluters, were not, however, contemplated (Brañes, 1994, cited in PROFEPA, 2000a). At the same time, environmental protection was switched from the health sector to the urban development secretariat, whose functions would include sanitation, natural resources, preservation of ecological equilibrium, environmental protection of land and aquatic wildlife, forestry protection and water management. In practice, the unification of urban and environmental issues under the roof of a single secretariat did not see any significant merging of the green and brown agendas. While the urbanists went ahead with establishing a national urban planning system to better cope with the brown agenda, environmental regulation was sparsely administered by a separate under-secretariat, completely unrelated to regional development and housing policy, or land-use planning. Until the 1980s, environmental regulation by the Mexican government has been described as 'weak' (Hogenboom and Alfie, 2003, p17) while remaining totally disconnected from national development strategy (Carabias and Provencio, 1994, cited in Hogenboom and Alfie, 2003).

Further constitutional reform in 1987 directly assimilated the 'preservation and restoration of ecological equilibrium' to the concept of 'public interest', thus strengthening the State's responsibility for environmental regulation.[6] At the same time, Congress was enabled to legislate specifically on environmental protection, which it did the following year with the passing of the first General Law of Ecological Equilibrium and Environmental Protection. Unlike its predecessor, this law was not limited to regulatory aspects, but encompassed a wide range of aspects, such as planning, ecological criteria in development strategies, education and research, environmental impact assessment, information and vigilance, not all of which were implemented (González and Montelongo, 1999, pp36–37). One important effect of the 1988 law was to oblige the state governments to pass local environmental legislation, thereby establishing the basis of decentralizing environmental policy (PROFEPA, 2000a).[7] In spite of (and often because of) ensuing local and federal legislation, the legal attributes of each level of government concerning environmental regulation remained highly ambiguous. In practice, the major tasks, relating to what might be considered the hard-core green agenda, remained (and still remain) in the federal government's hands.[8] State governments enforce federal legislation concerning atmospheric pollution from fixed and mobile sources, except contaminating industries and interstate transport; they also have some attributes for creating and protecting parks and ecological reserves, as well as public transport regulation. Policy implementation on softer, or less regulatory aspects of the green agenda, such as environmental education, vigilance, research and community participation is not limited to any level of government, while it is the municipalities which face the traditional brown agenda tasks of water and sanitation, solid waste collection and disposal and, since 1993, land-use planning and property taxation. This functional division has done little to help bridge the gap between the green and brown policies, which are implemented independently of, and often in contradiction with, each other.

The elimination of these administrative ambiguities, which were seen to hamper the decentralization of environmental policy, was a major concern which prompted reforms to the General Law of Ecological Equilibrium and Environmental Protection in 1996 (LGEEPA, 1996, 'Exposición de motivos'). But, by that time, the concerns raised by the Brundland report (WCED, 1987) and the Rio Declaration on Environment and Development (UNEP, 1992) were also incorporated into the federal legislative agenda: 'sustainable development', 'biodiversity', 'citizen participation', access to information and the right to sue on environmental grounds, environmental economic accounting and environmental impact assessment were all incorporated into the revised law, as was the recognition of the rights of the indigenous and other communities to preserve and exploit their natural resources. These precepts have been very unevenly put into practice. The exception is the regulation of industrial and vehicle emissions, especially in Mexico City, where legislation and implementation have advanced following international developments (see below).

After Río: The largest developing world mega-city in the international agenda

> *Over the next 30 years, a further 2 billion people are expected to be added to the cities of the developing world. This massive urbanisation will cause an exponential growth of the volume of resources consumed and of pollution created... Mexico City exemplifies this twin threat: it has the dubious distinction of being the largest and most polluted city in the world.* (Rogers 1997, pp27–28)

After the mid-1980s, Mexico City was often pointed to in admonishment of the dreadful urban future that awaits the world. Yet the precise nature of its dreadfulness has changed, even over the last decade or two. Although the preliminary results of the 1990 census were published that same year, they took some time to filter through to the international literature, which continued to quote estimates of a 25 million population well into the decade. So it was widely thought – erroneously in both cases – that Mexico City and São Paolo were overtaking New York and Tokyo as the largest cities in the world (Fernández Durán, 1993, p69; Haughton and Hunter, 1994, p33). This in itself inflated the magnitude of the green agenda. Lack of precision about the quality of habitat that unauthorized urban development produced led to further overestimations of the brown agenda. For example, the frequent affirmation that two-thirds of Mexico City's population live in 'shanty towns' without regular access to water and sanitation (Haughton and Hunter, 1994, p33) is simply not true. There is a very real sanitation problem, affecting maybe as much as a quarter of the population, but this does not amount to two-thirds of 25 million.

It is for its combination of brown and green agendas, however, that Mexico City was most widely cited in the mid-1990s. After the 1992 World Bank report, it was commonly believed to be 'notoriously the most polluted city in the world' (Goldemberg 1996, p99), having 'in the early 1990s, the worst air of all megacities' (McNeil, 2001, p81). The more recent study headed by the Mexican chemist and Nobel Prize-winner, Mario Molina, also recognizes Mexico City as 'one of the worst pollution problems in the world' (Molina and Molina, 2002, p2), although there is a time lag of five to seven years between the data source and the study's publication date.[9] Meanwhile, various historical and technical studies brought international attention to Mexico City's hydrological situation, having a water and drainage system that is the 'epitome of regional environmental displacement' (McGranahan et al, 2001, p62).

Although most of Mexico City's 19 million inhabitants suffer to a greater or lesser extent from these and other environmental problems, few would wholly share these catastrophic appreciations of their city. The exodus of mostly middle-class people to smaller cities that began in the late 1980s did not increase significantly during the following decade, which may be interpreted to mean that Mexico City is still a better option for most of those who can afford to choose. Statistically, Mexico City still provides better living conditions to a greater proportion of its residents than most other cities in the country, and certainly ensures healthier, longer lives than the Mexican rural environment. As neither Mexico City residents nor the politicians who curry their votes are particularly concerned about their city's impact on near or distant regions, and in the absence of a broad-based environmentalist lobby, 'environmental displacement' hardly impinges on the local environmental agenda, notwithstanding numerous excellent studies on the subject by anthropologists and historians.

Mexico City: A mine of local solutions?

> *Getting the community involved: Children from Chalco (a vast irregular settlement) plant trees as part of a large-scale reforestation project.* (Giradet, 1992, p139)

> *Going up by the self-help staircase: This house near the Santa Fe refuse dump in Mexico City ... as is usual with the first homes of these people, it will soon be improved.* (Giradet, 1992, p131)

> *Mexico City's congestion and pollution would be much worse were it not for the dynamic and wide-ranging transportation system that has evolved over the years in response to explosive growth...*
> *Mexico City's free enterprise, open-market approach to providing supplemental transit services has relevance beyond Third World megacities.* (Cervero, 1998, pp380, 395)

> *The improvement in servicing levels, the more sensitive and realistic housing policies, the emergence of a planning structure and process, and the vastly extended and improved transportation system, greater ecological awareness, and so on, have all had a real and positive impact upon the lives and life chances of Mexico City citizens.* (Ward, 1998, p284)

Although Mexico City is usually portrayed as a portent of global ecological disaster, it is not without admirers for its localized solutions, as these four quotations illustrate. Like the apocalyptic descriptions of the previous section, these rosy pictures are perhaps more effective on the pages of the textbooks than in convincing Mexico City residents on how to improve their environment.

The colour snapshots of Mexico City's slums and tree-planting children included by Giradet to illustrate what can be done to 'cure the city' (Giradet, 1992) might cause scepticism; the real problems of deforestation and soil erosion will not be solved by planting a few pine trees in the schoolyard. In fact, the urban areas of Mexico City, as most other urban oases in this arid country, is fairly well-endowed with trees, much better in fact than 50 years ago; but this has little to do with deforestation. Neither will the chronic environmental risks and problems associated with irregular settlement automatically disappear with consolidation. It is true that these settlements do improve and that they have provided an acceptable housing solution to millions of people, especially those upwardly mobile masses who participated in Mexico's unprecedented economic development from the 1940s up until 1982. But not all the houses improve, and some of the settlements will never have water, drainage, paved roads and garbage collection, due to their precarious location on hillsides, riverbeds and legally disputed areas. As the neighbourhoods become more densely populated, the badly built houses become overcrowded and more dysfunctional; illumination, ventilation and the ill-planned public services all deteriorate. Education no longer guarantees a job, and more people go out to work for less, especially women, who now have less time, energy and space to perform the self-help miracle.

Mexico City residents would be extremely surprised by the eulogistic account of the 'free market' para-transit system offered by Cervero (1998, pp379–399). These microbuses, which daily move 12 million people around the city, are more generally considered to be a threat to public safety, associated with road accidents and violent muggings, corruption and political backwardness, as well as traffic jams and air pollution. However, in a city increasingly organized around the needs of the private car, the microbuses, together with the over 100,000 taxis do provide a better degree of mobility for the 80 per cent of the population who do not possess an automobile, than the previous publicly owned bus network.

Finally, it is doubtful that most Mexico City residents would now agree with Peter Ward's well-informed favourable prognosis, written in 1998 just after the election of a first head of government for the Federal District. Democratically elected local government for this half of Mexico City has not turned out to be a magic wand capable of making the smog and all the other problems vanish into the thin altiplano air. The other half of Mexico City has always been governed by elected state and municipal administrations, but the opening up of electoral politics and the resulting party divisions between neighbouring governments has not helped metropolitan coordination on environmental issues. Eight years on, Ward's optimism regarding service provision, housing, planning, transport and, especially, 'ecological awareness', although well-founded at that time, needs to be revised in the light of more recent developments. These include the 2000 census results enabling a better understanding of social and demographic trends, more emphasis on poverty in public and private discourse, the emergence of new forms of housing and infrastructure provision, changes in local planning policy and improved information on almost everything, not least on air pollution and other forms of contamination. Where Ward would continue to meet widespread agreement is in his comments on the lack of public security. Perceived as far more hazardous than emissions to air, soil and water, muggings, burglaries and kidnappings daily threaten Mexico City residents from all walks of life.

Water and air: Two competing agendas?

While historians and anthropologists unearth past evidence of accumulative water-related burdens and conflicts, many experts agree that water still constitutes the most serious environmental problem for present-day and future inhabitants of Mexico City, rather than air pollution. Why, then, do we have a much more coherent agenda and government policy for clean air than for water? This question was one of many raised by Lezama (2000, pp14–15) in his excellent critique of the social construction of Mexico City's air policy. The answer he found, provided by the same experts, is that there is more public awareness of the threat to public health posed by the air pollution problem, supported by prolific data provided by the monitoring system and published daily on the internet and all major mass media.[10] Water is not perceived as an important health problem, at least not in the social milieu known to environmental experts, notwithstanding medical evidence of high mortality and morbidity from gastrointestinal infections. Public opinion is thus shown to be a major determining force in establishing agendas and policy priorities.

Public opinion is, however, rather difficult to identify and measure. It would be even more difficult, in the Mexican case at least, to convincingly argue that public opinion has a direct bearing on government policy decisions. How, then, have the environmental agendas and policies for water and air pollution been defined?

Water: A long history of engineering and environmental displacement

Mexico City was founded on a hydraulic agenda. When the nomadic Mexica (Aztec) tribe took refuge on the rocky islet in the middle of a lake around 1325, the whole future urban civilization would depend on water management. For building and agriculture, land had to be reclaimed from the lakes, using the chinampa system of 'floating gardens': beds of earth held up by stakes. The freshwater lakes to the south of the valley, fed by springs, had to be separated by dykes from the saline Texcoco surrounding the new capital of the emerging Mexica empire. Dykes were also used for transport, and so were the lakes themselves, and to facilitate this function in the dry season, channels were dug. These also served to drain the growing city-island, periodically devastated by unusually heavy rains in the wet season. Aquatic transport was vital for trade with the surrounding tribes and communities, and was also instrumental in their speedy conquest and domination by the Mexica. Potable water was lacking on the island, and had to be brought from a nearby hill by aqueduct. When the Spanish arrived in 1520, they found a prosperous city-region with between 200,000 and 300,000 inhabitants: if not the 'largest city in the world', much bigger and grander than anything in Spain.

Water undoubtedly helped the Mexica defend themselves from attack by discontented dominions, but it only delayed the final fall of Tenochtitlan to the Spanish conquistadors and their allies after a long siege in 1521. When Cortés founded the capital of New Spain on the ruins of Tenochtitlan, he adopted its hydrological problems, but with an agenda immensely complicated by the environmental revolution brought about by the Spanish Conquest (Crosby, 1993). The introduction of Old World crops and weeds, horses, cows, pigs and sheep, as well as smaller breeds and deathly viruses, caused irreparable alterations to the indigenous ecology: not least, the decimation of the local human population from overwork, starvation, measles and smallpox. This native population had prized and nurtured their water resources as a means of transport, defence and source of animal protein. The Spanish replaced canals with dams, causeways and roads for their horses and other draught animals, and preferred red meat to reptiles and wildfowl. This in itself made them immensely more vulnerable to the effects of flooding, but as they were also even more prodigious builders than the Aztecs, the threat from excess water was greater. Meanwhile, the lakeside slopes and hills surrounding the city were stripped for timber and firewood, agriculture and pastureland. Subsequent soil erosion and reduced filtration increased runoff, accelerating the millennial silting up of the lakes, thus reducing their capacity to retain water. Much of the city's hydraulic infrastructure had been destroyed during the siege, leaving it unprotected from the inevitable floods.

By the beginning of the 17th century, Mexico City flood protection occupied a prominent position on the Spanish imperial agenda. Foreign experts were called in; the Dutchman Adrian Boot, believing the major threat to come from the lakes fed by freshwater springs to the south, advocated a containment policy similar to the indigenous system. However, the resident official cosmographer, German-born Enrico Martínez, had other ideas. He understood perfectly how deforestation and erosion had caused the lake beds to rise, while the built-up area had sunk under its own weight, thus increasing risk from flooding in the rainy season (Martínez 1980 [1606]). For Martínez, the only feasible solution was to make an artificial outlet to the north of the valley, and to this end he had built a tunnel in 1608, 6.6km long with a cross sectional area of $10.5m^2$. This was too narrow to be totally effective, so it was widened and converted into an open channel: a project that would take 165 years to complete, at a cost of more than 7 million pesos and 200,000 lives in forced labour. This colossal engineering feat would, however, only partially save the city from flooding. Although the lakes receded throughout the whole colonial period, the pestilential Texcoco Lake remained at only a metre below central city floor level. From 1774 onwards the drainage of this lake, by means of a second artificial channel and tunnel to the northeast, became increasingly accepted as the only viable solution to this problem. The fact that this would mean drying up the whole valley was not deemed important in an agenda still dominated by the flooding problem, although water supply for domestic use and irrigation also provided interesting challenges to the colonial engineers. For example, the city's drinking water was mainly provided by an impressive aqueduct fed by springs in the western hills. The historical archives are full of entries about conflicts over the appropriation and use of water in the Mexico Valley, in which local communities usually lose, to the benefit of the haciendas and the city. The entire history of Mexico City's water agenda is a prime example of spatial displacement of environmental problems.[11]

The environmental displacement of Mexico City's water agenda during the Colonial Period is small compared to later developments. The definitive drainage of the Mexico Valley could not be achieved during the last years of Spanish rule, for want of finance and labour, problems that continued during the violent years following independence. However, the master plan for the project was completed by 1856. But in the latter half of the 19th century, a new water agenda compounded the existing flood problem. Following the example of European and North American cities,[12] it became necessary to replace the open channels, remnants of the pre-Columbian transport system, which slowly shifted the city's detritus into the Texcoco Lake. The new urgency for draining the lake coincided with a period of relative political stability and economic prosperity, under the dictatorship of Porifirio Díaz. This guaranteed the ways and means – foreign debt to finance British contractors – to build the necessary 40km of canal and tunnel that still remove sewage and stormwater combined, out of the valley into the Tula River to the north, and eventually to the Gulf of Mexico. The immediately affected area, originally a semi-desert, was thus provided with an irrigation system, which would at the same time create increasing risks from water and soil pollution. Meanwhile, the sanitary drainage system could be installed in Mexico City, mostly in the better areas to the west of the city. And to fill the drains, new modern aqueducts had to be built, bringing water from springs to the south, thus contributing to the depletion of the Xochimilco lake system. The system was completed in 1913.

Although Mexico City expanded considerably in area during this period, population growth was still only moderate, having increased from 300,000 around 1884 to 345,000 in 1900, reaching 471,000 in 1910 (INEGI, 1994). The new urbanizations and hydraulic infrastructure built at the beginning of the 20th century did, however, provide the material basis for Mexico City's subsequent explosive population growth. They also laid down the principles which would govern the environmental agenda for water, virtually unchallenged, throughout the century.

The first principle concerns who defined the agenda: in this case, it would be the engineers and contractors who decided what the problems were and how to solve them. And the two dominating problems have inevitably been flood prevention and the need for an ever-increasing potable water supply. An underlying supposition was that water and drainage provision are public services; if necessary (and inevitably it was necessary), they should be subsidized. The solutions have almost always entailed grandiose schemes to move growing volumes of water in and out of the valley, proud monuments to Mexican engineering and government achievement.[13]

Thus, when the original drainage tunnel proved insufficient, causing severe floods in 1937, another one was built. The flooding continued, due to the dysfunctional main drainage system in the sinking city. Sewage then had to be pumped up to the level of the Gran Canal, which, by the 1960s, was not only higher than central Mexico City, but it was insufficient to remove additional stormwater after heavy rains. A monumental gravity-fed deep drainage system was built from 1967 to 1975, and continues to be extended. At present it consists of a 50km main collector fed by over 150km of deep sewers. The outfall is into the Tula River north of the Mexico Valley. Although initially designed to remove only stormwater during the rainy season, allowing for routine maintenance and repairs during the rest of the year, Mexico City's deep drainage system now removes sewage year-round. By 2002, inspection and reparation of the system had been impossible for 12 years, so its precise state was unknown. Although subsequent and ongoing investment in additional pumping stations and other measures have partially alleviated the problem, it is estimated that the system is operating at between 50 and 70 per cent of its capacity, due to blockages. Some experts consider that catastrophic flooding, due to a total failure of these outlets, still poses the greatest environmental threat to Mexico City (Legorreta, 2004; Delgado, 2005; Meléndez, 2005).

As Mexico City grew, an increasing number of wells further depleted the aquifer as the spring water brought in from Xochimilco proved insufficient by the 1930s. The solution envisaged was to import water into the valley. The first aqueduct to bring water from wells at the source of the river Lerma, 80km away, was built in the 1940s and extended during the following decade, affecting local agriculture and aquiculture (Romero Lankao, 1993). This did not prevent further wells being bored within the Mexico Valley, not only to meet the demand generated by the accelerated city growth, but also to increase coverage. The city centre continued to sink unevenly, breaking water and drainage mains and contributing to leakage, contamination of both water mains and aquifer and, of course, exacerbating the problem of flooding. More distant sources were tapped; the Cutzamala system, bringing water over a distance of 127km was started in 1976 and continues to expand.

It is useful to reflect on some agendas that were not addressed or not given priority by the hydraulic engineers during the crucial, explosive, phases of Mexico City's growth. Displaced environmental impact was one, especially in the case of Xochimilco, though in Lerma the detrimental effect on the neighbouring city of Toluca's aquifer was considered to be one reason for seeking water elsewhere (Sahab Haddad, 1988, p71). Demand management and raising general consciousness about the scarcity of the water are other missing agendas, at least until the 1980s. Repairs, substitution and rationalization of an increasingly leaky and obsolete distribution network were also not given high priority, in spite of the fact that an estimated 30–40 per cent of the water supply is lost in leakage. An exception was the enforced introduction of the 6-litre tanks for all water closets sold after 1980. Reuse and conservation water within the valley was not an important feature on the agenda either. Early efforts in the 1930s to contain water within the valley by means of a series of dams were mainly motivated by flood protection. Dual drainage was never seriously contemplated, so recycling of stormwater within the valley entails extensive treatment. The first plant to treat residual water was built in 1956, to supply an artificial recreational lake in Chapultepec Park. Additional plants were built thereafter, capable of recycling water for irrigation and agricultural uses, but not for domestic or economic purposes. By the mid-1990s, there were 69 plants in the Federal District and a further 22 in Metropolitan Mexico State (Merino, 2000). Although these have a combined capacity for treating 24 per cent of Mexico City's wastewater, only half this amount is, in fact, processed (SMA, 2000, p164), due to inadequate design, soil mechanics and, significantly, insufficient demand for treated water (Mazarí-Hiriart and Noyola, 2000, p458).

Over the last decade, the water agenda in Mexico City has been affected by changes at a national level, partly in response to international pressures. The constitutional concept of public ownership of the nation's water has been reinterpreted in terms of water as a scarce natural resource. Symptomatic of this shift was the demotion in 1976 of the Hydraulic Resources Secretariat from ministerial level to a dependency of the Agricultural Secretariat (Aboites, 2002, pp30–34). In 1989, this was substituted by the National Water Commission (CNA for its initial in Spanish), whose new mission included financial self-sufficiency and decentralization of water management (Bitrán, 1999, p7). In the early 1990s, the conceptual shift from water as a 'scarce natural resource' to water as a 'scarce commodity' (Castro, 2004) was institutionalized in a number of ways, including the programmes to 'build institutional capacity' and privatization of delivery and charging operations (Downs, 2001).

In the case of Mexico City's Federal District, this resulted in the creation of a single commission to handle the distribution, measurement and charge of consumption of water, which took responsibility for the privatization of the service. A first phase of privatization occurred in 1992 when four international consortia, combining foreign water companies with Mexican contractors, won contracts for updating and digitizing the water register and distribution network, as well as meter instalment in four zones of the city (CMIC, 2000, p44). A second phase, begun in 1995 and programmed to finish in 2004, would extend the implication of the private sector control to metering, registry and emission charges. A final phase would have completed the privatization process. However, the triumph of the left-of-centre Democratic Revolution Party candidate in the Federal District's first-ever elections for mayor in 1997 brought about a change of agendas, reflecting a reconceptualization of the problem. While efforts continue to improve and extend the register, thus increasing cost recovery, the issues of social justice, 'environmental relevance' and the 'equitable distribution of a public good' also inform decision making (CADF, 2002, pp4–5). It is also recognized that the large consumers constitute a highly inelastic demand and continue to use water inefficiently even in the face of high prices, or just do not pay. A more promising source of revenue is to extend the registry of paying consumers and collecting outstanding debts (CADF, 2002, pp5–6). Meanwhile, the functions of water supply provision, assignation, distribution and administration continue to be institutionally divorced. One effect of this is hidden subsidies; although the water rates are progressive, the immense investments behind water and drainage provision effectively subsidize mainly the big consumers.

Another effect of the institutional segregation of water management is the equally segregated vision of water as an environmental agenda. Rarely is Mexico City's water problem seen in a national context, except for recognition of its hydraulic imbalance and relatively low rainfall compared to the coastal strips (Domínguez, 2001). Recently, however, new voices are heard expressing concern for regional-level water-related environmental problems, which have inevitably had some impact on the agenda and policies. Some of these voices come from the localities affected by the extraction of their water for Mexico City. Organized opposition in Michoacán and Guerrero has effectively halted the Temascaltepec extension of the Cutzamala aqueduct. There is also increased awareness that the other cities within the central region, most of which have more severe water problems than Mexico City itself, are competing for the same sources. Toluca, capital of the State of Mexico, where Cutzamala and Lerma are located, needs water from these sources: an important issue affecting political relations between the governments of this state and the increasingly autonomous Federal District.

Unlike the clean air agenda, which always includes citations of medical literature about the impact of atmospheric pollution on health, water-related diseases are seldom included in discussion about the water problem. Vital statistics and reports on medical research into gastro enteric and other water-transmitted diseases are simply not part of this agenda. Some known health risks associated with water are kept out of the public eye, such as the effects of chlorine used to make water potable, or irreversible contamination of the aquifer by gasoline spills and other sources. The water agenda, whether it is set by territorial struggles, efficiency in marketing a scarce commodity, or meeting urban demand for water and sanitation, is still almost exclusively a task for civil engineers.

The air pollution agenda: Technically fixing the automobile

In contrast to the centuries-old water agenda, green and brown, the air pollution agenda dates back only three decades. Sometime towards the end of the 1970s,[14] Mexicans woke up to the loss of the 'the air's most transparent region'.[15] As the snow-capped volcanoes disappeared from view, air pollution became to be perceived as *the* environmental contamination problem, the two expressions being used almost synonymously, but differentiated from 'ecological degradation'. Reduced visibility and probable, but as yet unproven, health hazards were the perceived effects of 'smog'; excessive population and industrial concentration, were invariably mentioned as the main cause. A group of international experts was convened to study the problem in 1978 (Lezama, 2000, p196). Following its recommendations, the first Coordinated Programme to Improve the Air Quality in the Mexico Valley (PCMCA, after its initials in Spanish) was drawn up in 1979 by a joint commission headed by the Health Secretariat. As Lezama (1997, p325, 2000, pp88–89) has pointed out, this programme outlined the basic agenda that would be adopted in all the subsequent phases of Mexico City's clean air policy: this, in spite of the fact that both the scientific understanding of the problem at that time was extremely limited, and that the nature of the problem would change radically over the following decades.[16]

The agenda starts with the diagnosis. The precise nature, causes and effects of atmospheric pollutants in Mexico City were virtually uncharted. But it was known that the thermoelectric plants and oil refineries were responsible for large amounts of sulphur dioxide, particles and carbon monoxide. It was also suspected that lead levels were high[17] (Ezcurra, 1990, p81). It was obvious that the dried up lake bed to the east of the city generated dust storms in the dry season, bringing with them the additional hazard of air-borne faecal matter generated by the then largely unserviced irregular settlements. The dust storms periodically arising from ploughed up lands around the city were equally noticeable, although agriculture was seldom blamed for environmental problems (all things green tend to be considered 'ecological'). There was also awareness that Mexican industrial and vehicle emissions standards were well below those introduced in the US and Western Europe after the 1972 UN Conference on the Human Environment in Stockholm. Finally, it was easy to observe, and explain, that the resulting 'smog' was particularly bad on cold winter mornings when thermal inversions suppressed the already constricted natural dispersal of pollutants.

The smog was explained by the combination of physical aspects, such as Mexico City's altitude, climate and bad quality fuel, with an excessive industrial and demographic concentration, as well as the upsurge of cars (Lezama, 1997, p359). Why and how this urbanization came about are not seen as part of the environmental problem, a limitation that remains today. For example, the energy inefficiency of the private car was recognized, but the only relevant causes mentioned in relation to its 'immoderate use' were the inadequate public transport system and traffic management systems. Urban expansion into protected areas was lamented for its effects on deforestation and soil erosion, but the reasons why this occurred in spite of existing legislation to the contrary were not considered. Patterns of domestic and industrial energy consumption, nurtured by a lifetime of subsidies, were not questioned. In general, the relationship between the prevalent model of industrial and urban development, energy consumption and resulting atmospheric emissions was not addressed. The general idea was that it was not the development, per se, that was the problem, but its 'excessive' concentration in the Mexico City valley.

The resulting agenda concentrated on coping with 'emergency situations', reducing 'unsatisfactory conditions' to 10 per cent in the medium term and eliminating them in the long term. The solutions envisaged for emergencies included improvements to atmospheric monitoring, which then was limited to a manual system installed in 1967 measuring total particulate matter, sulphur dioxide and formaldehyde (SMA, 2006), and the setting up of a public alert system (Lezama, 1997, p356). The medium- and long-term reduction of pollutants was to be achieved by a series of measures that, for the purpose of this analysis, may be grouped into four categories. First, and most effective, a series of technical solutions to vehicle and industrial emissions were aimed at bringing Mexico in line with standards in the US. These included:

- the substitution of diesel for natural gas in the thermoelectric plants and public transport;
- reformulation of gasoline and diesel;
- higher emissions standards for new vehicles and periodic testing on those in circulation;
- tighter standards and control of industrial emissions, including training programmes;

- the requirement of environmental impact studies for new industries and the revision of the municipal solid waste incinerator, among others.

(Lezama, 1997)

The second type of measure, the closure and transferral of industries that could not be cleaned up and the prohibition of new contaminating plants,[18] may be classified as frankly 'environmentally displacing'. Like the technical fix solutions, this would prove to be effective. In contrast, it would be difficult to argue that the third type of measure, purportedly directed at the causes of pollution, has had much effect on the quality of Mexico City's air. This measure concerns regional planning, land-use regulation, traffic management and public transport. Industrial decentralization, promotion of urban sub-centres and heterogeneous land-use patterns to dissuade car use, synchronized traffic lights and more municipal trolley buses were some of the suggested actions at this stage. As Lezama (1997) pointed out, these measures were totally unrelated, either to each other or to any causal hypothesis. They were also out of the hands of environmental policy makers. Some of these ideas were already being promoted independently of air pollution concerns; 'sub-centres' already featured on urban planning maps, but no one showed how they might reduce traffic.

The fourth type of measure concerned research, information and environmental education. These clearly form a vital part of the environmental agenda at any moment in time, while having a decisive effect on its subsequent evolution. The direction of scientific research, the information available and the ways the problem is presented to the general public and in schools are all prime determinants of future environmental agendas. To my knowledge, there has been no systematic evaluation of this kind of measure.

During the 1980s, much of the above technical agenda was implemented by the federal government via legislation regulating industrial emissions and emissions standards for new vehicles (see below) and other measures. The state-owned oil company, Petróleos Mexicanos (PEMEX), also forged ahead with fuel reformulation, introducing low-lead and unleaded gasoline in 1986 and 1989, respectively. In 1986, the two thermoelectric plants started to convert from fuel oil to natural gas (Lacy, 1993, p82). A federal government environmental programme in 1987 called '100 necessary actions' included all these measures in its agenda, adding other steps such as relocating steel plants and prohibiting further extensions to the oil refinery (Lezama, 2000, pp176–177). In line with these federal measures, in 1988 the Mexico City (Federal District) government introduced mandatory emissions testing for automobiles (Ezcurra, 1990, p87). The following year, the 'day without a car' programme was introduced as an emergency measure, taking one-fifth of all private vehicles out of circulation on weekdays.

Meanwhile, research financed by federal government and PEMEX, with international sponsorship, began to reveal what was in Mexico City's air. Early findings showed that, in 1983, this received 153,800 tonnes of particles,

3,720,000 tonnes of carbon monoxide, 525,000 tonnes of hydrocarbons, 411,000 tonnes of sulphur dioxide and 132,000 tonnes of nitrogen oxides: 5 million tonnes in all. Most of which (75 per cent), with the exception of sulphur dioxide and particles, was attributed to vehicles (Bravo and Torres, 1986, p45). In addition, an average of 2.6µg/m³ of suspended lead particles were measured in the winters of 1981 and 1982, caused by leaded gasoline (Bravo and Torres, 1986).[19] To gain a more accurate description of these pollutants and their evolution, a computerized monitoring system was set up in 1985 and has been constantly improved since then.[20]

Since 1986, the Federal Urban Development and Ecology Secretariat has published the monitoring system's output into an air quality indicator that can be understood by the general public. The 'metropolitan air quality index' (IMECA for its initials in Spanish) was adapted from the Thom and Ott (1975) index, based on prevailing standards in the US or NAAQS (National Ambient Air Quality Standards) for 'unacceptable' and 'significantly dangerous' levels of exposure to the following pollutants: particles, sulphur dioxide, carbon monoxide and ozone (Ezcurra, 1990, pp89–90).[21] The two levels of exposure, or cut-off points, represent 100 IMECAS and 500 IMECAS respectively, and the scale is calibrated by interpolating between these two points. Thus, 200 IMECAS of ozone means that there is twice the acceptable level of this pollutant. The average IMECA is calculated for each substance in five zones into which the metropolitan area was divided (the city centre and the four sectors of the quadrant) and reported to the general public via the internet, radio and press.

The IMECA system has been criticized for being subjective and because the cut-off points are arbitrarily based on standards that are inappropriate to Mexico City's specific climatic and topographical conditions (Mugica and Figueroa, 1996, p164). Ezcurra (1990, p90) also points out that the standards on which the IMECA is based are substantially lower than those applied in other countries: in Japan and the US (California), for example. This author also criticizes the fact that the public is only informed about the levels of the worst contaminant – usually ozone or particles – thus minimizing the risk of prolonged moderate exposure to other substances in the atmosphere (Ezcurra, 1990, p92). Lezama (2004, p254) assumes that official data underestimates the magnitude and effects of Mexico City's atmospheric pollution. Other criticisms point out that the IMECA does not indicate the real risks of pollution to different sections of the population in different locations in the city (Tavera, 2002). In spite of these (and other) defects, the IMECA and the monitoring system on which it is based contribute powerfully to establishing the air pollution agenda. The IMECA alerts to emergency situations, but also provides the criteria for evaluating the effectiveness of past measures, while setting the targets for ongoing environmental policy.

Apart from monitoring and information, another major priority area for research was to determine the causes of Mexico City's polluted air. The preliminary calculations by PEMEX (Bravo and Torres, 1984) were replaced by a first emissions inventory, published in 1989, which confirmed that 77 per cent of all pollutants were generated by vehicles (PROAIRE, 1995, p74). A second official emissions inventory was devised in 1994, with a breakdown of pollutants by type of industry and vehicle, and a further, more detailed version was published in 1998 (GDF/SEMARNAT, 2002).[22] This inventory included respirable particles and, significantly, distinguishes between the polluting capacities of different aged vehicles.

The emissions inventory is clearly a basic reference point in defining the air pollution agenda. In spite of constant improvements, however, it is only as good as its inputs. Emissions of informal economic activities, by definition, are difficult to estimate. The assumptions behind the estimates of transport emissions also lack credibility. Reliable information is lacking about the number of taxis and microbuses in circulation, as well as federally controlled heavy and medium goods vehicles, all of which are high polluters. There has been no original-destiny survey in Mexico City since 1994. Trip generations and modal splits are calculated by extrapolating the 1994 study, on the assumption that transport behaviour is determined by demand, basically by population and income. Radical changes in transport supply, demographic composition, the labour market and urban structure since the early 1990s have not been taken into account in estimating the polluting capacity of vehicles used in public and private transport. The confident charts and diagrams showing the causes to the problem and indicating the agenda to follow are less objective than they appear. Once the monitoring system, standards, the IMECA system and emissions inventory had been set up, Mexico City's air pollution policy could proceed along the tracks defined earlier, but now with greater scientific backing. This policy has been outlined in three successive programmes drawn up by the federal government and the two local governments involved (the Federal District government and the Mexico State government): the Integral Programme against Atmospheric Pollution in the Mexico City Metropolitan Zone (PICCA for its initials in Spanish); the Programme for Improving Air Quality in the Valley of Mexico 1995–2000 (PROAIRE, 1995) and PROAIRE 2002–1020 (PROAIRE, 2002).

The 1990 PICCA (DDF, 1990) incorporated into a single programme the policy measures initiated in the previous decade, reaffirming them on the basis of a much better understanding of Mexico City's atmosphere (Lezama, 2000, p200). Measures such as improved gasoline and diesel fuels, substitution of fuel oil and diesel for natural gas, higher standards and control of industrial and vehicle emission, were all to be continued and reinforced. Obligatory catalytic converters on new vehicles were introduced in 1991. Financial and technical aid for achieving this was provided by the World Bank, the Japanese Eximbank and the Japanese Overseas Development Fund (Lacey, 1993, p58). The explicit criteria for selecting the precise technical measures included their proven efficacy, commercial availability, cost, minimal impact on 'urban life and institutional activities' and significant effect on reducing the more toxic pollutants (Lacey, 1993). Included in these measures was the reduction of sulphur emissions from the oil refinery, a task that proved impossible, so it was closed down in 1991. Reforestation, both in and around Mexico City, was also on the agenda, equally assisted by World Bank and Japan. Research and information were afforded high priority, the results of which have already been noted. Among other things, improved information has provided mostly positive evaluations of the implemented measures. Important reductions in lead, sulphur dioxide and, to a lesser degree, carbon monoxide, may be directly attributed to improved fuel, catalytic converters and fuel substitution (PROAIRE, 2002). Inhalable particles then became a more important problem. But most of all, and partially as a result of the elimination of lead as a detonator in gasoline, ozone was increasingly identified as the most dangerous ingredient of air pollution in Mexico City. Extreme episodes of both particle and ozone pollution have been controlled by three-way converters introduced in 1993, improved diesel and tighter industrial standards in the same year (PROAIRE, 2002).

In all events, not all the results of the 1990–1994 PICCA programme are unanimously considered to be successful. The 'day without a car' programme, which had been introduced as an emergency measure, was permanently instated: with the result that the number of vehicles in circulation increased as car-dependent families rushed to buy their second, third or fourth automobile, often older, more polluting models, or intensified the use of their car on the days they were allowed (Ezcurra, 1990). Although the PICCA recognized the need to improve public transport, no concrete measures were suggested, beyond extending the metro and substituting the fleet of taxis, buses and microbuses. Ironically, as a result of this programme, the number of taxis in circulation escalated to its present-day estimate of over 100,000. There was an increase in number of concessions for new taxis in the Federal District, while the old ones continued to operate in the metropolitan municipalities in the State of Mexico. (It is said that there are more taxis per capita in Mexico City than anywhere else). It also gave unprecedented impulse to the other low-capacity high-energy public transport, the microbus (so eulogized by Cervero), whose estimated participation in the total transport leapt from 6 per cent to almost 60 per cent between 1986 and 1988 (PROAIRE, 2002, pp2–22).[23]

A change of government at the end of 1994 saw a new air pollution programme for Mexico City, PROAIRE 1995–2000, based on a burgeoning scientific understanding of its atmospheric chemistry. A list of 95 'instruments, actions and projects' was drawn up, mostly following the same course as the previous programmes. These were aimed at specific targets, such as reducing the emission of ozone precursors by tightening standards and new combustion technology. One innovation was the tentative introduction of environmental accounting: fiscal and price incentives to private enterprises converting to anti-pollution technology and the introduction of environmental cost into the price of vehicle fuels. This kind of measure was also incorporated into the Federal Environmental Law (LGEEPA), enacted in 1996. The 'day without a car' programme was given a new function: to motivate people to buy new cars, as those factory-fitted with catalytic converters would be exempt from the programme. Emissions standards and testing methods were tightened up further in 1998, aimed at eliminating pre-1985 models while favouring new cars.[24] A programme to trap vapour in petrol stations was set up, after the discovery of this important source of emissions. The level of pollution necessary for an emergency situation was also lowered: to 240 IMECAS (2.4 times the acceptable norm) of ozone, or 175 IMECAS of inhalable particles, or 175 IMECAS if ozone is higher than 225 IMECAS.

In general, the technical measures have been successful in reducing this type of contingency, the last recorded event being in 1999. However, high levels of ozone are still frequent, for example, in 2000 it was over twice the standard (200 IMECAS) on 19 days (PROAIRE, 2002). Interestingly, the sharpest dip in all the pollution curves, but especially that of ozone, is in 1995: the year of Mexico's devastating economic crisis after the so-called 'December mistakes' (PROAIRE, 2002).[25] The shape of these curves is notoriously similar to that of national car sales: a coincidence totally ignored in the later official diagnosis of the air pollution problem.[26]

Like the other programmes, PROAIRE 1995–2000 also contemplated modernizing public transport and a bundle of measures termed 'ecological recuperation', many of which are directly handled by environmental authorities. Regarding transport, the actions were either already programmed, such as the extension of Line B of the metro, or were totally unfeasible, such as private investment in high capacity bus transit. Ecological recuperation includes such diverse measures as paving the streets in irregular settlements and recuperation of the Texcoco Lake. These were already underway independently of the environmental agenda.

The next federal government came in on the wave of profound political changes. The Revolutionary Institutional Party (PRI), which had governed Mexico by a mixture of clientelism, corporatism and repression for more than half a century, lost the 2000 presidential election to the conservative and catholic National Action Party (PAN). The same year, the left-of-centre Democratic Revolutionary Party (PRD) won the election for the Federal District head of government, for the second time since 1997, when the first elections were held here since 1929.[27] These political events are recorded here only because it might be thought that different political parties could draw up alternative environmental agendas and pursue opposing policies. This is emphatically not the case in Mexico;[28] the version of the clean air programme (PROAIRE 2002–2010) drawn up by the new governments, contains essentially the same agenda as its predecessors. One difference is a new emphasis on simplifying the red tape facing the private sector in fulfilling their environmental obligations. Otherwise, as before, the programme's major line of advance is the incorporation of a better understanding of the chemistry of pollution and the better technology available to decrease emissions.

Even so, the targets are unambitious. It is now not thought realistically possible to reduce emissions of ozone precursors by the 70 per cent necessary to keep levels permanently below the admissible limit of 0.11 ppm (100 IMECAS). The goal for 2010 is to eliminate episodes of 200 or more IMECAS while reducing to an unspecified amount the days with 100–200 IMECAS (PROAIRE, 2002). Similar goals are established for respirable particles and the other pollutants.

The strategies for achieving these goals are listed and described in detail in the programme.[29] As before, most effort, scientific justification and resources are directed at technical and legal mechanisms for reducing vehicle and fixed-point emissions. Agriculture is still on the agenda under the rather vague heading of 'conservation of natural resources', together with land-use planning, forest fire prevention, restoration of open green spaces in the city, among other items. Lezama's criticisms (2000, p206) of earlier programmes, about lack of hierarchical structuring in causal relations and their relevance for the proposed measures, are still valid. In particular, there is a stark contradiction in the policy measure aimed explicitly at encouraging people to buy new cars by applying more stringent requirements for exemption from the 'day without a car' programme, and the much more nebulous proposition of improving public transport by exactly the same formulas that have generated the existing state of affairs. The lack of any attempt at a scientifically based diagnosis of the transport problem, the paucity of technological innovations in this field and the anachronistic methods adopted to calculate 'demand corridors', strongly contrast with the confident calculations by petrochemical expertise.

Since 1999, this petrochemical expertise has been greatly enhanced by the ambitious multidisciplinary research project set up in 1999 and directed from Massachusetts Institute of Technology by the Mexican chemist and Nobel laureate Mario Molina,[30] whose preliminary results (published in Molina and Molina, 2002) were also taken into account in PROAIRE 2002–2010. The Molina programme and its ramifications are now a decisive influence in the determination of Mexico's air pollution agenda, not only that of Mexico City, but its other major cities as well. This influence can be identified at various levels.

First, the pre-eminence of scientific enquiry in the determination of both the problem and the solution is reaffirmed. This is hardly surprising, given the Molina team's specialization in atmospheric chemistry, but may not have been intentional. The original objective of the programme was to go beyond the chemistry, meteorology and medicine of air pollution. The 'integrated assessment' involved a much wider set of considerations, such as institutional problems or the relationship between land use and mobility. In fact, the above criticisms of the assumptions and knowledge behind the emissions inventory were also raised in the first publication to come out of the Molina case study on Mexico City. Some of these criticisms and corresponding recommendations have been taken on board in the subsequent improvements to the inventory, while others have not. Those that have been ignored include questions outside the realm of the natural sciences: the inadequacy of transport data of all kinds, the lack of general understanding of mobility, and the absence of any kind of calculation of pollutants emitted by informal activities (Molina and Molina, 2002). These topics have been afforded scarce attention in recent years. In contrast, the diagnosis and recommendations concerning the chemistry of Mexico City's air pollutants, their interaction with meteorological conditions and health implications are prominent on the agenda.

Following this, the second level is the definition of the air pollution problem: the identification and measurement of pollutants themselves. It was recognized that the two most harmful pollutants are respirable particles and ozone. Of the former, ultra-fine particles (PM 2.5) have been shown to have a higher impact on human health (Molina and Molina, 2004), and these are generated both directly, by combustion in transport, and indirectly, by similar photochemical reaction to those producing ozone. So attention is now focused on the direct measurement and modelling of substances involved in ozone and PM2.5 formation. In response to this, the monitoring system has been expanded (see note 20). Since 2000, Mexico City's emissions inventory is now revised biannually and includes data on ultra-fine particles (PM 2.5), total and volatile organic compounds (TOCs and VOCs), ammonia (NH3) and Methane (CH_4), in line with recommendations of the Molina report (Molina and Molina, 2002).

The third way the Mexican air pollution agenda has responded to the Molina programme is its increased adoption of US scientific, technological and normative criteria. Although this has been the case right from the start, the convergence of the Mexican and North American air pollution agenda, especially the adoption of US norms and technology, is clearly favoured in the Molina report. This is all the more evident when the fourth level of influence is considered: an increasing preoccupation with the regional and global impacts of air pollution emitted by Mexico City, both for itself and as a case study for other mega-cities in the world. It therefore becomes important to understand the transport and evolution of pollutants, both within the Mexico Valley and, significantly, on a wider scale. As a step in this direction, the 2002 Mexico City emissions inventory, published in August 2005 (SMA, 2005a), incorporates the spatial and temporal distribution of pollutants. In addition, recent developments arising out of the Molina programme include major integrated research programmes called the 'Megacity Initiative: Local and Global Research Observations' (MILAGRO),[31] comprising four separate projects. The Mexico City Metropolitan Area – 2006, led by the Molina team, gathers data from ground-based monitors to examine emissions and boundary layer concentration within and around the Mexico Valley. This is jointly funded by Mexican institutions, the US National Science Foundation (NSF) and the US Department of Energy (DOE). The other three are funded by the US. At a somewhat expanded scale, the 'Megacity Aerosol Experiment in Mexico City', led and funded by the DOE, looks at the regional effects of Mexico City's aerosol plume. The 'Megacity Impacts on Regional and Global Environments', funded by the NSF, examines the evolution of the plume at wider regional scales. Finally, the 'Intercontinental Chemical Transport Experiment', led by the National Aeronautics and Space Administration (NASA) and co-funded with the NSF, studies the evolution and transport of air pollution at global scales; here the geographic coverage specifically includes the US Gulf states. The MILAGRO programme also contemplates meteorological analysis and studies on the health impact of fine particles in Mexico City.

Finally, while Mexico City's air pollution agenda is augmented to accommodate environmental impacts at regional and global scales, the immediate priority for Mexico City itself recommended by the Molina team would seem to have shrunk to technically fixing the automobile. This is evident from a report from a workshop coordinated by Mario Molina in 2004, entitled 'Proposal for Cleaning Mexico's Air in 10 Years' (Molina, 2004). The proposal contemplates three measures: reducing the sulphur content of diesel fuel and gasoline to a minimum; increasing emissions standards for all vehicles sold in Mexico to US levels; and implementing policies to accelerate vehicle turn-over, thus ensuring the substitution of ageing vehicles by new ones. The costs of these measures are estimated in the report and are shown to be less than the derived benefits.

It should be clear from the above that Mexico City's air pollution policy has been dominated almost exclusively by the technical agenda to clean up the automobile and clean up, or move out, industry. The policy has been successful in as much pollution has abated, or at least changed its composition. But the major scientifically recognized cause of air pollution, transport, remains undiagnosed and untreated. Not surprisingly, there is a corresponding lack of public awareness regarding the centrality of transport in the air pollution agenda; even less is the general public inclined to blame the private car.[32]

One serious effect of the centrality of technically fixing the automobile in the air pollution agenda is that it distracts attention from other problems related to the private car and other forms of motorized transport. Emissions are portrayed as the only drawback to cars and other energy intensive forms of transport. Other risks associated with the use of these vehicles, such as accidents, noise, obesity and stress are simply not on the agenda. The profusion of raw data, reports and research projects on air pollution sharply contrasts with available information on, say, motor accident casualties and deaths. Published statistics on the subject do not even distinguish between victims who were inside and outside the vehicle. Neither has there been attention paid to the specific environmental impacts of vehicle production and servicing. The question of how the increasingly car-orientated urban structure is affecting access to public spaces, goods and services by different sectors of the population is simply not part of the official environmental agenda.

As a final comment on the PROAIRE programme and the recent evolution of Mexico City's air pollution agenda, it is worth mentioning some contradictions with real environmental and other policies implemented. One example is the construction of a second level to parts of the major urban freeway serving the affluent southwest quarter of the city. This illogical piece of civil engineering goes in the face of the now-accepted wisdoms of transport studies and traffic engineering[33] as well as environmentalist concerns.[34] But this is now being completed in the name of the environment and by the Secretariat for the Environment of the Federal District Government. A complete ban on new dwellings anywhere except the central demarcations in the Federal District has also been enacted in the name of conserving natural resources on the periphery and promoting non-motorized mobility by increasing central densities. This may have contributed to the recent construction boom in middle- and lower-middle income housing within the central areas, but there is no evidence to show that this will reduce either motorized mobility or the use of the private car. The majority of new houses being built in the central areas has at least one or two parking places.

Outside the Federal District, and on the outskirts of practically all the medium and large cities in the country, Mexico City included, the landscape is changing in a way that can only worsen air pollution. The recent federal policy to reorganize mortgage finance favouring the emergence of a new housing industry has caused the massive construction of miniscule housing units on cheap outlying land. These new developments, which are usually unconnected to existing urban centres, generate an immediate dependence on low-capacity motorized transport, especially car ownership, for those that can afford it. Meanwhile cars have become more affordable. New car sales are boosted by an abundance of credit schemes fostered by national financial and economic policies. But that by no means eliminates older, more polluting cars from the market.

Another national policy that goes against the air pollution agenda, especially that of technically fixing the automobile, is the liberalization of imports of second-hand vehicles. There have always been periodic reprieves granted to cars and trucks illegally imported by migrants returning from the US. The legislation enacted in October 2005 unilaterally anticipates Mexico's obligations under the North American Free Trade Agreement by eliminating import duty and otherwise facilitating the legalization of vehicles between 10 and 12 years old originating in the US and Canada. Although it is thought that most of these vehicles will not circulate in metropolitan areas, and those that do will have to comply with emissions testing, this measure will clearly increase the amount of pollutants due to outdated combustion technology.

Conclusion

This chapter has illustrated how Mexico City's environmental agenda has evolved over the past 50 years. Hopefully, it will be clear from this analysis that the relationship between environmental 'burdens', 'agendas' and 'policies' is far from linear. The policies tend to generate their own agendas and the representation of the burdens in the agendas is, at best, selective. Perhaps one important conclusion is the need to address, not what is on the environmental agenda but what is not, and also to ask whose burdens never appear on any agenda.

Notes

1 Enrique Beltrán Castillo (1903–1994) wrote and taught extensively on biology, zoology, marine biology, protozoology, conservation of natural resources and

history of science in Mexico, and had at least 14 species named after him. In 1952, he founded the first environmentalist non-governmental organizations (NGO) in Mexico (Instituto Mexicano de Recursos Naturales Renovables).
2 Opposition to the LagunaVerde nuclear power plant was described as the 'backbone' of the Mexican ecologist movement in the mid-1980s (Muñoz, 1989, p65)
3 A constituent group of the UPREZ (Emiliano Zapata Revolutionary Popular Union).
4 For instance, the Union de Pueblos del Eje NeoVolcánico.
5 The 'Ley Federal para Prevenir y Controlar la Contaminación Ambiental', *Diario Oficial* 23-03-1971. Three subsequent legislations were derived from this law: 'Reglamento para la Prevención y Control de la Contaminación Atmosférica Originada por la Emisión de Humos y Polvos' (1971); 'Reglamento para el Control y Prevención de la Contaminación de las Aguas' (1973) and 'Reglamento parta Prevenir y Controlar la Copntaminación del Mar por Vertimiento de Desechos y Otras Materiales' (1979).
6 Article 27 of the Mexican Constitution is the basis of all public intervention in private property, including agrarian reform, expropriations and land-use planning. ('The Nation will have at all time the right to impose on private property the modalities dictated by public interest...')
7 The state-level governments of Mexico City, Mexico State and the Federal District passed their environmental laws in 1991 and 1996, respectively. The Federal District was later in legislating as it was only after constitutional and other political reforms in 1994, that it had an elected assembly with legislative faculties. Previously, the Federal District had no legislative governing bodies and was only granted an elected executive in 1997.
8 Under the revised General Law of Ecological Equilibrium and Environmental Protection, the Federal Prosecutor for Environmental Protection enforces legislation concerning toxic and biologically infectious wastes, atmospheric emissions from contaminating industries, major environmental impact studies, control of new vehicle emissions, forestry regulation, fisheries regulation, regulation of imported animal and vegetable species and derived products and protection of endangered species (PROFEPA, 2000b). Water resources and emissions to rivers, lakes and sea are handled by the National Water Commission. See González and Montelongo (1999) for a critical analysis of Mexican environmental law concerning the distribution of attributes between different levels of government.
9 Using World Bank Indicators published in 2001, Molina and Molina (2002) put Mexico City as the second largest city, after Tokyo, having also the third-highest emissions of total particulate matter ($279\mu g/m^3$, in 1995, compared to 415 for Delhi and 377 for Beijing) and sulphur dioxide ($74\mu g/m^3$ in 1998, compared to 129 for Río de Janeiro and 90 for Beijing). Mexico City is also listed as having the highest emissions of nitrogen dioxide, with $130\mu g/m^3$ in 1998.
10 In a later, more detailed, study of the social construction of environmental problems, Lezama (2004) found that only the representatives of political parties considered air pollution to be the 'only major environmental problem' of Mexico City, while government officials and ecological activists alike regarded air pollution to be extremely important, along with other hazards related to water, drainage, toxic waste and soil erosion.
11 An excellent historical overview from this perspective is provided by Musset (1992). The recent reorganization of the *Archivo Histórico del Agua,* which has its own bulletin, has also inspired renewed interest in the subject – for example,

the fine series published by the Centro de Investigación y Estudios Superiores de Antropología Social '*Biblioteca del Agua*'.

12 Sewer systems, including filtering and settlement tanks, began to be built in British cities from the mid-19th century (Briggs, 1968; Rolt, 1988) but these were not generalized in the US until after 1880 (Melosi, 2001). By this time, in Mexico, public health professionals (doctors, engineers, politicians) were aware of the relation between lack of sanitation and high mortality rates, which were more than double those registered in Europe by the turn of the century. For example, infant mortality in Mexico City in 1900 was 323 per 1000; life expectancy around 1880 was 24 years, compared to 46 in Paris (González Navarro, 1957, pp48–52).

13 The importance of hydraulic engineering (and hydraulic engineers) in Mexican national development policy, principally for irrigation schemes, is reflected in the existence of a Hydraulic Resources Secretariat from 1946 to 1976 (Aboites, 2000, 2002). The same kind of priorities are also reflected in the hierarchical and budgetary ranking of the two departments concerned with water – 'Water and Sanitation' and 'Hydraulic works' – in the Federal District government, over the same period.

14 Although the first scientific paper on Mexico City's air pollution was published two decades earlier (Bravo and Viniegra, 1958, cited in Molina and Molina, 2002).

15 The extraordinary clarity and luminescent effect of the altiplano atmosphere around Mexico City has been noted since the 16th century chroniclers. The much quoted epigraph '*la region más transparente del aire*' was immortalized in 1915 by the great writer, Alfonso Reyes (1889–1959) in his essay *Visión de Anáhuac*. It was later adopted by Carlos Fuentes for the title to his major novel, published in 1958, a portrait of the city in the throes of rapid modernization. Invariably cited in any pronouncement about the urban environment and associated sense of loss, the epigraph perhaps contributes to the idea that air pollution is the worst problem associated with Mexico City's growth (e.g. see Ezcurra, 1990).

16 According to Molina and Molina (2002), in the 1970s the problem was defined in terms of smoke and dust, reflecting an understanding air pollution comparable to that of the US in the 1960s.

17 Ezcurra (1990) citing his own findings relating to 1968, found an average value of 5mg of lead per cubic metre.

18 The substitution of maize cultivation in and around Mexico City for horticulture was also suggested.

19 This is substantially less than the $5.1\mu g/m^3$ recorded in Mexico City in 1970 (Bravo 1987, quoted in Ezcurra, 1990).

20 A Philips monitoring network with 22 stations had been installed in Mexico City since 1973. However, 24-hour systematized data were only generated after the system overhaul in 1985. The number of stations was increased to 32 in 1992. (A further five were added in 1995 but have not been incorporated into the computerized system.) This system measures ozone, carbon monoxide, nitrogen oxides, sulphur dioxide and particles (total PM and PM10). In 2003, eight more stations were added to the automatic system and continuous monitoring of PM2.5 was initiated. In addition, three differential optical absorption spectroscopes were installed to continuously measure ozone, nitrogen oxide, sulphur dioxide, nitrous acid, formaldehyde, ρ-xilene, benzene and toluene. By the end of 2005, a manual network with 26 remote bases samples to detect presence of particulate matter (differentiating between total particles, PM10 and PM2.5), heavy metals, nitrates and sulphates (Ramos, 2002; SMA, 2006). Acid rain has been monitored since

1987 by 16 remote stations (INEGI, 1999; SMA, 2006). The weather monitoring system now also measures ultraviolet radiation, wind speeds and temperature gradients up to 350m above ground level (Ramos, 2002; SMA, 2005b).
21 The norms were re-issued in 1994, when tolerated particle exposure was slightly reduced, as in the following table:

Pollutant	Length of exposure (hours)	Cut-off point of acute exposure = 100 IMECA		Chronic exposure 1994
		1985	1994	
Ozone O_2	1	0.11ppm	0.11ppm	
Sulphur dioxide SO_2	24	0.13ppm	0.13	0.03ppm annual mean
Nitrogen dioxide NO_2	1		0.21ppm	
Carbon monoxide CO	8	13ppm	11ppm	
Total suspended particles PMT	24	275µg/m³	260µg/m³	75µg/m³ annual mean
Inhalable particles PM_{10}	24		150µg/m³	50µg/m³ annual mean
Lead				1.5µg/m³ mean over 3 months
Nitrogen oxides NO_x	1	0.66 (= 200 IMECA)		
Interaction PMT/SO_2	24	24.5ppm (= 200 IMECA)		

Source: 1985: Ezcurra (1990, pp89, 19); SEMARNAP (1995, p17)

22 In addition to these official inventories, three versions were elaborated by international agencies: by the Japanese International Cooperation Agency in 1988; by the German Reinland Group in 1992 and by the World Bank in 1994 (CESPEDES, 1998). The Mexican Petroleum Institute also produced an emissions inventory in 1996, showing a substantially smaller contribution of hydrocarbons from domestic gas consumption and distribution than the 1994 PROAIRE inventory (CESPEDES, 1998).
23 The impact of this increase of taxis and microbuses on the mobility of Mexico City's inhabitants, in the context of an increasingly car-orientated urban structure, has yet to be studied.
24 Since 1998, new cars are exempted from emissions test for two years and can circulate any day of the week; cars with functional catalytic converters that pass the emissions test are also free to circulate any day. Other cars are banned from the streets one day a week, or two days, at times of environmental contingencies.
25 This refers to the government handling of the financial crisis, specifically to the leaking of signals about an imminent currency devaluation, provoking a massive outflow of capital and further devaluation. Generalized bankruptcies ensued in all sectors whose increased foreign currency obligations had been fostered by a decade of trade liberalization policy.

26 Vehicle sales are themselves a strong indicator of the state of the economy in general, so that it would be simplistic to just derive a simple causal relation between car sales and pollution, but rather, the whole relationship between development and environment should be addressed. On this point, the present under secretary for the environment, Fernando Tudela (1997) noted some years ago the reluctance of Latin American governments to recognize their development models as major determinants of environmental degradation.
27 The State of Mexico government, in whose jurisdiction the other half of Metropolitan Mexico City lies, has a different electoral calendar. Here, the PRI has never lost an election for governor, but about a third of the municipalities are in the hands of the PRD, while another third is PAN.
28 The lack of an environmental agenda in the major political parties was noted by Quadri and Provencio (1994) in relation to the 1994 presidential elections. The 'green' party, virtually commandeered by a single family on an entrepreneurial basis, competed in alliance with the PAN in the 2000 presidential elections and professes no specific environmental agenda.
29 Available freely via various Mexican government internet sites.
30 The Integrated Programme on Urban, Regional and Global Air Pollution, Mexico City Case, involves the participation of an interdisciplinary group of researchers from academic institutions and government bodies, as well as private consultants, in both the US and Mexico. These include the US-Mexico Foundation for Science, the Asociación Mexicana de la Industria Automotriz and the World Bank.
31 Information on the MILAGRO programme and its components is available at its website http://mce2.org/megacities
32 Of 3626 respondents to a survey undertaken in 2002 by researchers at the Instituto Mexicano del Petróleo, 32 per cent considered that 'industry in general' was the major source of Mexico City's air pollution; 28 per cent attributed this to public transport; 18 per cent to 'garbage and waste'; 16 per cent to private vehicles; 4 per cent to combustion vapours and 2 per cent to gas leaks (Adapted from Melgar et al, 2002).
33 As outlined, for example, in Goodwin et al (1991).
34 For a good criticism of the '*segundos pisos*', see Quadri (2002).

References

Aboites, L. (2000) 'Optimismo nacional: Geografía, ingeniería hidráulica y política en México 1926–1976', in B. von Mentz (coord), *Identidades, Estado Nacional y Globalidad: México, Siglos XIX y XX*, Centro de Investigaciones Sociales y Estudios Superiores en Antropología Social, México DF, pp95–152

Aboites, L. (2002) 'Fin de un sueño. Notas sobre la Extinción de la secretaría de recursos hidráulicos', *Boletín del Archivo Histórico del Agua*, Año 7, enero-abril, AHA/CIESAS, México DF, pp20–34

Barba, R. (1998) 'Participación de organismos no gubernamentales', in R. Barba (coord), *La Guía Ambiental*, The John D. and Catherine MacArthur Foundation, México DF, pp699–717

Beltrán. E. (1958) *El Hombre y su Ambiente. Ensayo sobre el Valle de México*, Fondo de Cultura Económica, México DF

Bitrán, D. (1999) *México: Inversiones en el Sector Agua, Alcantarillado y Saneamiento*, Serie Reformas Económicas no. 21, United Nations, Santiago de Chile

Brañes, R. (1994) *Manual de Derecho Ambiental Mexicano*, Fondo de Cultura Económica, México
Bravo, H. (1987) *La Contaminación del Aire en México*, Fundación Universo Veintiuno, México DF
Bravo, H. and Torres, R. (1984) *Niveles de los Contaminantes en la Atmósfera de la Ciudad de México. Alternativas de Control, Ciclo Ecológico y Desarrollo Industrial*, Gerencia de Protección Ecológica e Industrial, Petróleos Mexicanos, México
Bravo, H. and Torres, R. (1986) 'Problemas ambientales originados por el uso de los derivados del petróleo y gas natural', *Memoria del Simposio Energía y Ambiente*, Universidad Nacional Autónoma de México-Secetaría de Desarrollo Urbano y Ecología, México DF, pp35–50
Bravo, H. and Viniegra, O. G. (1958) 'Informe preliminar acerca de la polución atmosférica en la Ciudad de México', *Memorias de la XII Reunión Anual de la Sociedad Mexicana de Higiene Industria*, Puebla, México
Briggs, A. (1968) *Victorian Cities*, Pelican Books, Harmondsworth
CADF (2002) *Metodología para Analizar los Ingresos por los Derechos del Suministro de Agua Potable*, Unpublished report, preliminary edition, 25 June, Gobierno del Distrito Federal, Comisión de Aguas del Distrito Federal, México DF
Calvillo, A. and Velázquez, M. T. (1985) 'Dónde quedó la varilla', *Casa del Tiempo*, Universidad Autónoma Metropolitana, Azcapotzalco, vol 52, no 52, pp25–29
Camposortega, S. (1992) 'Evolución y tendencias demográficas de la ZMCM', in *CONAPO La Zona Metropolitana de la Ciudad de México. Problemática Actual y Perspectivas Demográficas y Urbanas*, Consejo Nacional de Población, México DF, pp3–16
Carabias, J. and Provencio, E. (1994) 'La política ambiental mexicana antes y después de Río', in A. Glender and V. Lichtinger (eds), *La Diplomacia Ambiental, México y la Conferencia de las Naciones Unidas sobre Medio Ambiente y Desarrollo*, Secretaría de Relaciones Exteriores y Fondo de Cultura Económica, México
Castillo, H. (2000) 'Basura', in G. Garza (coord), *La Ciudad de México en el Fin del Segundo Milenio*, El Colegio de México/Gobierno del Distrito Federal, México DF
Castro, J. E. (2004) 'Urban water and the politics of citizenship: The case of the Mexico City Metropolitan Area (1980s-1990s)', *Environment and Planning A*, vol. 36, no 2, pp327–346
CENVI/UCISV-Libertad (1986) *El Programa de Vivienda de El Molino. Una Experiencia Autogestiva de organización Popular*, Centro de la Vivienda y Estudios Urbanos AC, México DF
Cervero, R. (1998) *The Transit Metropolis. A Global Enquiry*, Island Press, Washington, DC
CESPEDES (1998) *Ciudad de México. Respirando el Futuro. Evalución del Porgama para Mejorar la Calidad del Aire en la Zona Metropolitana de la Ciudad de México 1995-2000*, Centro de Estudios del Sector Privado para el Desarrollo Sustentable/Cámara Nacional de la Industria de la Transformación, México DF
CMIC (2000) *El Desafío del Agua en la Ciudad de México*, Cámara Mexicana de la Industria de la Construcción/Centro de Estudios del Sector Privado par el Desarrollo Sustentable, México DF
CONAPO (1982) *México Demográfico*, Consejo Nacional de Población, México DF
CONAPO (2001) *La Población de México en el Nuevo Siglo*, Consejo Nacional de Población, México DF
Connolly, P. (1999) 'México City. Our Common Future?' *Environment and Urbanization*, vol 11, no 1, pp53–78

Crosby, A. (1993) *Ecological Imperialism*, Cambridge University Press, Canto edition, Cambridge

DDF (1990) *Programa Integral contra la Contaminación Atmosférica. Un Compromiso Común*, Departamento del Distrito Federal, México DF

Delgado, M. (2005) 'La crisis del drenaje profundo', *Reforma*, vol 10, no IV-2005, Sección Ciudad y Metrópoli

Domínguez, E. (2001) 'Agua: escasez y vulnerablilidad en la Zona Metropolitana de la Ciudad de México', Paper presented at the Día Mundial de Agua Conference, Mexico City, 22–23 March

Downs, T. (2001) 'Making sustainable development operational: Integrated capacity building for the water supply and sanitation sector in Mexico', *Journal of Environmental Planning and Management*, vol 44, no 4, pp525–544

Ezcurra, E. (1990) *De las Chinampas a la Megalópolis. El Medio Ambiente en la Cuenca de México*, Fondo de Cultura Económica, México DF

Fernández Durán, R. (1993) *Explosión del Desorden*, Editorial Fundamentos, Madrid

Garza, G. (1985) *El Proceso de Industrialización de la Ciudad de México 1821–1970*, El Colegio de México, México DF

GDF/SEMARNAT (2002) *Inventario de Emisiones de la Zona Metropolitana del Valle de México 1998*, Gobierno del Distrito Federal/Secretaría de Medio Ambiente y Recursos Naturales, México DF.

Giradet, H. (1992) *Ciudades. Alternativas para una Vida Urbana Sostenible*, Celeste Ediciones, Madrid

Goldemberg, J. (1996) *Energy, Environment and Development*, Earthscan, London

González, J. J. and Montelongo, I. (1999) *Introducción al Derecho Ambiental Mexicano*, Universidad Autónoma Metropolitana, Azcapotzalco, México

González Dueñas, M. (2000) 'Las organizaciones no gubernamentales ambientalistas (ONGAs) en la Ciudad de México: ¿Entre el Estado y la sociedad?' Unpublished thesis, Maestría en Desarrollo Urbano, El Colegio de México, México DF

González Navarro, M. (1957) *La Vida Social*, Daniel Cosío Villegas, *Historia Moderna de México. El Porfiriato*, Editorial Hermes, México DF

Goodwin, P., Hallet, S., Kenny, F. and Stokes, G. (1991) *Transport: the New Realism*, Report 624, Transport Studies Unit, University of Oxford, Oxford

Haughton, G. and Hunter, C. (1994) *Sustainable Cities*, Jessica Kingsley, London

Hogenboom, B. and Alfie, M. (2003) *Cross Border Activism and its Limits. Mexican Environmental Organization and the United States*, Centre for Latin American Research and Documentation, Amsterdam

INEGI (1994) *Estadísticas Históricas de México*, vol I, Instituto Nacional de Estadística, Geografía e Informática, México DF

INEGI (1999) *Estadísticas del Medio Ambiente del Distrito Federal y Zona Metropolitana*, Instituto Nacional de Estadística, Geografía e Informática, México DF

Lacy, R. (1993) *La Calidad del Aire en el Valle de México*, El Colegio de México, México DF

Legorreta (2004) 'Imperceptible para los capitalinos, la ciudad se ha hundido 10 metros', *La Jornada*, vol 8, no XVIII-2004, www.jornada.unam.mx/2004/07/08/02an1cul.php?origen=cultura.php&fly=2

Lezama, J. L. (1997) 'La construcción gubernamental de la contaminación ambiental: La política del aire para la Ciudad de México, 1979–1996', *Economía Sociedad y Territorio*, vol 1, no 2, pp317–362

Lezama, J. L. (2000) *Aire Dividido. Crítica a la Política del Aire en el Valle de México*, El Colegio de México, México DF

Lezama, J.L. (2004) *La Construcción Social y Política del Medio Ambiente*, El Colegio de México, México DF
LGEEPA (1996) *Ley General del Equilibrio Ecológico y la Protección al Ambiente*, Poder Ejecutivo Federal, México
Martínez, E. (1980 [1606]) *Repertorio de los Tiempos y Historia Natural de Nueva España*, facsimile edition of the 1606 publication, Centro de Estudios de Historia de Mexico CONDUMEX, Ciudad de México
Mazarí-Hiriart, M. and Noyola, A. (2000) 'Contaminación del agua', in G. Garza (coord.) *La Ciudad de México en el Fin del Segundo Milenio*, El Colegio de México/Gobierno del Distrito Federal, México DF
McGranahan, G., Jacobi, P., Songsore, J., Surjadi, C. and Kjellén, M. (2001) *The Citizens at Risk. From Urban Sanitation to Sustainable Cities*, Earthscan, London
McNeil, J. (2001) *Something New under the Sun. An Environmental History of the Twentieth Century*, Penguin Books, London
Meléndez, V. (2005) 'La Ciudad de México y su sistema de drenaje profundo', *Oncenoticias*, Canal 11
Melgar, E.M., Ruiz, M.E., Martínez, S., Yáñez, G. and Ceballos, J. E. (2002) *Valoración Contingente sobre la Calidad del Aire. Caso de Estudio Ciudad de México*, Consejo Nacional de Ciencias y Tecnología/Instituto Mexicano del Petróleo, México DF
Melosi, M. (2001) *Effluent America*, University of Pittsburgh Press, Pittsburgh
Merino, H. (2000) 'Sistema Hidráulico', in G. Garza (coord), *La Ciudad de México en el Fin del Segundo Milenio*, El Colegio de México/Gobierno del Distrito Federal, México DF
Molina, L. and Molina, M. (eds) (2002) *Air Quality in the Mexico Megacity. An Integrated Assessment*, Kluwer Academic Publishers, Dordrecht
Molina, L. and Molina, M. (2004) 'Improving air quality in megacities. Mexico City case study', *Annals of the New York Academy of Sciences*, vol 1023, pp142–158
Molina, M. (coord) (2004) *Propuesta para Limpiar el Aire de México en 10 Años. Reporte del Taller sobre la Contaminación del Aire en México*, Instituto Nacional de Ecología, México DF
Mugica, V. and Figueroa, J. (1996) *Contaminación Ambiental. Causas y Control*, Universidad Autónoma Metropolitana-Azcapotzalco, México DF
Muñoz, J. (1989) 'Laguna Verde y el movimiento antinuclear', *Ciudades*, no 1, enero-marzo, pp65–68
Musset, A. (1992) *El Agua en el Valle de México. Siglos XVI-XVIIi*, Pórtico de la Ciudad de México/Centro de Estudios Mexicanos y Centroamericanos, México DF
Nacional Financiera S. A. (1981) *La Economía Mexicana en Cifras*, Nacional Financiera, México DF
Pezzoli, K. (1997) *Human Settlements and Planning for Ecological Sustainability. The Case of Mexico City*, MIT Press, Cambridge, MA
PROAIRE (1995) *Programa para Mejorar la Calidad del Aire en el Valle de México*, Departamento del Distrito Federal, Gobierno del Estado de México, Secretaría de Medio Ambiente y Recursos Naturales, Secretaría de Salud, MéxicoDF
PROFEPA (2000a) *Discrepancia, Consenso Social y Unanimidad Legislativa. Crónica del la Reforma a la Ley General del Equilibrio Ecológico y la Protección al Ambiente, 1995-1996*, Procuraduría Federal del Protección al Ambiente, México DF
PROFEPA (2000b) *Informe 1995-2000*, Procuraduría Federal del Protección al Ambiente, México DF
PROAIRE (2002) *Programa para Mejorar la Calidad del Aire de la Zona Metropolitana del Valle de México 2002-2010*, Comisión Metropolitana del Medio Ambiente,

Gobierno del Distrito Federal, Gobierno del Estado de México, Secretaría de Medio Ambiente y Recursos Naturales, Secretaría de Salud, México DF
Quadri, G. (2002) *Un Segundo Piso a Vialidades Troncales en la Ciudad de México. Riesgos y Conjeturas*, Centro de Estudios del Sector Privado para el desarrollo Sustentable, México DF
Quadri, G. and Provencio, E. (1994) *Partidos Políticos y Medio Ambiente*, El Colegio de México DF
Ramos, R. (2002) 'Actualización del sistema de monitoreo atmosférico de la Ciudad de México y su zona metropolitana', Foro Calidad del Aire para México Conmemorativo del Día Interamericana de la Calidad del Aire, Instituto Nacional de Ecología, www.ine.gob.mx/cenica/forocalaire.html
Rogers, R. (1997) *Cities for a Small Planet*, Faber and Faber, London
Rolt, L. T. C. (1988) *Victorian Engineering*, Penguin, Harmondsworth
Romero Lankao, P. (1993) *Impacto Socioambiental, en Xochimilco y Lerma, de las Obras de Abastecimiento de la Ciudad de México*, Universidad Autónoma Metropolitana-Xochimiclo, México DF
Sahab Haddad, E. (1988) 'Abastecimiento de agua potable a la Zona Metropolitana de la Ciudad de México', *Revista Mexicana de la Construcción*, no. 406, septiembre, pp69–76
Schoijet, M. (1985) 'Claves para el debate nuclear en México', *Casa del Tiempo*, Universidad Autónoma Metropolitana, Azcapotzalco, vol. 52, no 52, pp20–24
SEMARNAP (1995) *Programa para Mejorar la Calidad del Aires en el Valle de la México (1995–2000)*, Secretaria de Medio Ambiente, Recursos Naturales y Pesca, México DF
SMA (2000) *Tercer Informe de Trabajo-2000*, Secretaría de Medio Ambiente, Gobierno del Distrito Federal, México DF
SMA (2005b) *Inventario de Emisiones de la Zona Metropolitana del Valle de México 2002*, Secretaría de Medio Ambiente, Gobierno del Distrito Federal, México DF, www.sma.df.gob.mx/sma/modules.php?name=AvantGo&file=print&sid=322
SMA (2005b) *Quinto Informe de Trabajo-2005*, Secretaría de Medio Ambiente, Gobierno del Distrito Federal, México DF
SMA (2006) '20 años de monitoreo atmosférico continuo en la Ciudad de México', Secretaria de Medio Ambiente, www.sma.df.gob.mx/sma/modules.php?name=AvantGo&file=print&sid=332, 5 January
SPP (1985) 'Programa de Desarrollo de la Zona Metropolitana de la Ciudad de México y de la Región Centro', October 1983, in *Antología de la Planeación en México 1917-1985. Vol. 15 Planeación regional e institucional (1982–1985)* Fondo de Cultura Económica, México DF
Tavera, L. (2002) 'Los impactos diferenciados de la contaminación atmosférica en la Zona Metropolitana de la Ciudad de México', Unpublished thesis, Maestría en Planeación y Políticas Metropolitanas, Universidad Autónoma Metropolitana-Azcapotzalco
Thom, G. C. and Ott, W. R. (1975) *Air Pollution Indices: A compendium and Assessment of Indices Used in the United States and Canada*, Council on Environmental Quality and the Environmental Protection Agency, Washington, DC
Toledo, A. (1985) 'Modernidad y Ecologismo', *Casa del Tiempo*, Universidad Autónoma Metropolitana, Azcapotzalco, vol 52, no 52, pp30–33
Tudela, F. (1997) 'Diez tesis sobre desarrollo y medio ambiente en Américo Latina y el Caribe', in G. López (coord), *Sociedad y Medio Ambiente en México*, El Colegio de Michoacán, México, pp59–70

UNEP (1992) *Rio Declaration on Environment and Development*, Emitted following the The United Nations Conference on Environment and Development, Rio de Janeiro 3–14 June, www.unep.org/Documents.multilingual/Default.asp?DocumentID=78&ArticleID=1163

Wallace, D. (1995) *Environmental Policy and Industrial Innovation. Strategies in Europe, The US and Japan*, Earthscan, London

Ward, P. (1998) *Mexico City*, John Wiley, New York

WCED (1987) *Our Common Future*, Report of the World Commission on Environment and Development to the United Nations General Assembly, a/42/427 New York, 4 August 1987

World Bank (1992) *The World Development Report*, The World Bank, Washington, DC

Yeager, P. C. (1991) *The Limits of Law. The Public Regulation of Private Pollution*, Cambridge University Press, Cambridge

11

In Pursuit of the Sustainable City

Graham Haughton

Problematizing affluence and sustainable urban development

This chapter provides an overview of environmental problems and affluent cities. This immediately raises some interesting problems of definition and analysis. What is an affluent city, and is it a meaningful category for analysis? The overall affluence of a city, for instance, can mask major internal inequalities in terms of uneven exposure to environmental risks and contributions to creating environmental problems. Moreover, many of the environmental problems and risks which we might associate with cities are actually felt outside the city and are rooted in social and economic processes that operate without regard to urban boundaries.

It is for these reasons that rather than a conventional exploration of issues such as levels of air and water pollution, car congestion and so on, this chapter steps back a bit to focus on the ways in which our understandings of environmental problems are constructed and on the political economy of environmental risk displacement. Although there is a focus on more affluent cities, this is situated within a critique of the dangers of over-emphasizing the 'urban-ness' of environmental problems. In particular, we need to be wary of reifying or privileging 'cities' as an analytical category for understanding environmental problems, since the underlying social and economic processes are worked out across scales, from the neighbourhood to the global.

As the title of this book suggests, urban environmental impacts need to be examined at various scales: we need to recognize that the causes and consequences of environmental problems associated with a city may well not be found within the city boundaries. The external environmental impacts of cities and nations have sometimes been referred to as their 'shadow ecologies' (MacNeill et al, 1991), sometimes as their 'ecological footprints' (EFs). In addition, studies of urban metabolism have long examined how cities draw in much of their material needs from external areas (Newcombe, 1984; Girardet, 1992). Such notions can help us to appreciate how urban consumption habits can lead to resource depletion from distant source areas, and also how urban pollution can impact on local, regional and increasingly global levels. Recent work by McGranahan et al (2001) has added a further dimension to such concerns, highlighting the ways in which different types and scales of environmental burdens can be identified for different cities, arguing that more affluent societies generate larger-scale burdens that are more likely to be externalized beyond their boundaries. This chapter sets out to extend some of the political economy aspects of such work, using recent work on risk and environmental cost-shifting, and the emergence of neoliberalist policy regimes to examine how policies for the urban environment are created that have facilitated the ability of more affluent people and more affluent areas to displace their environmental burdens in a variety of ways.

As part of the analysis of environmental burdens, this chapter emphasizes the processes by which particular sets of urban environmental 'problems' come to be acknowledged as suitable areas for policy actions. The argument here is that policy agendas are not created simply through neutral scientific exploration of facts and the development of rational, apolitical responses to discovery of 'problems': rather there is tremendous selectivity at work, as scientists focus on some potential problems rather than others, often guided by the availability of research funding, while policy makers can select from a wide variety of identified environmental problems, explanations of the causes of these problems, and often widely differing prescriptions for how best to address problems. Defining environmental problems and inserting them into policy discourses is almost invariably a highly politicized activity, even when cloaked in the guise of scientific objectivity. Paying attention to the social construction of knowledge and its associated power dynamics involves taking a more critical look at how the objects of policy come to be chosen, and especially how readings of specific 'problems' will reflect particular power-knowledge configurations, discursive practices, scientific understanding, technical possibilities, and the very real and variable predilections and preferences of particular disciplines and professional groups (e.g. economists, planners, ecologists), and political decision makers. Put in this way, the choice of an 'urban environmental problem' for policy to focus on requires a sophisticated reading of the emergence not simply of actual crises but also the ways in which crisis myths are constructed as mechanisms for influencing policy formation. Put another way, sometimes there may be more similarities in the environmental issues facing different types of cities than might first appear, somewhat disguised by the ways in which different policy discourses are constructed. Alternatively, policy discourses on possible solutions can sometimes gloss over different underlying causal processes and problems. For instance, water stress (shortage, drought and so on) and water access are often conflated in policy discourses in order to argue for changes in policy or greater state investment or aid, yet often the two are very different and unrelated problems. The important point to remember is that while environmental problems do vary in nature between different types of city, their nearly universal feature is that they disproportionately impact on the poor.

Building on this theme of the need to challenge overly simplistic interpretations of urban environmental problems, the chapter begins by highlighting some of the common fallacies that have pervaded some of the urban and environmental literature, in some cases for well over 200 years. The next section highlights the dangers of dualistic patterns of thinking and of accepting uncritically some of the metaphors used to describe urban growth. This is followed by a brief review of some of the literature on environmental risks and burdens. Before concluding, the chapter examines two of the more influential critiques of urban environmental problems, EFs and urban linear metabolism.

Dubious dualisms and muddled metaphors: Unpicking the rickety rhetorics of sustainable cities

The past 30 years have seen a burgeoning interest in the urban environment, often emphasizing the problematic nature of rising urban consumption habits associated with increased wealth, and also the rather different environmental problems confronting poorer cities, in particular the connected issues of poverty and poor public health. Cities, it seems, are somehow always seen to be a bit of a problem, whether because they are too 'rich' or too 'poor'. But this is too easy a reading, reflecting a series of unhelpful dualisms which have often contributed to over-simplistic understandings of environmental problems.

In order to make complex issues quick and easy to understand, academics, lobbyists and policy makers often show a tendency to contrast opposing views and experiences, which run the risk of creating 'overburdened dualisms', better at developing polemical contrasts than actual understanding of complex processes of change (Sayer 1989, p666). Both heuristic and rhetorical dualisms operate by relegating all possibilities into two more or less all-encompassing and mutually exclusive categories, in the process inevitably over-simplifying complex concepts, or over-exaggerating what are in practice more nuanced positions or understandings. For the researcher, the concern is that over-reliance on binary opposites can create a tendency 'to exaggerate differences, confound descriptions and prescription, and set up overburdened dualisms that miss continuities, underplay contingency and overstate the internal coherence of social forms' (Wacquant, 1996, pp124–125, in Graham, 2000). In other words, by focusing on dualist 'categories' there is a danger of over-simplifying our understanding of problems and how to analyse them.

One dominant form of dualistic thinking pervades much work on environmental policy and politics, where society and nature are often portrayed as in perpetual opposition, rather than as interdependent, that is complexly inter-related. Critical analyses in this vein emphasize how environmental campaigns have long been fought on the basis of selective understandings of nature, which in turn have been based on selective use of available knowledge and interpretations of facts (Katz and Kirkby, 1991). The growing bodies of work on political ecology and the 'social construction of nature' have both helped highlight how environmental campaign groups have often tended to base their case for conservation in terms of seeking to protect 'pristine' wilderness, 'saving' them from the encroachment of human influence (Katz and Kirkby, 1991; Braun and Castree, 1998; Macnaghten and Urry, 1998). Yet many of these 'pristine' landscapes are in truth the result of centuries of human intervention, including animal or forest husbandry. Conservation groups have typically been influenced in their choice of types of landscape for preservation by dominant societal and artistic aesthetics. Indeed, much of the early history of conservation planning has been the assertion of élite views of nature, dictating how and where it should be preserved, for example, at what stage of a landscape's evolution, and for whose benefit. Rather than posit a false nature:society dualism then, it is preferable to think of nature as socially constructed and contested (Macnaghten and Urry, 1998; Braun and Castree, 1998), and to focus on questions about whose interests are being served by efforts to promote particular formations of social-nature.

For instance, should we be promoting much greater provision of public parkland and open space in cities, rather than restricting urban expansion ('sprawl') through urban compaction policies in order to preserve rural landscapes, even when these are factory-farmed as a form of virtual monoculture, maintained by massive applications of artificial chemicals? Whose nature is being protected here and in the interests of whom? And if we want more urban parkland and open spaces, whose interests are being served by particular approaches? There are real choices involved in how best to promote social-nature between, for example, formal, stylized flower gardens, community food gardens or 'nature reserves'. More to the point, there are major problems when policy makers decide that particular types of social-nature are no longer appropriate in cities, for instance attempts to reduce urban agriculture smallholdings as being inappropriate or undesirable, perhaps because they are seen to be insanitary or 'backward' for a city whose politicians seek to present a modern image.

In similar vein, town and country have often been drawn as in opposition, where the city is portrayed as a malign influence whose creeping extension needs to be curbed in order to protect the unique character of the countryside. This privileging of rural over urban aesthetics has been particularly influential since the 19th century industrial revolution in the West, inspiring a constant stream of 'back to nature' writing based on a set of beliefs built up around some kind of imaginary rural idyll. In consequence, and rather perversely, anti-urban sentiments have been a major influence in the management of cities, for example, inspiring planning and related policies to protect the countryside from urban 'sprawl'. Anti-urban bias is evident in other ways too. Many Western aid agencies have been reluctant to prioritize urban poverty and environmental problems, instead focusing their efforts on rural poverty and the rural environment. Rural environments, it seems, are deemed more worthy of investment than already 'spoilt' urban environments. The persistent town versus country dualistic way of thinking seems at best anachronistic, but at worst it betrays a misunderstanding of the ways in which town and country are interdependent, as rural dwellers rely on urban services and indeed often on selling their goods and services in cities, while rural–urban commuting is increasingly common. In many ways then the distinctions between town and country are increasingly blurred, and not just at the boundaries.

This is but a small part of the problem, as the town versus country dualism is often associated with a related tension between mega-cities and small settlements, where large cities are seen as somehow inherently undesirable, associated with social malaise and community breakdown, economic inefficiency as congestion sets in, and greater local environmental damage. Smaller settlements, by contrast, are often presented as a more desirable form of urban living, less congested and more socially cohesive. Up to 30 years ago such thinking could still be seen in the literature on 'optimum city size', policies to tackle 'primate city' dominance of the urban hierarchy, and so forth (Richardson, 1978). Drawing on ideas of environmental doom found in landmark publications such as the Club of Rome *Limits to Growth* report (Meadows et al, 1972) and Schumacher's (1974) *Small is Beautiful*, there was an early 1970s boom in 'urban crisis' literature, which wrongly foresaw the implosion of mega-cities such as New York and London (Blair, 1974). While the revival of fortunes of London and New York helped give the lie to notions that urban size of itself is a problem, there is still a distinct tendency for people to assume that the good life is one which is best lived in smaller communities. Indeed, ideas for improving cities still often return to the theme of how to make them somehow more village-like, a theme which pervades much recent British planning, for instance, and which can often be found in the promotional brochures of developers of large housing schemes (Keil and Graham, 1998).

There has to be a concern too about accounts of globalization and localization that posit them as polar opposites in terms of constructing policy. In particular, some environmental literature tends to posit globalization as an inevitably undesirable process and localization as its necessary antidote. Drawing on the 'small is beautiful' line of thinking, 'local is beautiful' too. Although there have to be some very real concerns with aspects of contemporary globalization, it is dangerous to over-simplify these processes as promoting greater economic and cultural homogeneity, in the process over-exaggerating their dominance and influence. Far from implying the 'end of geography,' what is commonly termed 'globalization' necessarily needs to be understood as a complex series of intersecting global-local processes, nicely captured in Swyngedouw's (1997) term 'glocalization'. In similar vein, it is dangerous to over-emphasize the redemptive power and potential of the 'local' as a source of opposition to the assumed imperatives of 'globalization', where the global is painted as malign, the local benign. As Marvin and Guy (1997) argue, there is a danger of placing too much faith in the abilities of under-resourced local communities to address problems whose sources are often well beyond their capacities or competencies to address. More than this, for communities without specific resource endowments promoting greater self-reliance might well be problematic for them. Likewise, poor communities seeking to improve their standards of living by exporting goods and services are more concerned to move towards fair trade rather than to revert to being a self-sufficient local economy, unable to buy a full range of internationally available goods and services. In this sense it is the terms of trade, or the quality of trade relationships which is the core issue, not a simplified view that globalization with its promotion of trade is a detrimental process, or that localism on its own can answer complex, multi-scalar problems of economic governance. As with previous dualisms, the argument here is not that 'globalization' and 'localization' do not exist as tendencies, but rather that they need to be seen as intersecting and interdependent processes.

Given the general tenor of these arguments, it is perhaps almost inevitable that a cautionary note needs to be sounded about assuming that the problems of 'affluent' cities are necessarily in some sort of opposition to those of less affluent cities. Within many affluent cities, there are increasingly large numbers of less affluent members of society, some struggling below the starvation line and without adequate shelter, while in less affluent cities there are frequently substantial pockets of affluence. This is not to deny that overall there are major differences in wealth and poverty levels in different types of city, but rather to highlight the fact that it is not enough to assume that any one type of city only has one set of problems, be these the problems of affluence and over-consumption, or poverty, under-consumption and under-provision of the basic necessities for human survival.

This leads directly to wariness about creating a false dichotomy between cities pursuing either 'green' or 'brown' agendas. As earlier chapters have noted, 'green' issues such as addressing ozone layer depletion and global climate change are a particular preoccupation in more affluent societies, which have largely contributed to creating these problems, while less affluent societies with high levels of poverty and ill-health still need to prioritize 'brown' issues, such as providing basic amenities and improving public health. Though a valuable distinction to make, it can actually be problematic to assume that 'green' agenda issues are the exclusive preserve of affluent cities and that 'brown' agenda issues are the sole preserve of less affluent cities. With the rise of neoliberalist policy regimes, the state role in redistribution policies has been diminishing in many Western societies, with inequalities consequently growing, perhaps most notably within some affluent cities. Moreover, as the environmental justice and environmental racism literatures have highlighted, it is the poor within affluent (and less affluent) cities who are disproportionately most likely to be exposed to environmental hazards, such as toxic waste facilities (Bullard, 1999). 'Brown' issues are far from absent from affluent cities then, even in the US.

This is where it becomes important to appreciate the ways in which policy domains are discursively constructed. In affluent cities, with near-universal provision of drinking water, sanitation, education and so forth, animating the environmental agenda requires different sets of problems to be brought to the fore if policy makers are to give them their attention. It is in this respect that the 'green' agenda of global sustainable development can be used to generate consensus for addressing a range of often quite localized environmental issues, from traffic congestion, urban sprawl and rural land-take, to health campaigns, refuse disposal, and reducing demand for energy and water. Arguably, many of these issues are being addressed in less-affluent cities too, but animated by brown agenda discourses around public health, poverty and so forth. As Brugmann (Chapter 13) notes, cities such as Curitiba in Brazil show the potential for merging 'green' and 'brown' agendas.

Looked at another way, poverty, most commonly associated with the brown agenda, is increasingly being used to legitimate particular sets of policies for sustainable development in more affluent societies. Support for socially disadvantaged groups has the advantage for corporations of providing seemingly altruistic reasons for retaining 'business as usual'. For example, raising fuel taxes for cars can be resisted by powerful multi-national corporate lobbies keen to avoid sales falls by arguing that the poorest will be most heavily hit by price increases, and drawing on the concerns of anti-poverty groups to provide support for this view. Anti-poverty group fears about increases in house prices and home rentals, especially when linked to homelessness, can likewise be used by house builders to argue against planning policies for urban constraint (Haughton and Counsell, 2004). In other words, as soon as we begin to look at the messy world of real politics, highly politicized approaches for discursively constructing the terrain of policy making begin to emerge, which make simplistic readings of 'affluent' versus less affluent cities, and green versus brown agendas difficult to sustain.

The reason for beginning to sketch out the problematic nature of creating these kinds of false dualisms is that they are drawn upon by lobby groups, policy makers, and others constructing narratives of crisis, as they seek to influence our understanding of what constitutes a 'problem' and what therefore needs to be addressed by policy makers. If the problem is preserving the rural environment, for instance, then the solution may be to stop the encroachment of the city into the countryside. But if the problem is lack of housing for the poor, maybe the solution is to encourage more housing development in rural areas of low ecological value and low environmental risk. The discursive creation of crises (and crisis-myths) is one of the ways in which policy agendas are created at both local and global scales. Indeed, there are large institutional bodies keen to promote very selective understandings of the nature of urban problems in order to push forward their own political agendas, from bodies like the World Bank and World Trade Organization (WTO), to international environmental groups such as Greenpeace, the World Wide Fund for Nature (WWF) and Friends of the Earth.

The rhetorical and polemical usage of dualistic ways of thinking often involves moralistic judgements, as one category is set against another. Table 11.1 sets out some of the dualisms already outlined, in a way which tries to highlight some associated dualist moral judgements about what is 'good' and 'bad' which have come to be associated with some of these positions. The concern is that there is a tendency in many readings of urban environmental

Table 11.1 *Dubious dualisms*

Society	–	Nature
Economy	–	Ecology
City	–	Countryside
Mega-city	–	Country town/village
Global	–	Local
Globalization	–	Localization
Affluent cities	–	Less affluent cities
Technocentric	–	Ecocentric
Working against nature	–	Working with nature
Polluted	–	Pristine
Parasitic	–	Self-reliant
Unbalanced	–	Balanced
High risk	–	Low risk
Exports risks and burdens	–	Suffers from (urban) risks and burdens
Draws in resources	–	Has resources taken out
Parasite	–	Host

problems not simply to rely on unhelpful dualistic antagonisms, but also to read vertically down each column to make associations between binary opposites, which are in fact very different in their nature. This becomes problematic when the rhetoric of some environmentalists becomes intertwined with strong anti-urban sentiments.

As if dubious dualisms weren't enough of a problem, simplistic assertions of the nature of urban economies and their environmental destructiveness have given rise to a range of muddled metaphors for cities, for instance, that they are said to be 'parasitic' (Odum, 1989, p17) or cancers on the planet (Friedman, 1984). Such highly charged metaphors do little justice to the actual complexities of the interactions between cities and their local and global hinterlands, since they present only the troublesome side of cities, and say little about the advantages which cities can bring, if managed well. More than this, such metaphors wrongly portray the problem as being simply 'urbanness', neglecting the messy reality of the complex social, economic, political and cultural conditions that influence people's behaviour patterns, and the multi-scalar dynamics through which environmental problems are generally generated and distributed.

Towards a political economy of cost transference

The central focus of this book is the multi-scalar nature of environmental burdens. In particular it draws on the work of McGranahan et al (2001) on environmental transitions, which highlights how cities in wealthy societies in effect transfer some of their environmental risks and burdens to external areas, increasingly on a global scale. The Urban Environmental Transition (UET) approach emphasizes that more affluent cities tend to generate substantial environmental burdens, because of their higher consumption levels, while also creating a more globalized distribution of environmental risks and burdens. Not surprisingly, given the considerable backlash against models that have sought to highlight development stages, this work comes with a set of powerful warnings about over-simplistic readings of it as a 'predictive' model. Peter Taylor's (1989) critique of the errors of 'developmentalism' in Rostow's model of development stages is useful in reminding us of the dangers of reading off future trajectories from past ones, and of basing policies on attempting to expedite such transitions. Drawing on the literature on uneven development, Taylor particularly highlighted the need to examine processes of interdependence between different economies, rather than assuming that there is some 'natural' path of development stages through which societies might pass on the road to 'modernization'.

Recent work on environmental risk and burdens highlights political economy approaches that might help our understanding of urban environmental transitions. Particularly helpful is Beck's (1992, 1998) work on the risk society, in which he focuses on the growing importance of the production, distribution and amelioration of risk, created by the unintended consequences of the drive to 'progress' stimulated by science, technology and economics. Environmental risks, he argues, are 'produced industrially, externalised economically and minimised politically' (Beck, 1998, p26). The value of Beck's work is in drawing attention to growing environmental risks, and the complex ways in which these are produced and distributed, requiring approaches which move beyond simple state regulation towards more complex negotiations of understandings of risk and how best to address them, involving a wide range of actors, including non-government organizations (NGOs) and businesses. The major societal transformations that this approach suggests have been frequently contrasted with the 'ecological modernization' interpretation of sustainable development (e.g. Blowers, 1997; Davoudi, 2000), which in some variants assumes that with relatively minor technical and regulatory reforms, business as usual is possible under existing capitalist structures.

An alternative reading of the growing concern over environmental risk is to see it as part of the wider process of the unfolding of capitalism's attempts to legitimate and formalize its appropriation of natural assets. From this perspective, capitalist development requires systems of uneven development centred on exploitation of human labour and nature, involving attempts to reduce production and reproduction costs by shifting them onto other sectors, other businesses, the state, different communities, the biosphere, or to the future generations (O'Connor, J., 1988; O'Connor, M., 1993, 1994). Taking this argument further, Martin O'Connor (1993) argues that there is a clear tendency over time to move from 'cost-shifting' towards 'the capitalization of nature', as the possibilities for cost-shifting become circumscribed by changing social and political values. In this view, capitalism reinvents a legitimization for itself, by using the market to price nature, supposedly leading to efficient and rational allocations of resources. Far from doing this, capitalizing nature (pricing externalities, such as paying for entry to national parks or higher prices for water) still imposes the capitalist logic of seeking to ensure and legitimate access to natural resources and 'environmental services' at the lowest cost, linking in to systems of territorial exploitation where space is first captured extensively, then capitalized intensively (O'Connor, 1993; Altvater, 1989).

In this view, global capitalism requires predatory searches for 'uncapitalized domains of nature and humanity', with contestations over 'predation and cost-shifting' (O'Connor, 1993, p17) taking place under the banners and rhetorics of sustainable management. The market effectively becomes the vehicle for further exploitation of nature, albeit in ways which seem to suggest some kind of market rationality of reciprocal trading of rights and benefits, but where unfair terms of exchange as part of cost-cutting strategies remain the norm. The globalization of capitalism allows increasingly mobile trans-national capital to be shifted away from areas that impose restrictions on the exploitation of nature or workers, towards other, less strictly regulated countries or regions (O'Connor, 1994).

While broadly accepting the validity of O'Connor's (1993) arguments about a move from cost-shifting towards the capitalization of nature as altering qualitatively the nature of cost-shifting activities towards a market-legitimated framework, the nature of cost-shifting has changed in other ways too. In particular, it is worth remembering that cost transference is not simply about financial costs, though clearly as nature is capitalized there is a shift towards this, but can also take in non-costed issues of altered social relations, impacts on nature, quality of life, resource access and so on.

Something of the flavour of these arguments can also be seen in Friedmann's (1989) comments about how mobile middle classes can subvert environmental priorities and pressures in various ways. Typically, these might range from mobilizing to remove polluting industry from higher-income neighbourhoods, to maintain their quality of life, to voting against financing necessary environmental measures (e.g. a rapid transit system or sewage works), preferring instead to move to areas that can offer immediate improvements in quality of life without in the short term having to pay higher taxes. In a whole variety of ways then, people individually seek to avoid paying for, or being adversely affected by, environmental degradation caused by people and businesses at the collective level. This is not a straightforward failing of capitalism that can be read off in similar ways in similar places, or as part of linear progression in terms of evolving approach. Rather, capitalism at differing periods and in different places comes to different accommodations about how to reconcile individual and collective responsibilities, and where the burden lies between the state, individuals, communities and businesses, in terms of responsibility, reducing detrimental impacts and remedial actions.

More than this, there is a fluidity and interdependence in the relationships between those in affluent areas who are effectively transferring their environmental burdens and those who are being adversely impacted by them.

Since the lifestyles of affluent societies depend to some extent on their ability to export their harmful impacts to other areas, when these areas find ways of reducing their exploitation it becomes in the interest of the affluent societies to either undermine these changes or to shift their attention to other, less protected areas. In urban terms, these interdependent relationships exist between rich and poor neighbourhoods in a city, and between a city and other areas, be these the poor neighbourhoods of nearby cities or villages, or places on the other side of the planet.

Returning to Beck's (1998) comments on how risks are externalized economically and minimized politically, it is helpful to think about how risk transference has been incorporated into economic and regulatory systems. The last 150 years have seen the emergence of more and more sophisticated state regulation over, for example, pollution levels, which, combined with the effects of deindustrialization, has reduced the scale of *localized* externalities imposed by industry in many affluent societies. The most polluting aspects of the industrial chain have tended to move out of the more prosperous urban areas, and are instead more likely to be found in distant, poorer regions or in the poorer neighbourhoods of rich regions. This 'virtual pollution' is very similar to debates about 'virtual water', which highlight how many water-scarce countries have survived better than initially expected by virtue of importing cereals and so on grown using the water of the exporting countries (Allen, 1998).

There is an important scale issue here too in the sense that many of the key contemporary environmental problems are global in impact. In part this links to the increasing globalization of the economy, with complex and large-scale movements of raw materials, finished and semi-finished products, plus a growing global trade in services. These trends have exerted major environmental influences at the global scale, not least in the massive energy costs of transferring physical goods and people around the earth on an increasing scale. But there has also been a major transformation in the nature of environmental problems, away from the more readily controlled point-sourced pollution and resource degradation of the past, to more dispersed, more mobile pollution sources (especially motor vehicles) and often ill-understood human made environmental problems (nuclear waste disposal, chlorofluorocarbon (CFC) manufacture). So, for instance, while it has been possible with adequate political will to control the old-style, essentially localized sulphurous smogs that used to characterize Western industrial cities, not least by introducing smokeless fuel zones, these sorts of problems have been superseded by the more far reaching and difficult to control petrochemical smogs associated with the rise of the motor car. Perhaps most noteworthy has been the rise in other forms of transfrontier environmental degradation, notably the rise in acid deposition, the increased incidence of ozone layer depletion, and the emergent problem of global warming associated with the rise of greenhouse gases in the earth's atmosphere. Pollution has become a global issue in a way that it never has been before, linked to the rapid rise in consumption of natural resources, in particular the (inefficient) burning of fossil fuels.

While some steps have been made to achieving global agreements on environmental issues, for instance, on protecting the ozone layer, in practice the reluctance of some nations to accept these has undermined others, for example, on global warming. The reluctance on the part of some Western nations has reflected an unwillingness to absorb the additional costs they might incur as a result of such agreements lest they undermine some of their economic interests, while some poorer nations have understandably been reluctant to jeopardize their own growth prospects by having to share the costs of addressing problems brought about principally by Western risk-taking and high consumption levels. Every country wants economic growth, but there is little agreement on how to reduce the environmental and social costs involved and how to share them more equitably.

With little agreement on how to regulate for environmental activities at the global scale, the importance of regulatory actions at other scales becomes increasingly important, for example, the role of the European Union (EU) in developing stronger environmental regulatory frameworks, the role of national and local governments, and a whole host of more voluntaristic agreements across sectors (business, voluntary, state) and scales. The emerging patchwork of regulatory reforms undermines efforts to raise uniform standards for all, yet gives scope for more varied localized experiments in how best to regulate for environmental improvements.

Ecological footprints and urban metabolism: A critique

At the local level, policy making still tends to focus on the politics and possibilities for reform within a given set of political boundaries, with little incentive for addressing external impacts, which in the face of weak regional, national or supranational regulatory frameworks can lead to a continuing tendency to externalize environmental burdens. An essential step in raising the profile of such issues is to improve our understanding of the nature, scope and extent of external environmental impacts. Recent years have seen the development of two particularly influential frameworks for looking at the impacts of urban areas across scales, ecological footprints (EFs) and urban metabolism. However, both are limited in scope, addressing environmental issues in isolation of wider social and economic dynamics, while only charting the 'negative' aspects of urban consumption, with little to say about the benefits which they might bring.

The EF approach is based on attempting to measure the amount of land which is required to feed a city's consumption habits, for instance, the amount of land needed to produce its food, energy and forestry requirements. As such it provides a proxy method for gauging how much productive capacity has to be appropriated from elsewhere in order to meet a city's consumption habits. The approach is widely used already and has influence in some policy circles (including a study of Greater London, see www.citylimitslondon.com; for other examples see Wackernagel and Rees, 1996; Jones and Comfort, 2005; McManus and Haughton 2006). It is especially helpful in drawing attention to the larger footprint of more affluent cities, which are generally more heavily reliant on external sources to feed their consumption habits (Wackernagel and Rees, 1996). The key finding here is the extent of the 'appropriation' of external resources which is required to feed affluent cities in particular, involving global flows of inputs. While useful as one starting point for analysing the shaping of the relationship of cities to wider environmental factors, it remains just that, a starting point. In anticipating criticisms of their approach, Wackernagel and Rees (1996, pp16–27) acknowledge some of the limits of their tool but nonetheless rightly defend the broad utility of the concept. From the perspective of the current analysis, the problems with the EF approach as a means for examining the multi-scalar impacts of urban consumption habits are essentially five-fold.

Firstly, it fails to adequately convey the geography of the complex links between cities and their hinterlands. The crude nature of calculating land equivalence for urban resource inputs is a useful rhetorical tool but of limited value in terms of shaping policy interventions. Such are the complex flows of resources in most cities that it is impossible to see where the EF is doing most damage. Secondly, it fails to account for the complex reverse flows, of how urban benefits might have beneficial external impacts – from reducing the amount of land needed to house people to producing technological advances that can improve the productivity and well-being of other areas. Thirdly, it does not examine the 'carrying capacity' either of cities or their hinterlands, in terms of capacity to neutralize the waste streams of cities. Indeed it is generally strong on inputs to cities, weak on urban waste streams.

Fourthly, it fails to take into account the nature of the local environmental tolerance levels and management practices of exporting economies. There is a world of difference between a sustainably managed forest resource and those logging activities where stock is not replanted or where the original forest resource is so unique as a whole ecosystem that it cannot be sustainably harvested on a large scale to provide resources for external areas. Lastly, it takes inadequate account of social relations within cities and between cities and their hinterland areas. In terms of relations between cities and their hinterlands, it does not interrogate how resource extraction and production impact (both positively and negatively) on people in source regions. In addition, it fails to account for the damage done by environmentally degrading activities within cities, an issue perhaps most evident in concern over environmental justice and the way in which neighbourhoods with high proportions of poor people are disproportionately likely to be affected by toxic waste or similar noxious facilities (Bullard, 1999). It must be re-emphasized that these criticisms concern problems that the EF does not set out to address – it helps identify issues not solve them, and that is useful in itself as one approach among many to help identify and address issues of cost transference.

A second, less widely used approach to understanding the relationships between cities and their hinterlands is the urban metabolism model. Girardet

(1992) has developed a sophisticated and helpful argument that cities have tended to become less sustainable over time as they have begun to organize their economies and ecologies in terms of what he calls a linear metabolism, rather than a circular metabolism. The basis of the argument is that cities have increasingly drawn on resource inputs sourced from distant points around the world rather than locally, and that, increasingly, their pollution waste streams tend to be sent outwards. In combination, the effect is that city residents and businesses have become increasingly divorced from the external damage their consumption habits have. He argues that cities should be rethought and reorganized so that every output can become an input in an integrated local system – for example, reusing sewage to fertilize farmland, reducing the need for fertilizer factories. Cities organized on these kinds of 'circular' lines will source more of their resources locally and deal more locally with their waste streams, reconnecting with nature. The essence of his argument is that:

> *The city that uses linear processes makes its presence on the planet felt over a vast area. Almost everything the city needs has to be transported from far and wide, using valuable energy. The city that uses circular processes has impact over only a small area – because its needs are met by itself and its own immediate hinterlands.* (Girardet, 1992, p24)

While a useful rhetorical and even diagnostic device, four of the criticisms of the EF model also apply to the urban metabolism model: it is non-specific in its geography; it fails to take into account beneficial reverse flows between cities and their hinterlands; it does not look at regional and local social and environmental carrying capacities and tolerances; and it is quiet on the nature of social relations within cities. It is, however, less vulnerable to the charge that it does not deal with urban waste streams, although it does this in a fairly crude manner which does not deal with local carrying capacities.

One problem shared by both approaches is that they assume greater local self-reliance is always a good thing and by implication external trade is something to be avoided. The problem with this is that it deals only with the quantity of trade without considering the quality of trading relationships – there is a real difference between 'fair trade' relationships and more exploitative forms of trading. We need to unpick the political economy of urban-hinterland relationships much more carefully if we are to truly understand their nature, which might, for instance, require looking at global trade agreements, national attitudes to competitiveness and welfare, in addition to looking at local-level actions such as promoting greater self-reliance or circular metabolism.

Conclusions

The central theme of this chapter is that non-sustainable processes are promoted by complex social, political and economic systems of environmental cost and risk transference. Sophisticated political and economic systems have evolved for transferring many of the (environmental and social) burdens of our activities to future generations, to other (usually less privileged) social or ethnic groups, to other areas (from city to rural area, rural area to city, rich nation to poor nation), and to other parts of the biosphere (e.g. the destruction of rainforests and associated impacts on biodiversity). Understanding who is avoiding costs, how and why, and who is bearing them has to be a central part of any analysis of urban environmental problems.

The analysis presented here has deliberately shied away from using masses of data to prove that 'affluent cities' are generating particular types of environmental problem and contrasting these with the issues faced by less affluent cities, in part because of the concern that this simple opposition may mask the fact that the cities will also share many problems, for instance, with land contamination, loss of public open space, traffic congestion. Similarly, rather than detailing the nature, scale and spread of environmental burdens which might be said to characterize affluent cities, this chapter has sought to analyse how these burdens are generated, adopting an approach that emphasizes economic, social and political dynamics working across scales. In other words, rather than blame affluent cities for creating problems that are now stretching out across the world, this chapter has sought to develop a more nuanced understanding of urban problems, which emphasizes a more complex set of multi-scalar dynamics.

Acknowledgements

While writing this chapter I was a William Evans Visiting Fellow in the Department of Geography, University of Otago, New Zealand. The work drawn upon here is part of a larger project 'Changes in regional planning: a new opportunity for sustainable development?' funded by Economic and Social Research Council (ESRC) grant R000238368. I would like to thank the editors of this book for their critical and supportive comments on this chapter as it has developed, while of course exonerating them from any responsibility for errors, omissions and misinterpretations.

References

Allen, J. A. (1998) 'Watersheds and problemsheds: Explaining the absence of armed conflict over water in the Middle East', *Middle East Review of International Affairs*, vol 2, no 1, pp49–52

Altvater, E. (1989) 'Ecological and economic modalities of time and space', *Capital, Nature, Socialism*, vol 3, pp59–70

Beck, U. (1992) *The Risk Society: Towards a New Modernity*, Sage, London

Beck, U. (1998) *Democracy Without Enemies*, Polity Press, Cambridge

Blair, T. (1974) *The International Urban Crisis*, Hart-Davis, London

Blowers, A. (1997) 'Environmental policy: Ecological modernisation or the risk society?' *Urban Studies*, vol 34, no 5/6, pp845–71

Braun, B. and Castree, N. (1998) 'The construction of nature and the nature of construction', in B. Braun. and N. Castree (eds), *Remaining Reality: Nature at the Millennium*, Routledge, London

Bullard, R. (1999) 'Dismantling environmental racism in the USA', *Local Environment*, vol 4, no 1, pp5–19

Davoudi, S. (2000) 'Sustainability: A new "vision" for the British planning system', *Planning Perspectives*, vol 15, no 2, pp123–137

Friedman, Y. (1984) 'Towards a policy of urban survival', in F. DiCastri, F. Baker and M. Hadley (eds), *Ecology in Practice II: The Social Response*, Tycooly, Dublin, and United Nations Educational Scientific and Cultural Organization (UNESCO), Paris

Friedmann, J. (1989) 'Planning, politics and the environment', *Journal of the American Planning Association*, Summer, pp334–338

Girardet, H. (1992) *Cities: New Directions For Sustainable Urban Living*, Gaia Books, London

Graham, S. (2000) 'Constructing premium network spaces: Reflections on infrastructure networks and contemporary urban development', *International Journal of Urban and Regional Research*, vol 24, no 1, pp183–2000

Haughton, G. and Counsell, D. (2004) *Regions, Spatial Strategies and Sustainable Development*, Routledge, London

Jones, P. and Comfort, D. (2005) 'Ecological footprints over Europe', *Town and Country Planning*, September, vol 74, no 9, pp271–273

Katz, C. and Kirkby, A. (1991) 'In the nature of things: The environment and everyday life', *Transactions of the Institute of British Geographers*, NS16, pp259–271

Keil, R. and Graham, J. (1998) 'Reasserting nature: Constructing urban environments after Fordism', in B. Braun, and N. Castree (eds), *Remaining Reality: Nature at the Millennium*, Routledge, London, pp100–125

Macnaghten, P. and Urry, J. (1998) *Contested Natures*, Sage, London

MacNeill, J., Winsemius, P. and Yakushuji, T. (1991) *Beyond Interdependence: The Meshing of the World's Economy and the Earth's Ecology*, Oxford University Press, Oxford

Marvin, S. and Guy, S. (1997) 'Creating myths rather than sustainability: The transition fallacies of the new localism', *Local Environment*, vol 2, pp311–18

McGranahan, G., Jacobi, P., Songsore, J., Surjadi, C. and Kjellén, M. (2001) *The Citizens at Risk: From Urban Sanitation to Sustainable Cities*, Earthscan, London

McManus, P. and Haughton, G. (2006) 'Planning with ecological footprints: A sympathetic critique of theory and practice', *Environment and Urbanization*, vol 18, no 1, pp113–127

Meadows, D. H., Meadows, D. L., Randers, J. and Behrenv III, W. W. (1972) *The Limits to Growth*, University Books, New York

Newcombe, K. (1984) 'Energy conservation and diversification of energy sources in and around the city of Lae, Papua New Guinea', in F. DiCastri, F. Baker and M. Hadley (eds), *Ecology in Practice II: The Social Response*, Tycooly, Dublin, and UNESCO, Paris

O'Connor, J. (1988) 'Capitalism, nature, socialism: A theoretical introduction', *Capitalism, Nature, Socialism*, vol 1, pp11–38

O'Connor, M. (1993) 'On the misadventures of capitalist nature', *Capitalism, Nature, Socialism*, vol 4, no 3, pp7–40

O'Connor, M. (1994) 'The material/communal conditions of life', *Capitalism, Nature, Socialism*, vol 5, no 4, pp105–114

Odum, E. P. (1989) *Ecology and our Endangered Life-Support Systems*, Sinauer Associates, Sutherland, MA

Richardson, H. W. (1978) *Regional and Urban Economics*, Penguin, Harmondsworth

Sayer, A. (1989) 'Postfordism in question', *International Journal of Urban and Regional Research*, vol 13, pp666–695

Schumacher, E. F. (1974) *Small Is Beautiful: A Study of Economics As If People Mattered*, Abacus, London

Swyngedouw, E. (1997) 'Neither global nor local: "Glocalization" and the politics of scale', in K. R. Cox (ed), *Spaces of Globalization: Reasserting the Power of the Local*, Guildford Press, New York, pp137–166

Taylor, P. (1989) 'The error of developmentalism in human geography', in D. Gregory and R. Walford (eds), *Horizons in Human Geography*, Macmillan, London, pp303–319

Wackernagel, M. and Rees, W. (1996) *Our Ecological Footprint: Reducing Human Impact on the Earth*, New Society Publishers, Gariola Island, BC

Wacquant, L. (1996) 'The rise of advanced marginality: Notes on its nature and implications', *Acta Sociologica*, vol 39, pp121–139

12
The Metabolism of Urban Affluence: Notes from the Greater Manchester City-region

Joe Ravetz

Introduction

In the average UK or European Union (EU) city, representing the wealthier 20 per cent of the population in the 'developed' world, there are supermarkets with 50,000 product lines, and a level of affluence of about 50–100 times that of the poorest 20 per cent in the 'developing' world. These are apparently different worlds. And yet it is clear that one world is an integral part of the other world – that urban affluence appears to be interdependent with urban poverty.

Much of this structure of wealth and poverty revolves around 'environment', in its many meanings. In sectors such as housing, energy, transport, tourism and so on, we can see similar patterns of displacements and transfers of environmental values. Such systems of 'expropriation' tend to reproduce themselves, between local and global, between urban and rural, and developing and developed worlds. Such systems appear to be globalizing in scale and accelerating in pace, and the central theme of this book, the Urban Environment Transition (UET) hypothesis, is one way to characterize these structural changes.

This chapter explores the UET phenomenon from the perspective of a relatively affluent city-region, Greater Manchester (GM) in the UK. We focus on one particular dimension of the UET – the environmental metabolism or throughput of materials and energy. From this we develop a more general picture of patterns and transitions in systems of production, consumption and information transfer in GM, as representative of many 'developed' cities. The implication is then the dependencies and pressures such affluence exerts on 'developing' cities. This potentially huge topic is divided here into five sections:

- scoping the agenda: the multiple dimensions of the 'urban environment';
- review of the case study – conditions and trends, with the example of the city-regional airport;
- outline of the UET hypothesis, from several perspectives, with some preliminary evidence;
- the urban environmental metabolism or throughput of energy and materials, from developing to developed cities;
- synthesis and implications for the 'sustainable city' – an agenda that is often fuzzy and contested. The evidence shows how the urban metabolism perspective can help to identify alternative parallel models for policy and analysis.

This chapter draws from the book *City-Region 2020* (Ravetz, 2000a), and the ongoing research programme at the Centre for Urban and Regional Ecology, Manchester, including the projects Eco-Region NW, Eco-Budget UK, and the One Planet Economy Network.[1] The author would also like to acknowledge the work of John Barrett and the regional sustainability team at the Stockholm Environment Institute at York, UK, as partners in these projects and suppliers of excellent data.

Scoping the urban environment

Firstly I aim to explore the multiple meanings and layers of the 'urban environment', as a context for the trends and transitions at work. Clearly, there are many different types of environmental stock, condition, flow and pressure, in various scales in space and time, in relationship to the 'city'. These are some of the most significant levels:

- conditions and pressures on the *local urban* environment in urban areas;
- conditions and pressures on the *city-region or hinterland* environment;
- pressures on the *regional or global* environment resulting from urban activity;
- pressures of extraction, harvesting and other activities on environmental/physical *resources* at local, regional and global levels.

The scope of what is the 'city', 'city-region' or 'urban' is likewise open to debate. Here I suggest a multiple definition, with four types of spatial organization:

- urban hubs with central business functions and a surrounding built-up area – the definition most commonly used by urban policy in the UK and EU;

- settlements and built-up agglomerations of various sizes, however these are defined;
- functional city-regions or urban hinterlands, which may be defined by travel to work fields, employment types and so on;
- a global socio-economic pattern and network of interactions, of which cities are the hubs, facilitators and gateways.

Looking more closely, there are further interactions between each of the above to consider. Again, as a working method we can use the DPSIR ('driving forces, pressure, state, impact, response') framework of the European Environment Agency (EEA, 1999):

- *underlying and driving forces* in urban systems that interact with the environment, such as lifestyles, cultures and market forces. For example, the pressure for increased personal space can be seen to lead directly to the spatial development of cities;
- *activities and sectors* in urban areas that place environmental pressures, locally and globally. For example, the desire of householders for garden leisure places demands on suppliers of gravel, charcoal and so on;
- *physical patterns and infrastructures* in urban areas that contribute to environmental pressures;
- *environmental pressures* themselves and the interactions between them, such as air, water and ground pollution;
- *responses in urban systems*, including policy, economy or other interventions, to mitigate between urban activity and its environmental pressures. The first level of analysis is to look at the institutional divisions, such as between the public sector; private sector; civic or third sector; and the citizen or consumer sector;
- *interactions* between urban environments, and urban or regional economies and societies.

Figure 12.1 provides a mapping of economic, environmental and social factors at various scales, so that the interactions between them can be seen in context. Some example issues are charted out, but there are many more. This shows how environmental issues may simultaneously act as causes, conditions, effects or outcomes of socio economic processes. As with any mapping, there are endless possible levels of detail, and the very general level shown here is similar to that of a route map of any large city or region. Even this shows how sectors, such as housing or transport, can meet multiple needs with multiple actions, driven by multiple pressures, and causing multiple impacts at multiple scales.

The state of the city-region

The case study here is the dynamic and problematic conurbation of Greater Manchester (GM) in northern England. At the centre, the City of Manchester

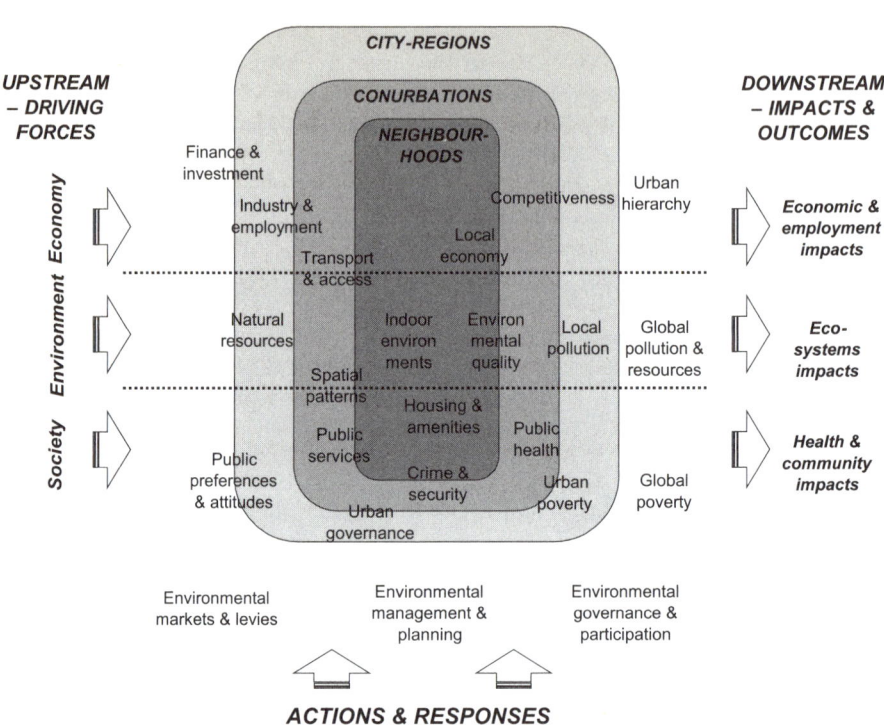

Source: Based on 'integrated assessment' framework as in Ravetz (2000a)

Figure 12.1 *Urban environment framework*

is an icon for style and sport, and a thriving centre for finance, media, education and culture. It is surrounded by the suburbs and exurbs of GM, a sprawling conurbation of 2.5 million people in a large urban core and a ring of satellites. A further 1.5 million people live in the 'metropolitan area' of a one-hour journey time, including Liverpool and other smaller cities. To the east and north are rolling hills surrounding an extended urban fringe, and to the west and south is mainly farmland with a patchwork of small towns and large suburbs. GM sits at a national crossroads, halfway between Scotland and London, and is also the gateway to the 'peripheral' North West region, and a playground for wealthy commuters and tourists. Its governance is divided among 10 autonomous municipalities, and many vital functions are devolved to the regional level. The location is shown in Figure 12.2, and vital statistics are shown in Table 12.1 in order of their growth trend, as a guide to where future pressures are coming from (Ravetz, 2000a).

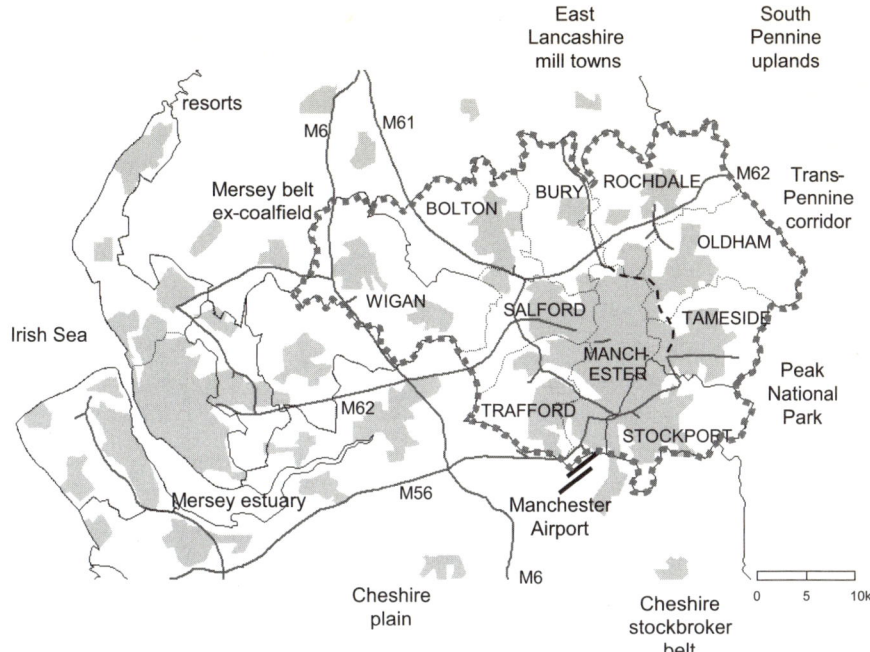

Source: based on Ravetz (2000a)

Figure 12.2 *Greater Manchester location*

Urban trends, past and future

GM a century ago was the 10th largest city in the world, and the classic industrial 'shock city' of free trade and enterprise. At various times it was the site of the world's first railway station, first free public library, first retail cooperative, first Trade Union Congress, and first stored memory computer. Here there emerged a unique combination of factors – access to iron and coal: a sea port for the British trading empire; a damp climate suitable for textile processing; and the non-conformist churches with their promotion of technical learning and innovation by the 'working classes' (Hall, 1998). A journey from east to west crosses many layers of this history, from the birth of the textile industry in the Pennine valleys, to the 'sunrise' business parks surrounding the airport (now self-styled as the 'world's best').

Current trends and projections are indicative of the prospects for many such city-regions in the developed world:

- 'globalization': integration of investment, production, trade and consumption;

Table 12.1 *Vital trends and statistics in Greater Manchester*

Conditions in the city-region	Annual growth factor (%)
'World's best' airport, 45,000 trips per day	8
GDP of £30 billion per year	2.5
about 1 million cars: 6 million trips per day	2
nearly 100,000 other buildings	2
derelict land on 6% of urban area	1.8
total waste arising: 11 million tonnes per year	1.5
about 1 million households	1
energy use: 90 billion kWh per year	1
CO_2 emissions: 32 million tonnes per year	0.7
population: 2.5 million people	0.2
urban area 55,000 hectares: 43% of total	0.15
700,000 bus trips per day, 70,000 local rail	0.1
8000km of roads: 152km motorways: 350km of railways	
10 autonomous local authorities: land area 1286km^2	

Source: Various data as of 1998, as compiled in Ravetz (2000a)

- 'connexity': global networks through information and communications technology (ICT), media, international travel (Mulgan, 1997);
- 'exclusion': new patterns of polarization and dependency for large sections of the population (Pacione, 1999);
- 'post-Fordism': dissolution of former economic, social and political structures which had a clearly defined logic (Amin and Thrift, 1995).

For several centuries, manufacturing activity was the basis of the industrial city – local and imported materials were processed to produce goods to send along water or railway corridors: economic 'advantage' could be defined in terms of the city-region's location and resources as a 'material processor' (Figure 12.3).

That model is now in transition to a more post-industrial 'city of flows' (Borja and Castells, 1997). The city-region now functions more as a node in a global 'hypergrid' – networks of motorways and airports for movement of people and goods, and networks of satellites and wires for movement of information and capital. Many patterns of urban activity and urban form are turning inside out, as the growth nodes of production and consumption migrate to the urban fringe or 'edge city' – retail, leisure and business parks with easy links to the hypergrid (Garreau, 1991). City functions now centre on services and consumption, and its cultural 'cachet' or branding now competes in a global hierarchy.

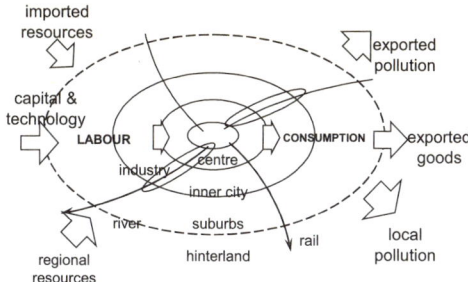

MATERIAL FLOW CITIES

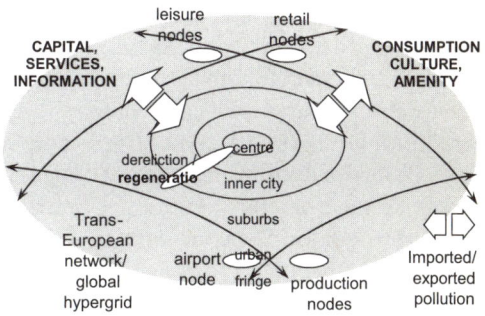

INFORMATION FLOW CITIES

Source: Based on Ravetz (2000a)

Figure 12.3 *From material flow to information flow*

There are many paradoxes in such a transition – in GM, for instance, there are 19th, 20th and 21st century cultures and economies, side by side, and often in competition and conflict. While production and consumption are globalized, there is a counter trend of 'localization' – a new kind of 'place advantage' gained through cultural amenity and attraction to mobile consumption and production (Dicken, 1998). In physical terms, edge cities are 'counter-urbanized', while historic centres are 'reurbanized' and industrial areas 'regenerated'. In social terms, 'uneven development' creates clusters of unemployment and exclusion. In environmental terms, the bulk of resources travels through the global hypergrid, which is increasingly privatized and deregulated, and where environmental management presents an even greater challenge than before.

Economic and social trends

A city-region such as GM displays both vulnerability and opportunity in the global hierarchy. With an industrial base which is partly obsolete, and partly booming with high-tech and tertiary activities, it is a major hub to a peripheral region, and at the same time saddled with socio-economic decline and dependency. The combined effect is the trend of segmentation – a 'sunrise' high-tech economy with global markets is surrounded by large areas on the threshold of dependency and decline.

In parallel are equally fundamental social dynamics. Demographic trends are changing age structure, gender balance, family structure, disposable time and income, and household organization. Cultural trends see a shift from former 'one nation' patterns towards self-identity and empowerment, and its counterpart of alienation and disorder. Former cultures of governance and welfare provision are replaced by one of 'enabling' in partnership with other organizations. The civic body is 'splintering', both culturally and physically, into countless fragments (Graham and Marvin, 2001).

Generally, there is a prospect of an ageing population, with rising disposable income and leisure time, chasing volatile employment and a diversifying set of lifestyle activities. Such trends may accelerate: many visions of the future, even without 'surprises' such as terrorism or sudden climate change, suggest structural conflict between cultures and corporate interests. Cities and urban systems have a continuing role not only as economic producers and consumers, but as arenas for creative conflict between the local and global, and between the corporate and the civic worlds.

Environmental trends
Environmental quality in GM shows the legacy of 200 years of heavy industry, and Manchester is still the 'pollution capital' of the UK (Ravetz, 2000a). The surrounding uplands are well over their 'critical loads' for acidity, river quality is only slightly less than toxic, and a tenth of urban land is potentially contaminated and unstable. A quarter of all households drink lead in their water, half are seriously disturbed by noise, and there is a 3 per cent annual growth trend in household waste.

In the longer run there are environmental dynamics that could enhance pollution hazards, whether actual or perceived. Economic growth will tend to increase material throughput, other things being equal, with new and more complex substances and processes. There will also be ever-tighter standards for health and amenity, and better evidence on environmental pathways, processes and impacts.

The result can be seen in the environmental history of the city-region as evidence of the UET model. In the first industrial phase of GM, starting about 220 years ago, the combined hazards of work, housing, diet and pollution resulted in an average life expectancy of 40 years (Ponting, 1992). In a second phase, the fossil fuels used in heating and transport dominated urban air pollution. In a third phase, many impacts have been displaced to a global level, and there are other, more insidious hazards now in the form of carcinogens, trace metals and genetic engineering. While life expectancies have doubled, public concerns on risks have multiplied (Beck, 1995). The risks of 'production' have now shifted to those of 'consumption' – transport, noise, waste, food chains, obesity and mental health. And in the modern 'risk society', social divisions are as sharp as ever. Pollution mapping shows the poor breathing the emissions of the rich; health mapping shows a seven-year lifespan difference between poor and rich areas; and 95 per cent of all major industrial polluters are in poor areas. An increasing proportion of the environmental burden of affluence is exported to poor countries overseas via resource extraction and climate emissions, even while the UK becomes gradually cleaner and greener.

With the general shift of urban activity from production to consumption, environmental management has likewise shifted from the 'dilute and disperse' approach of the industrial revolution, to integrated pollution control (IPC) for all media, as in the IPC system of the UK. More recently, environmental management has emerged as a driver for business opportunity and competitiveness, as in the principle of 'eco-modernization' (Weale, 1993). Taking this one step further, 'integrated chain management' coordinates all materials and processes (Wolters et al, 1997). The end goal is 'de-materialization', or de-linking of economic growth from material throughput, and this is the general goal of the UK strategy on 'sustainable consumption and production' (Jackson and Michaelis, 2003).

Case study: Airports and air travel
Here I focus on a very topical example, where economic, social and environmental goals collide head on, and where local, regional, national and global responsibilities are all entangled. The airport issue is now common to almost every city, developed or developing. It is commonly presented as a local environmental problem, and an incidental side-effect of economic growth. The much greater impacts of climate change emissions are generally conceived as 'someone else's' problem – the airlines, the international regulators, the travelling public, and so on. In the absence of a clear physical science threshold, the allocation of responsibility then becomes a political question, without as yet any clear answers.

Manchester Airport has put the city on the international map, and it now moves over 30 million passengers and 200,000 tonnes of freight per year. The expansion programme now in progress will provide for 42 million passenger movements by 2015, or half the current capacity of London Heathrow. Around 80,000 jobs are now related to the airport, possibly the most successful of any publicly owned enterprise in the UK. National and world projections are for a 5 per cent growth rate, or doubling of air traffic every 15 years (DFT, 2003). But urban airports such as Manchester are constrained by site area, access and noise limits, and future expansion will be difficult. Such a tension between supply and demand leads to several possible scenarios:

1   high-growth demand-led scenario, with traffic doubling between 2005 and 2020 and then beyond, using an offshore airport in the Irish Sea;
2   lower-growth 'business as usual' demand-led scenario, which would expand the existing site to the full, and network operations with other regional airports;

3   an environmental demand-management scenario, which would stabilize demand beyond 2005 and develop as an integrated regional transport hub.

The economic and social benefits of expanding air travel are huge, and have to be balanced with the environmental costs – but this is not simple, as any local or regional constraints could lead to traffic and trade diverting elsewhere (Caves, 1992). For both demand in general, and the supply via Manchester airport, there are many questions of scientific uncertainty and public controversy:

- how far air travel should be 'demand-led', or constrained by taxes or regulation;
- whether global impacts are the responsibility of the airport, the national infrastructure, the carriers, the passengers, or the travel industry;
- which local environmental impacts can be balanced against which economic benefits;
- how the airport could develop as an integrated multimodal transport hub, and the operator as a diversified service provider.

Questions on regional, national or global air travel management are linked – expansion of the UK regional airports will relieve congestion around London, and allow more efficient single leg flights from northern England to overseas (Logan, 1992). It is also likely that business and consumer demand will increase, whether or not the local airport expands – but as with roads, new infrastructure tends to increase both capacity and demand (Royal Commission on Environmental Pollution, 1994). Hence 'sustainable air travel' is not so much a local or regional matter, as a national, EU and global issue.

The economic benefits of the airport are huge – as the global gateway it could support indirectly over 5 per cent of the population. But while the North West contains the largest regional airport in the UK, it is still one of the poorest regions, and over half the passenger movements are holiday charters, of which the bulk are outward bound. This raises questions on the employment projections, the economic benefits and sustainability of overseas tourism, and the pattern of economic growth that excludes the costs of its external impacts. The airport's role as an economic generator is also crucial to development and property values across the region – already many nearby sites are under pressure for business premises and airport parking, and the challenge is to turn such pressures into opportunities for business and employment.

In reality, even modest levels of demand management will be controversial with users and operators, and appear to be ruled out by the recent UK aviation strategy (DFT, 2003). In contrast, a more creative approach to a 'sustainable airport' should aim to look at wider prospects on both demand and supply sides, and in particular the potential for diversified networks. This scenario might see the airport as the hub of multiple-transport modes, including high-speed and light rail, demand-responsive buses and minibuses, and advanced video conferencing. Road access and parking would be contained within local limits, while airport facilities would be networked across the region and the UK. Rising demand would be contained with a combination of taxation, mode substitution, and ICT. Travel intensive industries would be linked with airport operations for minimum impact and maximum added value through diversification into other service industries. The airport, and air travel in general, would be an integral part of a city-region transport–environment strategy, bringing economic and environmental pressures into balance.

Such an integrated scenario may be unlikely: while it aims at a precautionary approach of minimizing environmental risk, it may increase the political risk in the sense of extending responsibility beyond the remit of the institutions involved – that is, government, business and civic bodies. This crucial question of responsibility throws light on the more general analysis in the next section.

Sustainability initiatives

GM is also a kind of laboratory for initiatives and campaigns, building on a long history of urban reform and idealism. The word 'sustainability' was first quoted in the Global Forum 94 event in Manchester, intended as the follow-up to the Rio Summit (World Summit on Sustainable Development – WSSD). Some current examples of initiatives as of 2006 include:

- At the municipality scale is 'Manchester Green City' – a raft of local policies and networks, including purchasing renewable energy, active kerbside recycling, and large-scale urban tree planting. On a proactive front, www.Manchesterismyplanet.com is a current campaign and pledge system, which over six months has engaged 10,000 citizens, businesses and other organizations.
- There is more critical mass at the regional level, much of which is promoted through non-governmental organizations (NGOs) such as Sustainability North West (www.snw.org.uk). 'Responsibility North West' is a networking and promotions scheme for business Corporate Social Responsibility. The 'Climate Change Charter' for the North West builds capacity and offers assessment tools for larger organizations. 'Enworks' promotes environmental management among small- to medium-size enterprises and 'Enviro-link' promotes environmental technologies.
- Most large corporate bodies have sustainability appraisals and mission statements: these include many of the largest polluters in the region – for example, Shell UK, Manchester Airport, and British Nuclear Fuels.
- Most public policies are now subject to sustainability assessment and evaluation, and there is active experimentation in toolkits for 'integrated appraisal'. On close inspection, many of these are broad enough to quote the principles and then proceed with the original plans, showing that the real issue is not the appraisal method, but the definition of options, boundaries, trends and responsibilities (Ravetz et al, 2004).

A pragmatic and critical view of urban sustainability in GM sees much rhetoric about public transport, but 86 per cent of all journeys going by car. There is intensive research on climate change, alongside an airport that doubles in size every 10 years: many policies on fostering local communities, but where 10 per cent of local shops close every year. Possibly the largest single success is outside the conventional urban policy agenda – the 25 per cent per year growth of organic production and ethical food trading, with its 'responsible capitalism' model for global supply chains (Barrientos and Dolan, 2006).

Urban transitions in progress

The above sketch of a 'post-Fordist' city-region highlights the core themes of this chapter. First, that the phenomenon of the UET can be conceived as parallel and integral with other forms of urban transition – social, economic, technological, political, institutional, and informational. Second, that the common factor across these transitions is the changing nature of 'cities' themselves, as both local and global hubs; and the changing agenda for the 'environment' in its many forms. By implication, the UET can be seen as a re-structuring and reorganization of environment-resource systems in space and time. To follow this through we examine different types of transitions which overlap and interlink: including transitions in production, consumption, environments, resources, and sociotechnical-informational systems. Figure 12.4 shows, very roughly, the idea of these various transitions intersecting at local and global scales, in a complex tangle of cumulative causation.

We can also draw the conclusion that the UET is both cause and effect with new forms of 'accumulation' and 'division' on a global scale, for capital, informational, labour and positional goods. One way to highlight this accumulation is to compare international evidence on production, consumption and intermediation in the urban arena. This serves to inform the 'phase model' of the UET and the Environmental Kuznets Curve (EKC) approach (Bai and Imura, 2000; Rothman and de Bruyn, 1998). We illustrate this with work in progress on the environmental metabolism of GM, and its context in the UK and EU economy.

Finally, it is clear that there are alternative options for the future – modernization, material affluence and the hierarchies of developed and developing urban systems are not necessarily given and static cases. In order to 'invent the future' constructively we can bring the simplistic notion of the 'sustainable city' to the more complex reality of cities and regions in the global arena.

Transitions in productivity

Urban and economic analysis is generally focused on 'production', in the sense of applying the factors of labour, capital and resources to bring forward tangible goods or services for exchange in the marketplace. Thus, the analysis

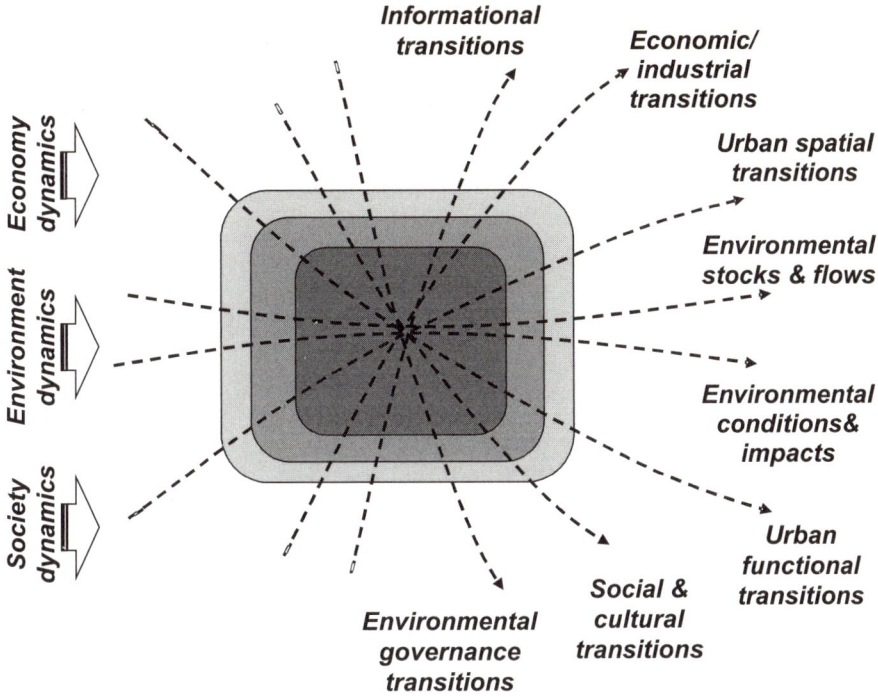

Source: Based on 'integrated assessment' framework as in Ravetz (2000a)

Figure 12.4 *Urban transitions and cumulative effects*

of 'resource productivity' is geared around various combinations of inputs to outputs – that is, materials produced per unit of capital or labour employed. Recent thinking has focused on 'ecological modernization' as an application of environmental management to quality management in business. This then leads towards 'dematerialization' as a means of delivering goods and services with diminishing amounts of material and energy consumption (Leadbeater, 1998). This corresponds with the conventional model of economic activity as production to meet the unrestricted demands of consumers. The 'greening of business' agenda places the firm at the centre of competing pressures both pulling and pushing (Figure 12.5).

There are apparent contradictions in this approach. The dematerialization of certain manufacturing sectors is dependent on physical resources, such as transport, waste or water. The dematerialization of the service sectors is dependent on abundant supplies of construction, transport, paper and computer hardware. There is also emerging conflict between demand management as an environmental goal versus equality of access as a social goal: the current debate on road pricing is one example. Meanwhile, there is increasing recognition

304 Scaling Urban Environmental Challenges: From Local to Global and Back

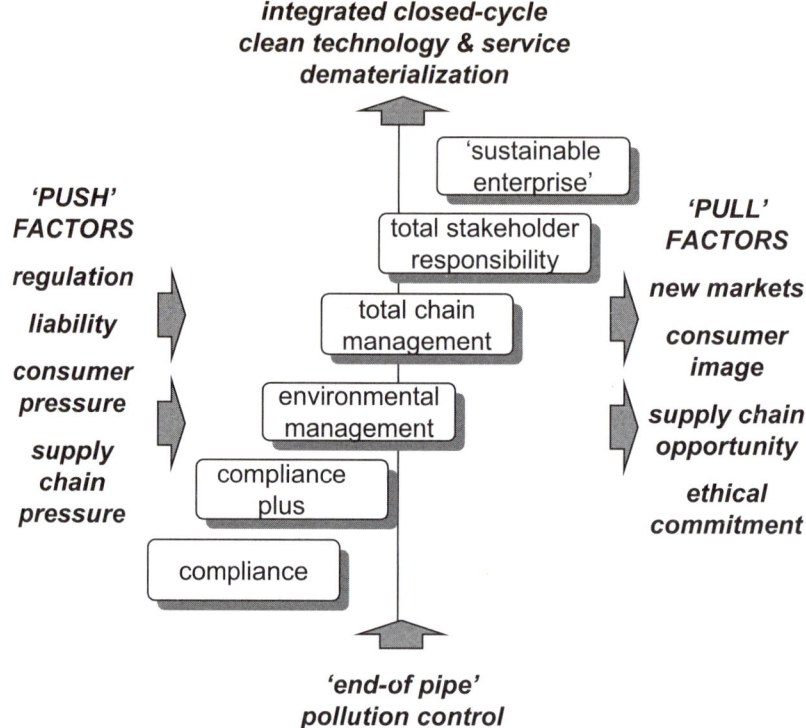

Push and pull motivation for shift from end-of-pipe to 'sustainable firm'.

Source: Adapted from Wood (1995); Gouldson and Murphy (1998)

Figure 12.5 *Transitions in production: The sustainable firm*

of the 'environmental economy' of pollution control, environmental and land management as a growing sector of the economy. This is also seen for its indirect effects on competitiveness and entrepreneurship, where city-regional environmental quality and positive image then generate inward investment.

Following this through suggests a more structural analysis of the economies of cities and urban systems, in terms of different intensities of materials/labour/information/capital. This could draw on recent structural studies of technological classifications (Pavitt, 1984; Green et al, 2002). These explore how economic development and technological innovation in different parts of the world may influence future trends in energy and resource use. This analysis identifies categories ranging from 'supplier dominated', such as agriculture, to 'scale-intensive informational', such as media, to 'specialized science-based', such as biotechnology.

In the context of the UET we can now suggest expanding this classification to a more comprehensive profile of different sectors within the productive economy (Ravetz, 2006). This suggests looking at business sectors not only as units, but also as components of overlapping supply chains, and not only as suppliers of material goods and product but of 'added value' to both producers and consumers. These can then be typed in terms of production 'frontiers' (Tyteca, 1996). The categories here are identified in terms of 'production intensities', which can be typed in order to build up sectoral benchmarks and profiles:

- *scale intensity* – the returns to scale of the industry/product;
- *informational intensity* – science/other content in terms of knowledge-based industries;
- *infrastructure intensity* – the added value derived from the presence of physical infrastructure, urban or other;
- *labour and capital intensity* – the conventional measures of labour or capital employed per unit of added value;
- *material intensity* – useful material input per unit of added value;
- *resource efficiency* – material input as a ratio of material output in the product;
- *energy intensity* – energy per unit of gross added value.

In addition, there are some analytic measures that aim to capture some of the more structural aspects of the business activity, in its context of supply chains, market shares and demand side profile:

- *positional intensity* – a view of how much the product added value is positional in relation to other products, such as in housing, fashion or media – and this may be very significant in terms of environmental quality and added value;
- *chain intensity* – a measure of how close the business or sector is to the critical path or central supply chain: for example, to capture the difference between a construction materials firm and an estate agent;
- *external intensity* – this is a prototype measure that aims to capture the ratio of external resources/impacts to internal resources/impacts: for example, the index of 'internalization' as in the direct/indirect resource ratio, shown in the next section.

Transitions in consumption

As a counterpart to production and 'productivity', we suggest a parallel analysis of consumption and 'consumptivity'. If consumer spending is shifted from material products towards (apparently) dematerialized services, then this can change radically the profile of resource use and hence the 'phase' of the UET trajectory. But what factors are involved in this shift?

Consumption in standard economic analysis is taken as the aggregate of a set of utility functions – for example, the larger/better the houses or cars that are 'consumed', the more utility and hence welfare of consumers. In economic policy terms the 'affluence' of consumption is seen as not synonymous, but a good proxy between economic throughput and social well-being. However, there are newly emerging approaches into consumption, starting from a human needs approach, and exploring the psychological, social and cultural dimensions of consumption (Van den Bergh et al, 2000). We can draw from this to outline alternative and overlapping modes of consumption:

- *physical based consumption* – the necessities of shelter, food, clothing and so on;
- *service-based consumption* – the higher order provision of health, education, and so on, generally, but not always, involving human content;
- *consumption as symbolic identity* – particularly seen with fashion or style, where material products have ever lower relative costs and ever-shorter lifetimes;
- *consumption as value accumulation* – based on the metrics of non-market values in social, cultural and environmental terms;
- *positional consumption* – new patterns of competition in late capitalist urbanization, often involving the 'positional goods' factors of environmental quality and location.

Such alternative perspectives also contribute to the new policy focus on 'sustainable consumption', at least in the UK: this focuses on the conflict between economic, social and environmental goals, even while in policy terms it is difficult to identify responsibility for such a problem (Jackson and Michaelis, 2003). In terms of the UET, the driving forces above can be separated into parameters on a material spectrum: not only the scale and type of material input, but the location in space and time, where delivery of simple products may involve long chains of energy-intensive and high-impact logistics.

One vital shift is on the boundary between production and consumption. A conventional economic perspective considers production of goods or services up to the point of final demand (i.e. purchase by consumers or government), which is assumed as the act of physical appropriation or, in the case of indirect services, legal appropriation. The distinction becomes less clear in many new service- and knowledge-based sectors, for instance, tourism, which identifies the locality itself as a commodity. Figure 12.6 is one way to visualize the merging of previous boundaries based on accumulation or spending of capital, and on positive or negative forms of experience.

This approach can be summed with the term 'consumptivity' – a counterpart to the 'productivity' ratios above (Ravetz, 2006). Consumptivity aims to be a multi-scalar representation of the degree of human welfare per unit of production, taking into account the various dimensions above, which may be entirely qualitative and non-additive. The generic metrics and benchmarks can then be summed up with three identities:

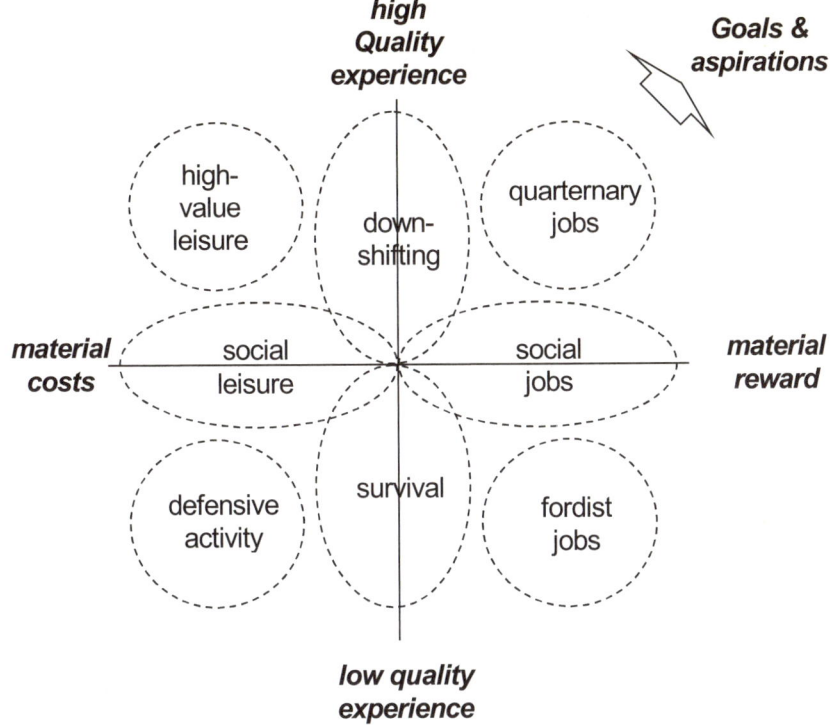

Changing balance of quantity of material reward and quality experience in work and leisure in post-industrial societies.

Source: Adapted from Handy (1995); Rifkin (1994)

Figure 12.6 *Transitions in consumption: Shifting boundaries*

- 'productivity' = inputs to production/outputs from production;
- 'consumptivity' = inputs to consumption/social outputs from consumption;
- resource effectiveness = productivity/consumptivity = inputs to production/ social outputs from consumption.

In terms of the 'consumptivity transition', there are topical questions on future directions. The broad measures of 'decoupling' in advanced economies such as the UK appear to show that direct environmental impact is static while economic growth continues (DEFRA, 2005). Meanwhile, there is evidence that social welfare is decreasing, although this clearly cannot be measured directly (Layard, 2005). The measurement of indirect impact contained in imports is as yet very rudimentary, but it is clearly significant and growing rapidly.

Meanwhile, it is clear that the role of consumer goods and services as identity-forming, positional and cultural commodities has increased greatly with the spread of affluence. It is also clear that most of this trend is based on material and energy-intensive goods and services. However, in many emerging aspects of post-industrial urban cultures, there are signs that some groups value quality over quantity, services over physical objects, and location over space. It may be that in a highly distributed and networked pattern of living and working, the conventional act of consumption as 'purchase, enjoyment and custody of physical items' will be shifted towards that of 'access to experience' in a time-share model.

Informational transitions

One of the main drivers of change in production, consumption systems and their environmental effects is ICT. Current trends show that services such as housing and education are increasingly distributed and networked, just as retail and employment are becoming now; in the near future the norm is likely to comprise a range of combined living/working/leisure locations at different distances around the world. One example in GM is the Manchester United Football Club, which has over 600,000 members worldwide who follow a completely multi-national team. Such alternative virtual identities to the city-region – a 'space of flows' rather than a 'place of home and work' – are likely to multiply as economic and social activity is globally networked.

This transition is very topical as it cuts across the conventional production and consumption boundaries, with new categories such as 'business to consumer' (B2C), 'consumer to consumer' (C2C), and so on. E-commerce can be seen as one dimension of structural change in the productive economy, which then involves markets, technologies, institutions and consumers (Wilsdon, 2001). It is raised here because there are direct and indirect linkages between the material flow of physical goods, and the dematerialized flow of digital information.

Much analysis of economic change through ICT/e-commerce tends to assume that markets, production processes, societies and so on will remain the same except for the e-commerce impact on speed and scale of activity. In contrast I would suggest that e-commerce is already instrumental in shaping much more fundamental and qualitative change, even while it is now available to a minority of people and businesses:

- qualitative change in economic and market structures – that is, instant/virtual markets, virtual distributed corporations, virtual stakeholder networks, consumer agglomeration markets, reverse auctions, consumer–consumer markets (C2C);

- qualitative change in institutional structures – that is, relations between governments and markets, transparency and accountability of corporations (consumer to business – C2B);
- qualitative change in industrial and technological processes – just-in-time production, outsourcing, multi-agent contracting (business to business – B2B);
- consequent qualitative changes in retail and distribution are likely to focus on new capabilities for products to be personalized and customized (B2C);
- in the social economy realm, there are many forms of trading, file sharing, open auctions and so on, which open up new possibilities in the consumer sphere (C2C). The most obvious internet application is in facilitating trading of goods for reuse or recycling, as in e-bay, and the non-profit equivalent, www.freecycling.com.

Clearly such possibilities have profound implications for the UET – but are they positive or negative? Information, communication and transport have historically been two sides of the same phenomenon – centred on the 'connectivity' of human organizations, economies and societies (Mulgan, 1997). Positive perspectives focus on the 'richness of cities' with creative open diversity, and large capacity for learning and resilience (Christie and Levett, 1999). Critical perspectives look at the digital divide and its effect on the 'splintered urbanism' of separate realities (Graham and Marvin, 2001). Evolutionary perspectives look at the potential for ICT-based connectivity with the aspirations for competitiveness and innovation in the 'learning region' (Morgan, 1997).

In summary, there are many possible linkages between economic, social and informational pressures and dynamics, and the environmental effects on the UET. This is conditioned by the different evolutionary stages of cities in different parts of the international order, in different functional relationships, different internal geographies and so on. First we summarize the key influences and linkages with the urban environmental agenda, from local to global (Table 12.2).

Transitions in spatial structures

While the city-region's functions are increasingly aspatial and globally networked, for a consumer-based society locational qualities are as important as ever, as the generators of economic value, competitiveness, social identity and quality of life. For GM, as for many post-Fordist city-regions, various trends are running in parallel (Champion et al, 1998):

- Thinning out – the reducing size of the average household means that the population of most existing areas is gradually reducing in density as demand rises for space, privacy, amenity and 'locational identity'. Similar pressures apply to industrial and commercial property.

Table 12.2 *Key transitions and implications for the UET*

	Individual/ household environmental health	Local-urban environmental conditions	Urban-regional impacts	Global environmental impacts	Supply chain and resources
Economic production transition	Basic domestic infrastructure	Rising efficiency of utility and infrastructure	Displacement of regional resources	Growing impact on climate, resources	Reorganization to global level systems
Productivity/ business transition	Customized business models	New institutions for infrastructure	New institutions for regional resources	Growth in corporate responsibility	Resource exploitation due to global finance
Consumption/ lifestyle transition	Consumer choice and access	Segmentation of housing and neighbourhoods	Segmentation of urban and regional communities	Growth and specialization of leisure and tourism	Increasing complexity and material turnover
Informational/ structural transition	ICT in personal profiles, diet, health care	ICT in monitoring and management	Counter-urbanization production and consumption	Functional specialization of global system	Rapid re-structuring from global logistics and new economy
Demographic/ health transition	Rapidly reducing mortality	Rapid reduction in pollution and communicable disease	New patterns of urban-regional migration	Rapid increases in material consumption	New risks in food chains and resource inequity

Source: Based on George et al (2007)

- Urbanization – the conventional spread of urban areas at their peripheries due to population growth, inward migration or demographic change. This may be more or less contained by Green Belt or similar planning policy.
- Re-urbanization – the return of populations with choice to city centres, inner cities and regeneration areas. In GM there are now 20,000 people living in and adjacent to the city centre, in contrast to less than 1000 only 15 years ago.
- Counter-urbanization – the wider distribution of urban populations across rural areas. Formerly measured by commuter flows, this trend might now be characterized by 'metropolitanization' of rural areas – that is, the shift of economic and social patterns by incoming or semi-retired social groups, new injections of housing finance, urban-centred lifestyles and so on.
- National-scale re-structuring – across the UK there is a continuing trend of inter-urban and regional migration towards the London and the South East, shire counties and coastal areas.
- Uneven development – the spatial polarization of growth and decline, opportunity and deprivation, security and risk. This occurs between

neighbourhoods, between core and satellite towns and cities, and between regional growth and decline.

The combined result of such trends might be characterized as *agglomeration* – a general scaling up of the urban functional system and merging of smaller into larger units. This can be defined in terms of physical residential or workplace location, although in most developed countries there are planning or zoning restrictions to contain this. It can also be defined in functional terms of social and economic patterns: or in terms of lifestyles, consumption patterns and behavioural factors such as choice of public services, local identities and so on. Cultural studies show an overwhelming desire for 'sustainable communities' with high levels of security, amenity and accessibility, and where such combinations are traded in the marketplace, the result is the exclusion of lower-value groups and activities. The trends above of economic and social fragmentation are likely to reproduce in the spatial polarization of urban and rural areas as the wealthy and mobile secure their private versions of the 'sustainable community'.

These spatial trends may each combine to form a broader 'spatial transition'. In the crowded territory of the UK this may be more subtle than in other more recent and faster-growing urban systems (United Nations Secretariat, 2001). It may be more concerned with the re-structuring and rearrangement of groups and activities within the city-region. It may also concern the 'metropolitanization' over a much wider area, of activities, financial and environmental metabolisms, as in Table 12.3.

Transitions in environmental management and policy

The typical city-region in the developed world shows a changing pattern of risks and opportunities. Many common environmental pollutants are being replaced with the more insidious and uncertain hazards of modern production and consumption – genotoxics, carcinogens and food chain viruses. As heavy industry migrates overseas, the clean-up of the urban environment shows gradual improvement, while rising affluence generates consumption of imported goods.

On the ground, local territorial conflicts are mounting over 'positional goods' such as amenity and location. Sectors such as housing, transport and waste management are each embroiled in controversies over environmental risk and justice, and these also define new social groupings and subcultures (Beck et al, 1994). New ways of managing such conflicts will be needed, whether or not 'sustainability' is on the agenda. In general the environmental agenda for post-industrial city-regions is marked by rising affluence and aspirations for identity-creating goods and lifestyles of all kinds. It will also be marked by the polarization of communities and social groups and networks, which focuses on access and environmental quality. Topical questions of market versus state in the distribution in quasi-market goods, such as housing, transport or waste management, may come to revolve around such polarization.

Table 12.3 *Spatial dynamics and environmental metabolism*

	Environmental flows	Environmental stocks	Environmental conditions	Environmental impacts	Environmental benefits
Urbanization	Direct increase in urban metabolism	Direct land-use change	Intensification of urban conditions	Transport demand growth	Transport and energy efficiency
Suburbanization	Shift in metabolism to suburban	Rapid land-use change: shift of biodiversity	Outward spread of urban conditions	Transport/ energy demand growth	Increase in domestic green space
Counter-urbanization	Shift to long-range commuting/ networking pattern	Rapid activity and community changes	Displacement of urban conditions	Transport/ energy demand growth	New rural–urban fringe landscapes
Re-urbanization	Shift to affluent urbanist metabolism	Intensification of urban land use	Intensification of urban conditions with greater affluence	Gentrification with loss of biodiversity on derelict land	Land and water reclamation
Functional agglomeration	Increasingly complex metabolism due to specialization	Specialization of land use and activities	Polarization of conditions due to increased fragmentation and segmentation	Transport demand growth	Increased investment due to economic growth

Source: Based on George et al (2007)

In terms of environmental processes the picture is complex. Material movements are 'trans-boundary', with long distances from origin to destination; 'trans-media' with many processes between gases, liquids and solids; and 'trans-generational', transferring impacts and responsibilities from present to future (Blowers, 1993). Assessment of hazard and risk depends on how the system boundaries are drawn – even detailed lifecycle analysis (LCA) of products or processes can easily underestimate total system impacts (Lave et al, 1995). Standards for such impacts can be set with 'thresholds' and 'critical capacities', for resource demands, industrial emissions, environmental themes such as acidification, and human or ecological health risk. However, these various kinds of standards do not often match, with large uncertainties between them.

Parallel to these are the transitions of environmental management and policy, which can be seen on another spectrum:

- first, the direct regulation of emissions, as in the UK Clean Air Acts of 1956;
- there follows a more negotiated dialogue with polluters as in the BPEO principle (Best Practical Environmental Option);

- then, the integration of all environmental pressures through integrated pollution control (IPC) and the EU Integrated Pollution Prevention and Control;
- then, a more structural foundation in terms of integrated datasets on pressures, conditions, benchmarks, LCA, and underlying material flows and environmental processes;
- an integrated assessment approach: a more open discursive and participative approach to complex problems, involving different stakeholders, different levels of time and space, different chains of production and consumption, and different perceptions of risk and values (Bailey, 1997; Ravetz, 2000b).

Generally, we can observe a growing displacement between environmental producers and consumers, between environmental causes and effects, and between environmental beneficiaries and victims. This shift is running in parallel to a qualitative change in policy and management. The implication is that the UET is not only concerned with the transition of urban environmental technologies and impacts, but with the institutional questions of environmental policy and management. This then, in turn, drives a structural shift in the type of information and policy instruments which are relevant and effective. Each of these is then most relevant and effective at different scales of spatial organization and urbanization, as in Table 12.4.

Metabolism of urban affluence

Each of the above transitions is combined in various patterns across the international urban system. Inevitably, the course of change is not smooth or predictable, the causes and effects are entangled, and the outcomes are open to debate. Bringing this back to our central theme of the UET, we focus on the 'environmental metabolism of urban affluence' – that is, the patterns of physical stocks, flows and impacts, through the cities and urban systems of the developed world. As in the second section above, such cities often contain their former functions as material processors, in parallel with emerging functions as information processors, even while the materials continue to flow.

To understand this metabolism better we might put the metabolism of GM alongside the UK, EU or world average; or we could relate the current conditions and trends, to alternative projections and possible targets. Also relevant is the distribution of environmental flows and resources between locations, social groups or economic types.

This all points towards the focus of this chapter – the 'metabolism of affluence'. Such an urban environmental metabolism can be seen as driven by the international urban hierarchy and division of functions. This in turn is a result of the role of urban systems as arenas and enablers of the international capitalist order and division of labour (Knox and Taylor, 1995).

Table 12.4 Urban hierarchy and environmental metabolism

	Resource use	Production	Services	Urban infrastructure	Consumption	External impacts
Global urban systems	S–N commodity transfers	International division of industry and labour: supply chains and CSR	International flows of capital, info, media: international travel	Global fixed capital investment: international travel	Global influences on consumption: tourism, etc	Climate impacts; Marine pollution; resource use:
National/ regional urban systems	Water, minerals, energy	Regional manufacturing	Functional specialization	Utility investment: inter–urban transport	Re-structuring of spaces for consumption, leisure, etc	Air/water pollution: resource use
Urban structures	Land-use competition	Urban manufacturing	City services	Water, energy, sanitation	Aggregate demand	Air/water pollution
Local structures	Land-use competition	Local production	Local services	Street and block structures	Local lifestyles	Environmental health
Households	Space and territory	Domestic economy	Household economy	Household technology	Consumer demand	Home hazards

Note: S–N = South to North; CSR = corporate social responsibility.

Source: Based on George et al (2007)

In practical terms, the role of a post-industrial city-region such as GM can then be understood as a kind of consumption and service sector 'bubble'. Its relative comfort and order is directly or indirectly supported by other larger workforces and city systems, which are generally more labour intensive, materials intensive, undercapitalized, disorganized and environmentally hazardous. There are of course many qualifications to this simplistic view:

- there are great local/urban/regional differentiations: within the city, the city-region and the wider functional region;
- there is also social/cultural/employment segmentation of social and economic groups and organizations;
- rather than a simple division of 'developed versus developing' nations and cities, there is a spectrum of affluence between and within nations.

A further question concerns the role of the city or city-region itself in the environmental metabolism. In one sense, a city may be nothing more than a location where the supply chains of production and consumption happen to be concentrated. But in practical terms it is clear that the urban system is the provider and enabler of many types of infrastructure and 'factors of production' – energy, transport, buildings, and the labour market itself. We might apply some of the quantitative evidence to identify between these urban 'factors' of buildings and so on, and the industrial supply chains of consumption and production. In more structural terms, the city system also provides the external benefits of agglomeration, with specialized facilities, economic districts and critical mass, as the enabler and facilitator of innovation and competitiveness. The implication is that there is not one simple boundary around what is an urban agenda and what is a purely economic agenda for production and consumption – each overlaps the other. The evidence below aims to reflect these multiple boundaries.

Resource flows and the urban metabolism

First we look at how consumption and production are represented in the urban metabolism. This shows three main measures, which are different but interrelated, and each providing a contrasting view on resource flows and their impacts.

- *Carbon dioxide emissions*: this is the most common and easily calculated resource flow, and the most topical as the largest human cause of climate change. Carbon dioxide is usually measured in terms of 'production' or territorial emissions: there is however recent analysis that reallocates the responsibility for carbon dioxide to consumption, which then includes imports. Carbon dioxide is mainly generated by energy production and consumption (energy flow), which runs in parallel to material flow.
- *Material flow:* this can be measured in Direct Material Consumption: the total amount of materials directly 'used' – that is, consumed by final demand in the region, excluding exports. There is also the larger Total Material Consumption, which includes the indirect or 'hidden' material flows generated, although the calculation here is less precise.
- *Ecological footprint* (EF): this includes the land area equivalent for the sequestering of carbon dioxide emissions, plus other impacts on 'bio-productive' land use, and is a simpler measure of total impact than LCA.

EF is allocated on the 'consumption responsibility' basis (i.e. measuring total impacts from all material flows implicated in the delivery of products to household and government final demand (Rees and Wackernagel, 1995; Simmons et al, 2000)). It is measured in a standardized area unit equivalent to a world average productive hectare or 'global hectare' (gha), and is usually expressed as global hectares per person (gha/cap) to permit comparisons between countries or regions. This is often divided into 'land footprint' and 'energy footprint':

- *Land footprint:* The land footprint includes the area required to produce all the crops, grazing land required to provide meat, forest land required to produce forest products, and fishing ground required to produce the fish and seafood products consumed by people living in a defined area.
- *Energy footprint*: An energy footprint represents the area required to sustain energy consumption. This includes the energy used directly by households and services in the region and the indirect energy to produce goods imported

and consumed within the region. The footprint is calculated as the area of forest that would be required to absorb the resulting carbon dioxide emissions, excluding the proportion that is absorbed by the oceans.

There are many critiques of the EF method. It is clear that the value lies as much in communications and awareness raising as in scientific analysis, which is prone to all the caveats of environmental accounting (see Chapter 11). One key point of reference is the 'Factor Four' concept (von Weizsäcker et al, 1997). This is all the more topical, now that the Factor Four concept is not only the idealistic dream of environmentalists: it is enshrined in the UK Energy White Paper, which sets out the long-term scientific target for climate change emissions, of 60 per cent reduction by 2050. While there is a difference between accounting for carbon dioxide and EF, they are very closely related. Current projects in the UK are aiming towards this 60 per cent target, but on a global scale, taking account of disparities in wealth and in emissions between developed and developing nations. This is the fundamental logic for the overall target of Factor Four, or 75 per cent reduction by 2050.

For each of these measures, there has emerged a standard accounting arrangement for consumption and production, as for instance in the Blue Book tables of UK national accounts (ONS, 2005). These show the contribution of imports, exports, capital investment, and the supply and demand from other industries, and a similar structure can also show the resource flows above – material, energy, carbon dioxide and EF. In the arrangement of the economic accounts, the total production equals exactly the total consumption, and this is also reflected in the 'mass balance' principle of the resource flow framework. As in Figure 12.7, this shows an outline of a whole supply chain, rather than single sectors of production.

The logic of this resource flow framework is to apply economic accounting practice to the environmental metabolism, from 'cradle to grave' – raw materials, energy, emissions, waste, recycling and so on. Obviously, this represents only a part of total environmental impacts, leaving out the effects of toxicity, ecosystems loss, landscape change and so on: but it does represent a common thread that links and summarizes more complex effects. Also, such a framework can only be as good as the data available: for instance, current UK waste data do not contain details of its material content, its industry source or its location of origin. So the framework shown here is greatly simplified compared to the reality, where many materials are used to make many products, at many intermediate stages, in many sectors, with many environmental effects. Recent analysis from the Ecological Budget UK project is enough to provide an outline five-stage resource flow model of the UK and regional economy (WWF-UK et al, 2006).

- The five stages correspond to the classification of economic sectors as primary, secondary, tertiary, demand and 'externalities'. Each of these stages shows a different type of material intensity: from the gross tonnages of minerals and agricultural sectors through to the service sectors, where

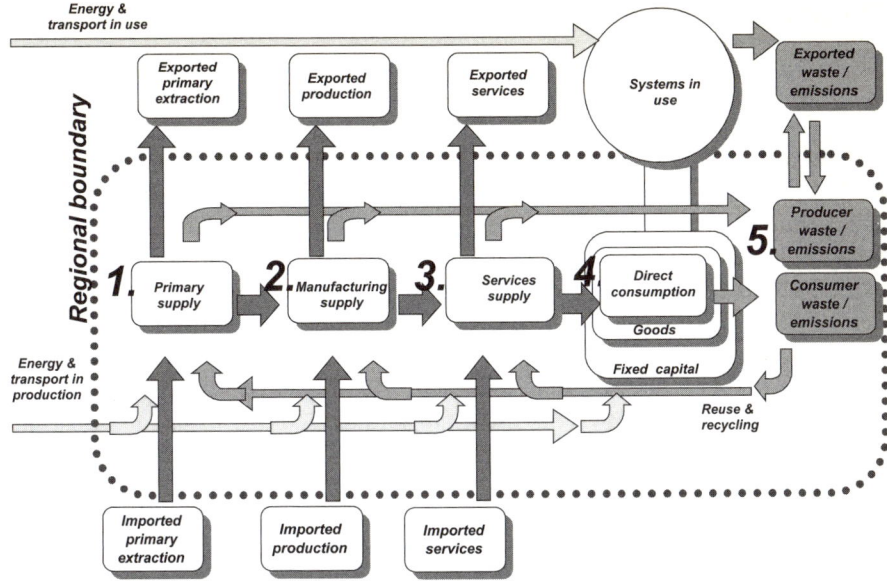

Source: Based on Ravetz (2006)

Figure 12.7 *Resource flow framework*

materials are involved but of little added value compared to that of labour inputs.
- Various kinds of waste stream are shown by the shaded boxes on the right-hand side of Figure 2.7, coming off each of the stages, and the recycling loops can be seen going back to each stage.
- Various inputs of the 'factors' of energy and transport are also shown at each of the five stages in Figure 2.7.
- On the consumption side is a simple breakdown: consumables are items with less than a year's life; durables are items with a life of more than a year; fixed capital are items such as buildings with an indefinite life span. For each of these types, a stock flow model can be constructed if needed. Many consumption types, such as vehicles or buildings, may have larger demands and impacts from their lifetime in use.

The implication of this more evolved scheme is that 'resource productivity', the useful outputs per unit of input, can be measured in different ways at each stage of the supply chain in terms of material inputs or outputs, energy, emissions or waste. Each of these can be indexed against the inputs, outputs or outcomes such as capital employed, labour employed, turnover, finished products, or value added.

Metabolism of the global urban order

Applying this scheme to the international urban order, we can identify how the resource flow supply chain is redistributed to different cities with specific roles to play. This then reinterprets Figure 12.3 – on the material and information processing functions – to a geographical context. Figure 12.8 shows a typical developing city as the hub and gateway of the typical developing nation (bearing in mind that all nations and cities are actually unique). There may be large volumes of primary extractions and harvests, a small service sector, large export volumes, and often a drastic under-investment in capital equipment and infrastructure. Pollution is often high due to rapid industrialization, and the external impacts of extraction are often very high in terms of forced displacement of ethnic peoples, destruction of ecosystems, and the downstream consequences in rapid and chaotic urbanization.

By contrast, the developed city in Figure 12.8 shows large import volumes, a slim manufacturing sector, and an overgrown service sector. The factors of energy, transport and infrastructure are large, as are capital investments. While emissions are relatively small, waste volumes are high due to the sheer throughput of material and advanced packaging systems. Some waste is recycled, and increasing amounts are 'repatriated' in return container loads, often with drastic impacts in the destination country.

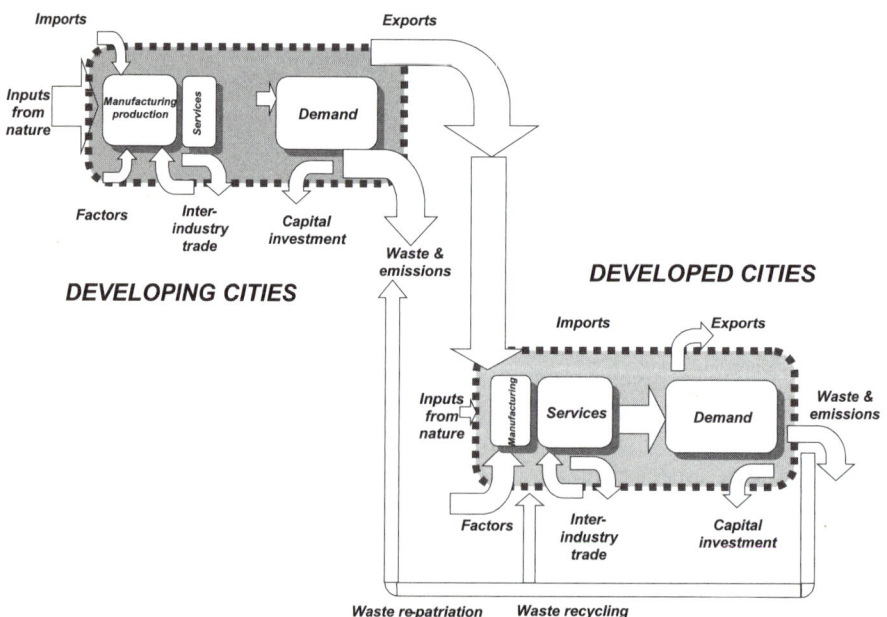

Source: Author's concept diagram

Figure 12.8 *The global urban environmental system*

The picture as a whole shows in a simple but graphic way how the metabolism of urban affluence is inextricably linked with the metabolism of urban poverty in developing nation cities. Further analysis would show how the developing cities are themselves acting as expropriators of the physical resources of their hinterlands.

Measuring the metabolism of affluence

At present, the evidence base for the overview above is rudimentary, and new research directions are emerging rapidly. The analysis of resource flows in cities and regions has been aided by a recent research programme on the mass balance and EF of sectors and territories in the UK and EU (Wuppertal Institute, 2001; Barrett, 2001; Simmons et al, 2000; McEvoy et al, 2004). Much of this was generated by the 'mass balance' programme of Biffaward, funded by the UK 'landfill tax credit scheme' (RSWT, 2006). Each of the studies highlights the fundamental difference between production and consumption as the frame of reference. Production-based analysis includes exports and locally generated pollution and waste. Consumption-based analysis includes imports and their impacts, which can be anywhere in the world, depending on how the 'responsibility' principle is used (Eder and Narodoslawski, 1999).

The example results shown here are generated by the Ecological Budget UK project and the REAP modelling system (WWF-UK et al, 2006). This is situated within the Global Footprint Network international methodology and accounting system (EEA, 2005). The results are then checked against a method of bottom-up calculation used previously for the North West region (Wiedman et al, 2004). The scenario data and cross-sectoral links are based on the 'integrated sustainable cities assessment method' and its comprehensive study of the GM city-region (Ravetz, 2000a). The key results are shown in Table 12.5, with a further breakdown of carbon dioxide emissions and EF in Figure 12.9.

- The total direct material input to the economy is 15.2 tonnes per person (also described as tonnes/capita or t/cap).
- Of this total, 26 per cent comes from imports; 21 per cent of all material production is exported.
- The average household directly purchases 2.5 tonnes a year and throws away a tonne of waste each year.
- 35 per cent of all material goes to capital stocks (new buildings, roads and other infrastructure).
- 37 per cent of all material used in the economy ultimately ends up as waste.
- The build-up of products in the regional economy and infrastructure is in the order of 9 million tonnes per year (4 tonnes for every person).
- The total carbon dioxide emissions from UK production is 10.8 t/cap. The total carbon dioxide emissions from UK consumption is 11.9 t/cap.
- The total carbon dioxide emissions in imports for consumption is between 10 and 30 per cent of the total UK emissions.

Table 12.5 Material flows in the UK average regional economy

NACE Rev I class A31		TOTAL FLOW dom+exp+imp	Total domestic material inputs	Total export materials	Total import materials	Total solid waste	Total recycled/ reused	SUBTOTAL intermediate flow	Final demand: households	GFCF - capital investment, stocks	Total exports of goods and services	Subtotal final demand	Import fraction of DMI (%)	Waste fraction of DMI (%)
AA, BB	Agriculture	2043	1749	100	195			921	700	25	104	828	11	
CA	Fuel extraction	6925	4628	1422	875			3029	47	20	1532	1599	19	
CB	Mining and quarrying	3931	3248	280	403			1601	6	360	1280	1647	12	
DA	Man.: food, etc	1788	1352	129	307	145	63	445	805	-2	104	907	23	11
DB, DC	Man.: textiles, etc	127	50	20	57	25	8	9	34		6	41	115	50
DD, DE	Man.: wood, paper, etc	770	392	93	286	80	43	550	55	3	33	91	73	20
DF	Man.: energy fuels	532	0	294	238	124	61							
DG	Man.: chemicals, etc	982	512	209	261	105	27	237	105	2	167	275	51	21
DH, DI, DJ	Man.: other materials	3353	2802	270	282	173	93	2839	135	282	441	857	10	6
DK	Man.: machinery, etc	724	210	429	84	19	9	61	28	59	63	150	40	9

Table 12.5 *Continued*

DL	Man.: electronics, etc	413	304	48	61	10	4	111	31	62	101	194	20	3
DM	Man.: vehicles, etc	322	185	47	89	30	18	49	57	26	54	136	48	16
DN	Man.: other	86	36	19	31	14	6	6	22	5	3	30	85	37
E	Electricity, gas, water													
F	Construction					1723								
G, H	Wholesale, retail					315	141							
I	Transport, comms					43	15							
J, K	General services					235	61							
	Private households					591	118							
	TOTALS	15468	3359	3170	3631	666	9858	2025	0	842	3888	6755		

Notes: Flows shown as kg/person/year; note that material flow data is not available for NACE sectors E-K, and Final demand by government; waste data not shown for primary industries.

dom+exp+imp = domestic + export + import; DMI = direct material input; GFCF = gross fixed capital formation; man. = manufacturing; comms = communication.

Source: Adapted from WWF-UK et al (2006)

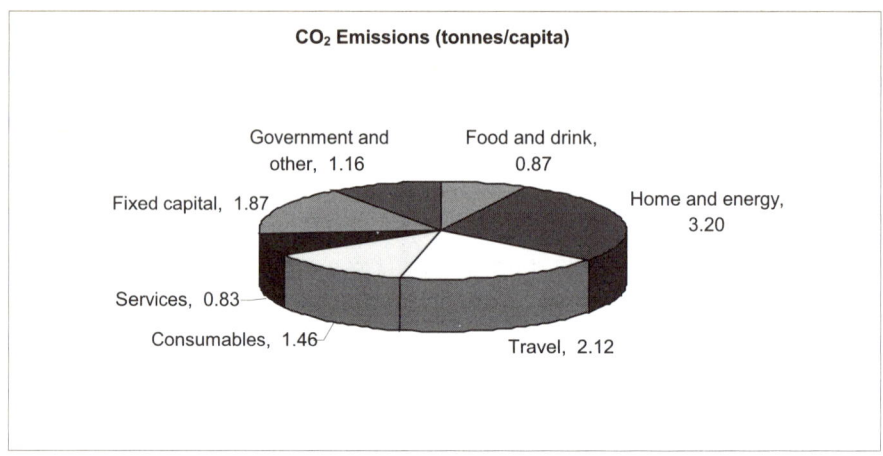

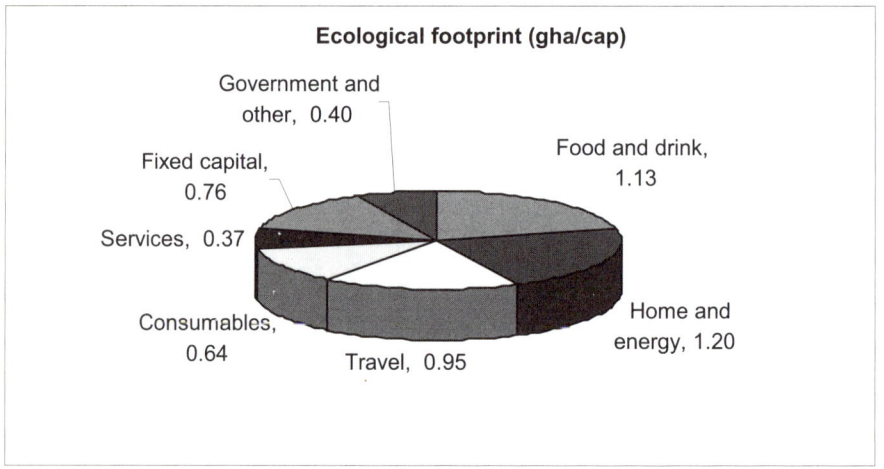

Source: WWF et al (2006)

Figure 12.9 *Breakdown of CO_2 emissions and ecological footprint*

- In terms of carbon dioxide emissions, the most resource intensive sectors are the cement industry followed by electricity generation.
- The most indirectly resource intensive sector is banking and finance.
- The consumption type with the highest impact is domestic energy consumption, followed by car use.
- The consumption type with highest impact per pound sterling spent is electricity generation.

Global footprint comparisons

The overall results can be summed up with the ET measure, as expressed in gha/cap: the notional bio-productive land area needed to supply resources and absorb emissions, in various places around the world. The total EF in GM is 5.4 gha/cap, at about the UK average. If this footprint was to be distributed evenly among the global population, the UK is currently overshooting that by a factor of three.

Full comparative analysis of the material flows in developed and developing cities will await further research. However, in the meantime, the recent databases on global footprint results act as a first order proxy, as in Table 12.6.

Table 12.6 shows the comparison of EF as the proxy for urban environmental metabolism, between GM and the highest- and lowest-income nations. These can be projected on a simple what-if basis, to show the effects of current trends to the year 2050. A further 'global equity' scenario shows the 'One Planet' goal of equality between nations within the global bio-productive capacity, which is itself expected to be reduced from current levels. At present, the total EF of the world's population exceeds the available land area or bio-productive capacity by 22 per cent. This 'overshoot' could possibly escalate by a factor of 10, as the middle-income countries rapidly catch up with Western standards of material affluence.

Conclusions

The main implication coming from such analysis of the 'metabolism of affluence' is to ask how could things improve? In other words, how could a city-region such as GM realize its aspirations of economic growth and quality of life, while improving its own environment and reducing its impact on the rest of the world; and meanwhile enabling other cities and regions to do the same? There are several directions for responses to this – alternative models for the UET; practical actions for city-regional strategy; and avenues for further research.

Alternative models for the UET

The outline of the 'global urban metabolism' in the previous section suggests that there may be alternative models for organizing urban management and policy to enable a more sustainable UET. This divergence often confuses many current attempts to frame the 'sustainable city' agenda (Marvin and Guy, 1997). From the perspective of the environmental metabolism, we can identify how these alternatives each focus on different components including production, consumption, informational-structural, and urban infrastructure. Each of these models represents an alternative focus within the overall urban environmental metabolism.

Table 12.6 *Global footprint comparisons and trends*

	Population 2005 (millions)	Total EF Current 2005 gha/cap	Land footprint gha/cap	Energy Footprint gha/cap	Total bio-capacity gha/cap	EF scenario Business as usual trend 2050 gha/cap	EF scenario One Planet target 2050 gha/cap
GM/NW-UK	2.5	5.4	1.7	3.6		8.5	1.5
GM/NW-UK growth trend/year	0.2%	1%				1%	–3%
Highest EF (US)	288	9.5	2.9	6.3	4.7		
High-income countries	920	6.4	2.1	4.1	3.4	9.6	1.5
Middle-income countries	2971	1.9	0.9	0.9	2.1	6.4	1.5
Low-income countries	2226	0.8	0.5	0.3	0.7	1.6	1.5
Lowest EF (Afghanistan)	23	0.1	0.1	0	0.3		
World total	6148						
World average EF		2.2	0.9	1.2		5.1	1.5
World available bio-productive land area					1.8	*1.5*	1.5
Overshoot factor		22%				240%	0

Source: Adapted from WWF-UK and Global Footprint Network (2005)

- *Production innovation: the evolutionary model:* The Factor Four approach to 'dematerialization' and 'decarbonization' of the economy would be a shift on a massive scale. It relies on businesses and organizations anticipating such shifts in their own terms over years or decades, and steering their innovation activity to turn potential problems into opportunities.
- *Service model*: This focuses on the producer/procurement side, where products are leased, taken back, re-manufactured or recycled, with huge savings in raw materials, processing energy and waste impacts; plus the consumers' facility is continuously updated. This model overlaps with the *Informational model*, which relies on the potential of ICT to enable the 'internalization of externalities'. This enables new and more efficient patterns of trading producer and consumer markets, and personalizes production and consumption in order to reduce overall impact.

- *Social and cultural economy model*: This focuses on the consumer demand side. In many cases there are opportunities to reduce material consumption while increasing human satisfaction, by community networks, social trading schemes, equipment and lift sharing, and many forms of social cohesion.
- *Integrated urban resource management model*: This comes back to the agenda for the city and city-region, and its potential for supporting low-impact infrastructure for production and consumption. Such infrastructure can be 'hard' pipes and wires, and/or 'soft' organizations and networks, and/or the processes of innovation, learning, participation, personalization and others.

City-region strategy for the UET

At the city-regional level there is often a strong correspondence and 'fit' between physical functions, social identity, economic units and political territories. Because of this the regional level brings opportunities to improve on the current state of fragmentation of policy objectives, and move towards the goals of sustainable development. In general, the city-regional level offers an opportunity to make new linkages for the sustainable development agenda, between the local and the national scale, where economic and urban policy is often in a greater state of flux. All these add up to huge opportunities for sustainability development strategy, with a much wider scope than the conventional focus on economic production. The policy questions are then how such opportunities could be taken, at what kind of risk, and whether the structures are in place to realize them. The evidence from GM on sustainability agenda shows much complexity and contradiction, not only in physical results but in the policy discourse, split as it is between the diverging models above (McEvoy et al, 2004; Ravetz et al, 2004). This is highlighted by some of the current 'sustainability' headline themes, which can be seen to mix communications, aspirations, strategies and tactics, in a post-modern display of mediums and messages:

- 'Sustainable consumption and production' is promoted by the UK Department of Environment, Food and Rural Affairs (DEFRA); while its more powerful neighbour, the Department of Trade and Industry (DTI), continues on a business-oriented material growth trajectory.
- 'Sustainable economic development' and 'sustainable regeneration' are widely sounded: on close inspection, the definition means the lowest-cost strategy that leads to economic recovery while reducing public sector funding.
- 'Sustainable communities' are promoted at the city level with the agenda for social inclusion and avoidance of dependency. The same title is promoted at the UK national level, as a plan to relieve housing shortages in the London hinterland, and revive failed housing markets in the North.

The urban development trajectory of GM over 200 years can now be observed in only 20 years in some Asian cities, and this enormous acceleration can create its own problems. However, the overlapping of different cultures, economies, spatial patterns and environmental problems in GM can be seen from both sides. It creates the problems of weak governance, segmented societies and fragmented urban form; but it also provides a melting pot for diversity, innovation, cultural creativity and new forms of social networks. Other cities with much faster rates of growth may also be able to turn such problems into opportunities.

From affluence to effluence

We can now come back to the starting point and ask, what has the 'green agenda' of the affluent northern cities in common with the 'brown agenda' of southern cities, separated as they are by huge income differences? At one end is the average US city with a per capita income of $35,000, where a private swimming pool is standard, and an EF of 10 gha/cap. At the other end is an average sub-Saharan African city, with a per capita income of less than $350, where clean drinking water is unattainable luxury, and an EF is only 0.5gha/cap. Ironically, the US is both the richest and most unequal nation in the north, and not by chance also has the lowest life expectancy of any developed nation. But the reality of developed and developing in practice is less of opposites, and more of a spectrum.

Firstly, many developing nation cities are on a trajectory that is heading rapidly towards the standards and conditions of developed nation cities. If they are not yet experiencing the problems of majority affluence, it is likely to be on the horizon. Secondly, the underlying contradictions and problems of the cities of the North – in their social economy, social cohesion, social movements, and so on – often resemble the more surface level activities of the cities of the South. The phenomena of privatization, fragmentation and exclusion affect both types of city equally.

Furthermore, there is no clear economic or social division between either end of the spectrum, as the cities of many transitional, Latin American and Asian countries are approaching rapidly the levels of the affluent North. It makes more sense to talk about a 'world urban system' than a binary dependency (Wallerstein and Hopkins, 1996). Meanwhile, there is increasingly direct interaction, through immigrants and migrant workers, where family and community networks are maintained and strengthened through international transport and ICT. The environmental agenda of the urban North is more concerned with property values than with drinking water, but the same underlying factors of dependency and expropriation may still apply. Within each kind of city there is a process of segmentation, exclusion, fracturing – and similar words to describe how the economic, social and physical space is divided and partitioned, generally to define and protect the lifestyles and environments of the wealthy.

Within each kind of city there is also a conflict between levels of governance and ownership – the question of whose city. Does it belong to the capitalists or the workers, the private sector or the public sector, the national capital or the local warlords, the foreign investors or the local professionals? Often, the real patterns of power and influence are complex and difficult to distinguish on the surface.

This also revolves around the question of what and where is the city. The partitioning process above may lead to a situation where most of the wealth is located outside the city boundary and pays no taxes or levies. Elsewhere it can lead to the opposite case, where the majority of the population lives in squatter camps outside the city boundary, receiving no services or rights of property or citizenship.

Finally, each urban system is clearly interdependent on the other – many of the consumer goods which pack the shopping malls and then the waste landfills of Northern cities, are generally produced by low-cost labour in Southern cities, often in extreme environmental conditions. By implication, the sustainable development agenda needs to work on both sides as part of a global whole.

Notes

1 Further details and working papers available on www.eco-region.org and www.ecologicalbudget.org.uk.

References

Amin, A. and Thrift, N. (1995) 'Globalization, institutional thickness and the local economy', in P. Healey et al (eds), *Managing Cities: the New Urban Context*, John Wiley & Sons, Chichester

Bai, X. and Imura, H. (2000) 'A comparative study of urban environment in East Asia: Stage model of urban environmental evolution', *International Review for Environmental Strategies*, vol 1, no 1, pp135–158

Bailey, P. (1997) 'IEA: A new methodology for environmental policy', *Environmental Impact Assessment Review*, vol 17, pp221–226

Barrett, J. (2001) 'Component ecological footprint: Developing sustainable scenarios', *Impact Assessment and Project Appraisal*, vol 19, pp107–108

Barrientos, S. and Dolan, C. (2006) 'Whither ethical sourcing', in S. Barrientos and C. Dolan (eds), *Ethical Sourcing in the Global Food System*, Earthscan, London

Beck, U. (1995) *Ecological Politics in an Age Of Risk*, Polity Press, Cambridge

Beck, U., Giddens, A. and Lash, S. (1994) *Reflexive Modernization: Politics, Tradition and Aesthetics in the Modern Social Order*, Polity Press, Cambridge

Blowers, A. (1993) 'Pollution and waste', in A. Blowers (ed), *Planning for a Sustainable Environment*, Earthscan, London

Borja, J. and Castells, M. (1997) *Local and Global: The Management of Cities in the Information Age*, Earthscan, London

Caves, R. (1992) 'Aviation policy', in J. Roberts et al (eds), *Travel Sickness: The Need for a Sustainable Transport Policy for Britain*, Lawrence and Wishart, London

Champion, T., Atkins, D., Coombes, M. and Fotheringham, S. (1998) *Urban Exodus: A Report for CPRE*, Campaign for the Protection of Rural England, London

Christie, I. and Levett, R. (1999) *Towards the Ecopolis: Sustainable Development and Urban Governance; Report no 12 of the Richness of Cities project*, ECOS Distribution, Leicester

DEFRA (2005) *Sustainable Consumption and Production Indicators: A Revised Basket of 'Decoupling' Indicators*, Department of Environment, Food and Rural Affairs, London, available 2006 at www.defra.gov.uk/environment/statistics/scp/index.htm

DFT (2003) *The Future of Aviation (White Paper CM 6046)*, Department for Transport, London, available at www.dft.gov.uk/stellent/groups/dft_aviation/documents/page/dft_aviation_031516.pdf

Dicken, P. (1998) *Global Shift: Transforming the World Economy*, Paul Chapman, London

Eder, P. and Narodoslawski, M. (1999) 'What environmental pressures are a region's industries responsible for? A method of analysis with descriptive indices and input–output models', *Ecological Economics*, vol 29, no 3, pp359–374

EEA (1999) *A Checklist for State of the Environment Reporting: Technical Paper No 15*, European Environment Agency, Copenhagen, available at http://reports.eea.eu.int/index_table?sort=Published

EEA (2005) *Europe 2005: The Ecological Footprint: Report by Global Footprint Network*, European Environment Agency, Copenhagen, http://reports.eea.eu.int/index_table?sort=Published

Garreau, J. (1991) *Edge City: Life on the New Frontier*, Doubleday, New York

George, C., Howe, J., Ravetz, J. and Roberts, P. (2007) *Environment and the City: Critical Perspectives on the Urban Environment Around the World*, Routledge, Abingdon (forthcoming)

Gouldson, A. and Murphy, J. (1999) *Regulatory Realities: The Implementation and Impact of Industrial Environmental Regulations*, Earthscan, London

Graham, S. and Marvin, S. (2001) *Splintering Urbanism: Networked Infrastructures, Technological Mobilities and the Urban Condition*, Routledge, London

Green, K., Shackley, S., Dewick, P. and Miozzo, M. (2002) 'Long-wave theories of technological change and the global environment', *Global Environmental Change – Human and Policy Dimensions*, vol 12, no 2, pp79–81

Hall, P. (1998) *Cities in Civilization – Culture, Innovation and Urban Order*, Weidenfeld and Nicolson, London

Handy, C. (1994) *The Empty Raincoat: Making Sense of the Future*, Hutchinson, London

Jackson, T. and Michaelis, L. (2003) *Policies for Sustainable Consumption*, Sustainable Development Commission, London, available at www.sdc.gov.uk

Knox, P. L. and Taylor, P. J. (1995) *World Cities in a World-System*, Cambridge University Press, New York

Lave, L., Cobas-Flores, E., Hendrickson, C. and McMichael, F. (1995) 'Using input–output analysis to estimate economy-wide discharges', *Environmental Science and Technology*, vol 29, no 9, pp421–426

Layard, R. (2005) *Happiness: Lessons From a New Science*, Allen Lane, London

Leadbeater, C. (1998) 'Welcome to the knowledge economy', in A. Hargreaves and I. Christie (eds), *Politics of the Future: The Third Way and Beyond*, Demos, London

Logan, M. (1992) 'Environmental capacity of airports: A method of assessment', in Roberts, J. (ed), *Travel Sickness: The Need for a Sustainable Transport Policy for Britain*, Lawrence and Wishart, London
Marvin, S. and Guy, S. (1997) 'Constructing myths rather than sustainability: The transition fallacies of the new localism', *Local Environment*, vol 2, no 3, pp311–318
McEvoy, D., Ravetz, J. and Handley, J. (2004) 'Managing the flow of construction minerals in the north west region of England – A mass balance approach', *Journal of Industrial Ecology*, vol 8, no 3, pp121–140
Morgan, K. (1997) 'The learning region: Institutions, innovation and regional renewal', *Regional Studies*, vol 31, no 5, pp491–504
Mulgan, G. (1997) *Connexity*, Calder and Boyars, London
ONS (2005) *UK National Accounts: The Blue Book*, Office of National Statistics, London, available on www.statistics.gov.uk
Pacione, M. (ed) (1999) *Britain's Cities: Geographies of Division in Urban Britain*, Routledge, London
Pavitt, K. (1984) 'Sectoral patterns of technical change: Towards a taxonomy and a theory', *Research Policy*, vol 13, pp343–361
Ponting, C. (1992) *A Green History of the World*, Penguin, Harmondsworth
Ravetz, J. (2000a) *City-Region 2020: Integrated Planning for a Sustainable Environment*, Earthscan, in association with the Town and Country Planning Association, London
Ravetz, J. (2000b) 'Integrated assessment for sustainability appraisal in cities and regions', *Environmental Impact Assessment Review*, vol 20, pp31–64
Ravetz, J. (2006) 'Regional innovation and resource productivity – New approaches to analysis and communication', in S. Randles and K. Green (eds), *Industrial Ecology and Spaces of Innovation*, Ashgate, Aldershot
Ravetz, J., Coccossis, H., Schleicher-Tappeser, R. and Steele, P. (2004) 'Evaluation of regional sustainable development – Transitions and prospects', *Journal of Environmental Assessment Planning and Management*, vol 6, no 4, pp585–619
Rees, W. and Wackernagel, M. (1995) *Our Ecological Footprint: Reducing Human Impact on the Earth*, New Society Publishers, Gabriola Island, BC
Rifkin, J. (1995) *The End of Work: The Decline of the Global Labor Force*, G. P. Putnam's Sons, New York
Rothman, D. S. and de Bruyn, S. M. (1998) 'Probing into the environmental Kuznets curve hypothesis,' *Ecological Economics*, vol 25, pp143–145
Royal Commission on Environmental Pollution (1994) *18th Report: Transport and the Environment*, Her Majesty's Stationery Office, London
RSWT (2006) *The Mass Balance Movement: The Definitive Reference for Resource Flows Within the UK Environmental Economy*, Royal Society for Wildlife Trusts, Newark, available at www.massbalance.org
Simmons, C., Chambers, N. and Wackernagel, M. (2000) *Sharing Nature's Interest: Ecological Footprints as an Indicator of Sustainability*, Earthscan, London
Tyteca, D. (1996) 'On the measurement of the environmental performance of firms – A literature review and a productive efficiency perspective', *Journal of Environmental Management*, vol 46, p281
United Nations Secretariat, Department of Economic and Social Affairs, Population Division (2001) *World Urbanization Prospects: The 2001 Revision*, United Nations, New York, www.un.org/esa/population/publications/wup2001/wup2001dh.pdf
Van den Bergh, J. C. M., Ferrer-i-Carbonell, A. and Munda, G. (2000) 'Alternative models of individual behaviour and implications for environmental policy', *Ecological*

Economics, vol 32, no 1, pp43–61

von Weizsäcker, E., Lovins, A. and Lovins, L. H. (1997) *Factor of Four: Doubling Wealth, Halving Resource Use*, Earthscan, London

Wallerstein, I. and Hopkins, T. K. (1996) *The Age of Transition: Trajectory of the World-System, 1945-2025*, Zed Books, London

Weale, A. (1993) *The New Politics of Pollution*, Manchester University Press, Manchester

Wiedman, T., Birch, R. and Ravetz, J. (2004) *A Preliminary Ecological Footprint of the North West Region*, Stockholm Environment Institute, York, available at www.ecoregion.org

Wilsdon, J. (ed) (2001) *Digital Futures: Living in a Networked World*, Earthscan, London

Wolters, T., James, P. and Bowman, M. (1997) 'Stepping stones for integrated chain management in the firm', *Business Strategy and Environment*, vol 6, no 3, pp121–132

Wood, C. (1996) *Trading in Futures: The Role of Business in Sustainability*, Wildlife Trusts, Lincoln

Wuppertal Institute (2001) *Resource Use and Efficiency of the UK Economy: A Report to DEFRA*, Department of Environment, Food and Rural Affairs, London, available at www.defra.gov.uk/environment/statistics/waste/research/download/mfaressum.pdf

WWF-UK, Stockholm Environment Institute and Centre for Urban and Regional Ecology (2006) *Counting Consumption: CO_2 Emissions, Material Flows and Ecological Footprint of the UK by Region and Devolved Country*, Worldwide Fund for Nature-UK, Godalming, available at www.ecologicalbudget.org.uk

13

Locating the 'Local Agenda': Preserving Public Interest in the Evolving Urban World

Jeb Brugmann

Introduction

The Urban Environmental Transition (UET) concept does not suggest the necessity of common transition patterns from city to city across different regions during specific historical periods. However, in the modern era we do, in fact, observe similar patterns of environmental transition in widely different cities of diverse societies. Large, low-income settlements, whether in 19th century London or late 20th century Lima, appear to suffer similar public health problems. Industrialization appears to correlate with patterned increases in certain ambient pollutants. High-wealth cities appear to displace increasing portions of their environmental burdens to other territories.

Similarities in the urban environmental transitions of our time can be viewed as arising, in large part, from common institutional factors. In parallel with economic and policy globalization, nations and their cities and towns increasingly share similar norms of production and wealth accumulation, technology preferences and engineering standards, regulatory and governance approaches. These shared norms have been and continue to be spread and institutionalized via global institutions and their political projects. The norms produce similar developmental pathways and, therefore, related ways of managing the environment, of generating and managing waste flows, and of distributing environmental benefits and risks. For example, a globalized 'green' agenda, operationalized through mechanisms such as international conventions and technical assistance programmes, will tend to reproduce very similar regimes of practice from city to city, especially since these mechanisms are dominated by Organisation for Economic Co-operation and Development

(OECD) countries and their established accommodations to industry, their technology preferences and so on.

This chapter explores the institutional preconditions for variations from the observed, dominant patterns of transition. Our particular interest is in transitions that better reflect dominant *local* aspirations, and that thereby more effectively address public interests such as equity, justice and sustainability in the heterogeneous landscape of local ecology, community and culture. The chapter considers the strategic requirements for advancement of such distinct, local public interest agendas.

These issues are explored through consideration of the case of the worldwide Local Agenda 21 (LA21) 'movement'. During the 1990s, this loose but broad federation of local planning and governance projects working under the banner of LA21 engaged more than 6400 cities and towns in some 113 countries in developing locally specific strategies for sustainable development. The LA21 case is of specific interest to the UET discussion because its principal objective has been to resolve the historical tension between global public agendas, which function through the localization of uniform global standards, and local public agendas, which struggle to advance distinct local development aspirations.

'Agendas'

The colloquialism 'agenda' denotes a political project, with an associated discourse, that promotes institutions and patterns of production and social relations that complement its political objectives. The term 'global agendas' is used here to refer to political projects that promote generalized propositions about the world, its needs, priorities and 'best' practices. Global agendas are characterized by their universalist and often modernist impulses. Whether championed by private sector organizations, governments, international civil society (international non-govenmental organizations, or INGOs), or internationalized professions (e.g. engineering), the international nation-states system, coordinated via the United Nations (UN) and Bretton Woods institutions, World Trade Organization (WTO) and myriad international standards organizations, has generally provided the vehicle for the advancement of current global agendas. 'Development', 'sustainability' and 'human rights' are examples of global agendas.

Global agendas advance coordinated global development pathways and transitions, employing largely uniform rights, codes of conduct, standards and priorities. As such, they are abstracted from place. Their relevance, definition and legitimacy are not contingent upon the specificities of local place. Global agendas have both led and increased in prominence with global economic and social integration; many argue that global strategies are needed now more than ever. Their intensifying reach into local society has, however, met with increasing resistance, whether because of their unanticipated impacts on local culture and power relations, because of the dependencies that they create, or because of the sheer inefficiencies associated with their monumental scale.

The term 'local agendas' is used to refer to the political projects of local community systems as they respond to their unique social, economic and ecological conditions, and also as they seek their self-preservation and self-determination. They are vernacular in character and cannot be meaningfully abstracted from place. Local agendas frequently do not apply or reflect the categories constructed by global agendas. For instance, Guha and Martinez-Alier have documented forms of local community action that at once integrate and defy the abstracted categories of 'human rights', 'green' and 'brown' agendas that are often presumed to be distinct, or even in competition, in global policy discourse (Guha and Martinez-Alier, 1997).

From its beginnings, a central objective of the LA21 initiative was to find a way to give localities a lead in (re)defining the meaning of the rejuvenated global development agenda, called 'sustainable development', for their unique contexts. This, its founders argued, might facilitate an important institutional reform of the global development project whereby localities would gain greater strategic control over their developmental pathways, and national and global institutions would support these strategies. LA21's objective was to make development more responsive to unique local conditions, more open to experimentation, and more transparent and legitimate to its presumed (local) beneficiaries. It proposed to build local capacity for managing the conflicts implicit in development's prioritizing of economic, social and environmental objectives. In short, it represented an attempt to reform the machinery of global agenda implementation, providing primary strategic control to localities for achieving global public objectives.

The result, in many instances, was the formation of local development strategies with strong vernacular characteristics. Thus, in the city of Betim, Brazil, a very typical LA21 effort involved the following activities. The LA21 process focused on strengthening the working relationship between the municipality and a low-income district, Citrolândia, a historically isolated and stigmatized community due to its former status as a leper colony. The development of this relationship focused on the rehabilitation of Citrolândia's riverfront for recreational uses, due to the community's complete lack of recreational spaces and its interest in achieving more equal social status with other city districts:

> *Over 3,000 square meters of sidewalks were paved, improving conditions in both the wet and dry seasons. The installation of 35 streetlights contributed to increased safety for local residents. Recreational benches and chess tables were installed. Water quality in the river is continuing to be checked by a monitoring group. The project also contributed to changes in the personal behavior of participants, not only in relation to environmental attitudes such as keeping the yards clean, but also in their personal and family relations.* (ICLEI, 2000)

Whether this activity aligns most with the UN's 'green' LA21, the 'brown' Habitat Agenda, or the 1995 Social Summit Programme of Action is impossible to discern – and locally irrelevant. LA 21 provided a mechanism for Betim to plan its own agenda.

Both district-level local agendas, like the above example, and broader city-wide agendas, as in the well-known case of Curitiba, Brazil, respond to the material, local world as it appears in local places: as an integrated reality. For this reason, unlike global agendas, which aim to shape social reality to abstracted categories and global institutional norms, local agendas were proposed to produce elegant, locality-specific solutions involving simple interventions to address multiple, related social, economic and environmental problems.

Given their different orientations, global and local agendas frequently advance conflicting concepts and norms for governance. The often noted 'distance between the urban poor and the donors' (Satterthwaite, 2001) describes the disconnection between a global agenda (development) and its related governance mechanisms, and diverse locally formulated agendas and their distinct governance processes. Global agendas work to establish and maintain systems of global social regulation, and therefore also generally privilege the operational needs of global institutions and their private and civil society partners. Local agendas work to maintain or to re-establish locally embedded norms and to reinforce local community systems.

This being said, the simple dichotomy of local and global, and the associated choice between 'top-down' and 'bottom-up' governance, does not sufficiently explain our world as it functions. This chapter proposes that local communities be considered as autopoietic systems (Luhmann, 1995), shedding further light on the 'distance' described by Satterthwaite. As Jessop (2001) writes:

> *The concept of autopoiesis (from the Greek for self-production) is used to denote a specific class of systems (whether natural, social, or artificial) that are concerned, at least in the first instance, with their own self-reproduction. Thus their operations are directed at maintaining their own existence rather than at serving the needs of other systems. In this sense such systems are self-constituting, self-organizing, and self-reproducing. These properties make them resistant to top-down internal management and to direct intervention from outside. They nonetheless co-exist and co-evolve in complex ways with other systems with which they are reciprocally interdependent. This poses in turn major problems regarding possible external steering (governing, guiding, managing) and/or strategic coordination.* (Jessop, 2001)

The concept of autopoiesis offers a theoretical framework for understanding the persistent difficulties faced by international development institutions in making successful interventions at the local level. More pertinent to this discussion, it also explains why local communities, in a globalizing world, are today concerning themselves more with governance and policy issues at national, regional and global scales. This concern is increasingly actualized by urban strategies that are designed to influence conditions at all scales.

Therefore, for example, a community system (e.g. a fishing town) which co-evolves within a bioregional system (e.g. Lake Victoria), quickly learns that engaging in regional governance is a meaningful element of any local agenda to reduce the adverse impacts of aggressive exotic species – for example, Nile perch and water hyacinth (Grossman, 1995). Thus, the LA21 process in the coastal city of Mwanza, Tanzania – one of whose major traditional livelihoods is fishing indigenous lake species but whose major industrial employers are fish-processing plants for exports of the invader Nile perch to Europe – closely links its focus on provision of basic water and sanitation services in hillside squatter settlements with a second focus on reducing pollution to the coastal ecosystem and its indigenous fishery (ICLEI, 1998; ICLEI et al 1998b). From a local Mwanza perspective, issues of slum upgrading, sanitation, pollution control, employment and ecological rehabilitation are integrated concerns that must be addressed together and at multiple scales of engagement. Similarly, to use another example, autopoiesis helps explain why more than 600 cities and towns in some 30 countries and 6 continents have joined together in a strategically coordinated 'Cities for Climate Protection (CCP) Campaign' to develop local strategies for mitigating and adapting to climate change.[1]

The concept of autopoiesis helps to clarify why thousands of independent projects to prepare very diverse local agendas also constitute, and are aligning with, a coordinated, strategic global agenda, which is here called the 'Local Agenda'. On the one hand, the suggested Local Agenda, like other global agendas, promotes universal concepts such as 'sustainability', 'equity' and 'participation' and proposes a universal institutional mechanism, a strengthened local state, for application of these concepts. On the other hand, the principal aim of the Local Agenda, as pursued through programmes such as LA21 and the CCP Campaign, is to increase the political space, and to build effective institutional mechanisms, for the development and assertion of vernacular local agendas and their implied, distinct local 'transitions'.

It comes as no surprise that this strategic project is most actively advanced by cities, both independently and in alliance through their national and international municipal associations. Cities themselves represent socio-economic systems that are increasingly operative at scales ranging from the street and neighbourhood to globally operative inter-urban networks and economically-connected distant hinterlands. A now vast body of literature has described and theorized this scalar expansion of urban systems (Lo and Marcotullio, 2000; Sassen, 2002).

This scalar expansion has added to the complexity of urban development processes by increasing the range of agents engaged in bargaining over local development priorities and resource allocation. The articulation and promotion of strategic interests in such multi-agency systems requires complex mechanisms of negotiation, facilitation, coordination, resource allocation – that is, 'governance'. Therefore, the Local Agenda, as a new global agenda, is challenged to address the governance challenges associated with achieving local agendas within a multi-scalar urban system.

Good governance in this complex environment requires an institution or institutional framework that can provide strategic coordination in favour of the local agenda (Brugmann, 1994). The institution charged with such strategic coordination must manage the interaction between local objectives and universal global agendas. As a programmatic initiative focused on establishing the Local Agenda, the LA21 initiative proposed the local state for this function, for three reasons. While local government in many countries was shaped by global colonial projects, it is generally articulated to the specificities of place and legitimated in the context of place. At the same time, as part of the nation state, local government can claim a formal and legitimate place within the international nation-states system that governs global agendas. Finally, in most countries, local governments govern and operate existing development assets: the infrastructure of roads, sewers, water supply, local markets, and services such as waste management through which future development pathways and environmental transitions can be pursued.

We now review in greater detail how the LA21 project unfolded in the decade following its endorsement by the 1992 UN Conference on Environment and Development (UNCED), and how it related to other global agendas seeking to localize their priorities during this same period.

Local Agenda 21's first decade: Discerning the institutional project

The origins of LA21 strategy and practice predated the UNCED and had separate roots from the emerging global sustainable development agenda, with which LA21 is most associated.

The early practice of what later assumed the label of LA21 planning represented classic local agenda formation efforts. These were distinguishable from other kinds of local planning practice *by the strategic and institutional reform orientation of their approach*. The International Council for Local Environmental Initiatives (ICLEI), which founded LA21 as a general planning concept and international programme, explicitly constructed its generalized LA21 approach from cases that pre-dated the UNCED and its Agenda 21 (ICLEI, 1996). Two frequently referenced cases were those of Cajamarca, Peru and Hamilton-Wentworth, Canada. In the early 1990s, the highlands Province of Cajamarca undertook a dramatic re-structuring of local government and used the resulting decentralized system to create its well-known, multi-stakeholder sustainable development planning process (ICLEI, 1995). In 1989, the Region of Hamilton-Wentworth, Canada, confronting the decline of its manufacturing sector, 'decided that new mechanisms were needed to improve the coordination between municipal budget decisions and policy goals and objectives' as well as the integration of the region's Official Plan and Economic Development Strategy (ICLEI, 1998, p79). From this motivation, the region in that year started its internationally recognized Vision 2020/Sustainable Community Initiative. Both Cajamarca and Hamilton-Wentworth were preoccupied less with being 'green' (as defined at the Rio Summit) than with ways to create greater strategic, *public sector* leverage over pressing, and inter-related, local social, economic and environmental concerns.

ICLEI itself was the offspring of the 1980s, grassroots 'municipal foreign policy' (MFP) movement. It had no origins or substantial support in the international environmental community. Driven by local elected officials, the MFP movement promoted the strategic interests of cities and towns in a variety of international relations areas, including refugee policy, local self-determination (e.g. in apartheid South Africa, in Sandanista Nicaragua), and international security (Shuman, 1986). ICLEI was but one of a number of new international organizations created by the local government sector in the 1984–1990 period – including the World Association of Major Metropolises and the United Towns Development Agency – to advance the sector's international interests. In 1990, ICLEI elaborated the global LA21 strategy to advance the interests of local agenda formation in the context of the emerging global sustainable development agenda.

To ICLEI's organizers, the upcoming UNCED provided a critical strategic platform for promotion of greater development planning capacities in the urban sector. ICLEI proposed the LA21 concept to the UNCED Secretary General Maurice Strong and his team in January 1991. With their support, the LA21 concept was endorsed by the nation-states community in a distinct chapter (Chapter 28) of LA21 at the UNCED (Hom, 2002).

Since that time, 'Local Agenda 21' has become the descriptor for a wide variety of local processes that reform local governance and planning approaches in order to address the primary social, economic and environmental challenges of a neighbourhood, town or city in a more strategic and integrated way. In 2001, ICLEI completed a worldwide survey of local authorities and local authority associations engaged in LA21 practice (ICLEI/UNDESA, 2002). The survey identified more than 6400 LA21-type processes in 113 countries, documenting a substantial increase from the 1800 LA21 processes identified in a similar survey undertaken in 1996 (ICLEI/UNDPCSD, 1997). Since that time the LA21 process has continued to expand in new countries and to evolve into new forms, as in the recent development of 'advanced local area management' as a key new community development planning concept in India's urban sector.

The 2001 data presented below is derived from the 633 survey responses received from local authorities (representing 9.9 per cent of the total identified LA21 processes) and from 146 responses received from local government associations. ICLEI tabulated results by country, region and gross domestic product (GDP) level. Where respondents from the same country reported different conditions, ICLEI undertook direct follow-up interviews with relevant respondents. In further analyzing the primary survey data, I made a further review of international and country-level documentation of LA21 activities, including the document archives of the United Nations Development Programme UNDP Capacity 21 Programme. Where ICLEI survey data did not permit characterization of LA21 activities in particular countries, I relied on other researches (e.g. Lafferty, 2001), on correspondences with leading LA21 actors in the relevant countries[2] as well as on my direct field experiences with LA21 activities in 17 countries.

Table 13.1 *Institutional origins of LA21 planning*

Region	Number of identified LA21 processes	LA21 processes motivated by local government development objectives and initiated by			Unknown or other
		Local gov't LGO	National/ international programme	Civil society/ NGO	
Africa (28 countries)	151	26	46	?	79
Asia-Pacific (17 countries)	674	351	267	22	34
Europe (36 countries)	5292	4206	485	254	347
Latin America (17 countries)	119	71	8	21	18
Middle East (13 countries)	79	?	?	?	79
North America (2 countries)	101	15	?	?	86
Total	6416	4669	806	297	643

Source: ICLEI (2002); interviews and correspondence with principal country-level LA21 experts

Analysis of the survey data reveals the formation of a distinctive institutional project that can be differentiated from much of what is frequently generalized today as 'local initiatives'.

Table 13.1 presents the survey findings according to institutional origin and motivation of LA21 activities. As can be seen, the primary instigators of LA21 activities on a country-by-country basis have been organizations motivated by explicit local government development objectives, often within the context of, or in reaction to, decentralization. Such organizations were the primary agents behind 5772, or 90 per cent, of the identified 6416 LA21 processes (in 113 countries). Seventy-three per cent, or 4669 of the identified LA21 processes, were first instigated by local government organizations (LGOs), such as national associations of local government. By comparison, approximately 1100 processes were instigated by national or international programmes that explicitly and actively promote and support local government development, (e.g. UNDP Capacity 21, Deutsche Gesellschaft für Technische Zusammenarbeit (ZGTZ)).

The LA21 survey data reveal a frequently *coordinated* effort by the local government sector, with both *institutional* and *political* dimensions, to strengthen the position of local agendas within a globalized development process.

Coordinated

The coordination of this effort is reflected by the substantial contribution of regional and national LA21 campaigns to the overall definition and growth of LA21 activities. These campaigns were primarily organized and coordinated by national or regional associations of local government or LGOs, which differ from INGOs through their direct, democratic accountability to a local government membership. The primary general purpose of these LGOs is to coordinate policy positions among their diverse municipal members and promote their common strategic interests at the national, continental and international level.

In 2001, LA21 national campaigns in 18 countries, under the management of national LGOs, accounted for 41 per cent of the global total of LA21 processes. On average, the national campaigns each involved 146 cities and towns. In some countries these campaigns succeeded in involving nearly all the country's municipalities in LA21 planning. As an indication of the centrality of these coordination mechanisms to the LA21 project, the countries without national campaigns had, in contrast, an average of 40 participating cities and towns.

The formation of national campaigns had been a central element of ICLEI's LA21 coordination strategy since 1994 (ICLEI/UNDPCSD), 1997). ICLEI, itself an LGO, served as the catalyst or implementation partner in the formation of diverse national and regional campaigns in Africa, Europe and Latin America.

In some regions the leadership of national campaigns joined with regional LGOs to create regional LA21 campaigns. Regional campaigns promoted and supported national campaigns as well as regional LA21 practice generally. In 2002, the European Campaign for Sustainable Cities and Towns counted 1650 cities and towns from 39 countries in its membership and coordinated work among 10 regional city networks. It organized projects, seminars and conferences, provided guidance materials and best practice resources, and undertook LA21 research with academic institutions. It also actively advocated LA21-related policy positions to the European Union (EU). Demonstrating its strategic orientation, in 2003 the campaign launched a new project, called Common Cause, to support global coordination among LA21 actors. The project's website posed the questions, 'How can a Local Agenda 21 Campaign look beyond the local level? How can it be linked up with others to work together on a world-wide level?'

Institutional

The ICLEI survey provides evidence that LA21 has been a local government-led institutional project. In 71 per cent of the reported LA21 processes, the local authority has been the responsible, lead agent for the process. Local authorities have directly managed the process and its budget in 60 per cent of reported cases. LA21 processes have been integrated into the official municipal planning and decision making systems in 59 per cent of the reported cases.

In addition to the specific projects and investments that were generated by these processes, the survey data indicated that LA21 activities were being used to reshape the form and functioning of the local state. In particular, LA21 was used to strengthen the working relationship between the local state and civil society. For instance, 67 per cent of the surveyed African LA21s reported an increase in municipal public consultation. Sixty-one per cent reported an increase in multi-stakeholder partnerships. More than 40 per cent reported an increase in interdepartmental coordination and municipal transparency as well as changes in formal decision making structures. More than one-third of the LA21 survey respondents from Africa, Europe and Latin America reported that LA21 planning has been integrated into formal municipal decision making, budgeting and/or statutory planning processes.

Political

Through ICLEI and the aforementioned campaigns, the self-described 'LA21 movement' actively promoted positions within international policy fora such as the UN Commission on Sustainable Development, the 1996 UN Conference on Human Settlements, the 2002 World Summit on Sustainable Development (WSSD), and the Asia Pacific Economic Co-operation (APEC) process. These positions consistently sought fuller recognition and support for the role of local government; called for direct flows of overseas development assistance (ODA) to local governments and for recognition of decentralized (city-to-city) cooperation as a mode of development assistance; and critiqued decentralization efforts that affected a withdrawal of the state from development responsibilities. In 1998, ICLEI went further to caution against the adverse impacts of privatization policies. It coordinated LGO efforts to oppose the proposed Multilateral Agreement on Investment in order to protect local powers to establish their own procurement preferences – that is, their ability to advance local agendas through their purchasing decisions.

The ICLEI survey data, along with a considerable body of case studies, indicate that the engaged local government community did succeed in establishing a coherent project of local state reform and rebuilding, aimed at increasing local capacities for the development and promotion of locally-distinct development strategies or 'local agendas'. Of course, in such an extensive undertaking, a range of practice can be found; some of the shortcomings of practice will be reviewed in the last section of this chapter. Overall, however, the documented LA21 activities are distinguishable from 'local initiatives' generally due to their emphasis on building local government as the lead strategic agent and facilitator for local development planning. In so doing, many LA21 efforts consistently approached the concept of 'participation' as a structured process of instigating community-based development projects through the engagement of local stakeholders *as citizens* in the reformed processes of local state priority setting, policy development and resource allocation. Along the way, in each locality, in each country, as well as in key international forums, the agents of the LA21 project confronted numerous challenges.

Firstly, the LA21 processes established in each city had to confront myriad local social, political and institutional impediments to the reform of development processes. For instance, the ICLEI survey data indicate limited success in attracting private sector participation to LA21 processes.

Secondly, the often arms-length LA21 stakeholder planning bodies established by many municipalities to coordinate LA21 planning had to manage resistance within local authorities themselves to participatory reforms in governance, operating procedures, and policy.

Thirdly, the LA21 movement, as a global project, had to confront resistance from alternative global projects for reforming governance and development patterns in cities worldwide. As will be described below, the proponents of the LA21 project naively assumed, on the basis of their initial positive reception at UNCED and the emergence of linked discourses on sustainable development, participation and decentralization, that ample international resources and national policy reform could be secured to help local practitioners deal effectively with the first two areas of challenge. To their surprise, the opposite would be the case.

Competing for the 'local': How the international community responds to Local Agenda 21

In preparing the endorsement of LA21 in a special chapter of the UN Agenda 21, the UNCED's last preparatory committee meeting delegated responsibility to 'UNDP, the United Nations Centre for Human Settlements (Habitat) and UNEP [the United Nations Environment Agency], the World Bank, regional banks' to mobilize resources for LA21 implementation (UN, 1992). The response was mixed at best. No international community support was made available until 1994–1995, when the UNDP and UNEP made small grants, together totalling less than US$100,000. Meanwhile, LA21 activities were started with local resources and, in a few countries such as the UK, with support from national governments. For this reason, LA21 planning first established prominence in Europe, where budgetary resources from national municipal associations, central governments and local authorities were quickly mobilized to establish national LA21 campaigns.

The Government of Canada made the first large investment to support LA21 activities in developing countries, reflecting Canada's prominent role in the UNCED. In the mid- to late 1990s the UNDP Capacity 21 Programme assumed a major role in providing support for the establishment of LA21 programmes in developing countries. The Capacity 21 support was augmented by some bilateral development assistance agencies, particularly from countries where LA21 planning had taken hold, such as the Netherlands. Excepting the sustained policy support for LA21 provided by the UN Commission on Sustainable Development, the Capacity 21 effort remained the only sustained, resourced effort of the international community to support LA21 activities on the ground.

The United Nations Human Settlements Programme (UN-Habitat) showed substantial resistance to the promotion of LA21 both leading up to and throughout the 1996 Habitat II Conference process. During the Habitat II preparatory process, sympathetic governments repeatedly proposed, to no objections, the inclusion of text that endorsed LA21 as an implementation mechanism for the Habitat Agenda.[3] But the texts subsequently produced by the secretariat repeatedly omitted these proposals. Even after eventual, explicit LA21 endorsement at the Habitat II Conference, UN-Habitat endeavoured to establish its own 'local Habitat Agenda' process in parallel (and in competition to) LA21. It was not until 1999, when the UN Commission on Human Settlements finally endorsed LA21 as a key Habitat Agenda implementation mechanism, that this resistance was largely put to rest.

At the request of the government of Colombia, the World Bank supported an LA21 capacity-building programme in that country. This was the limit of support provided by the Bank and the regional development banks, which generally pursued their urban investment programmes without any reference to LA21 planning, even when they were taking place in parallel with LA21 processes in the same city. When the World Bank decided to support participatory development planning on a large scale, it undertook a review of LA21 methodology and experiences. The outcome was the establishment of its own ambitious programme in cooperation with UN-Habitat, called the City Development Strategies (CDS) initiative. This initiative, launched in 1999 – the same year that the Unified Nations Centre for Human Settlements (UNCHS) finally endorsed LA21 – avoided association with LA21 and, as will be explored, generally promoted a quite different agenda for localities.

Preparations for the 2002 WSSD provided an opportunity for the international community to align more fully with the expanding LA21 movement. On the eve of the WSSD, the OECD Development Assistance Committee (DAC) issued a special set of guidelines for implementing the promises of Agenda 21, called *Strategies for Sustainable Development*. The document conspicuously focused on the limitations of both local government and LA21.

Representative of many international community documents in the post-UNCED period, the DAC strategy document highlighted problems with local government accountability, transparency and effectiveness – although recognizing that LA21 'can also become a means for promoting these qualities'. It highlighted the faults of a minority of cases in order to diminish the majority: 'While many [LA21s] have led to practical results and impacts, some may be little more than documents setting out goals or plans of government agencies developed with little consultation. They may, in other words, simply be conventional plans renamed.' Even in success, the DAC guidelines reported failure: 'Other LA21s developed in highly participatory fashion and resulting in well-developed action plans, have however foundered because of the limited capacity of city authorities to work in partnership with other groups' (OECD, 2001, p32).

The 'Local-level Strategies' section of the DAC guidelines does not once mention local government. It focuses instead on an alternative conception of local agency, using the notion of 'local communities' to highlight the central roles of community-based organizations, non-governmental organizations (NGOs), traditional fora, 'local groups' and 'user groups'. The section on 'Convergence and links between national, sub-national and local strategies' also does not mention local government. In a section on decentralization, the guidelines emphasize only the need for 'local level institutions for planning and decision-making' (OECD, 2001, p20). The ongoing role of local government as regulators, service providers, developers and managers of public infrastructure goes unnoted.

The DAC guidelines call for 'deep structural changes' to achieve sustainable development (OECD, 2001). But in the local context they appear to envision a deepening direct engagement of international institutions (directly or via INGO intermediaries) with local residents or community-based organizations (CBOs) – not with any locally legitimated public institution, via LA21 processes or otherwise.

Thus, when the parties gathered for the WSSD, the stage was set for an ambivalent position on LA21. On the one hand, governments and senior UN officials offered distinctive recognition to the worldwide LA21 movement. In his opening address to the summit, Mr Nitin Desai, the Summit's Secretary General, placed LA21 at the top of his list of sustainable development accomplishments during the last decade, stating:

> *Many assessments have been made, Mr. President, in preparation for this conference on how much progress has been made in meeting the Rio challenges...We know that there have been some successes – that there is heightened awareness, and that there have been many concrete achievements, particularly in communities which have established local [sic] Agenda 21s.* (Desai, 2002)

But the ceremonial recognition was not accompanied by any substantive support. As the plan's main reference to LA21 indicates, the international community avoided any explicit commitment to LA21's institutional project to shift strategic direction for development to localities via a strengthened and rejuvenated local state. The plan endorsed efforts to:

> *Enhance the role and capacity of local authorities as well as stakeholders in implementing Agenda 21 ... and in strengthening the continuing support for local [sic] Agenda 21 programmes and associated initiatives and partnerships.* (UN, 2002)

Careful drafting ensured that 'local authorities as well as stakeholders' are grammatically designated to provide the indicated 'strengthening' of support for LA21. The plan avoids any explicit or implicit statement of commitment on behalf of central governments or official development assistance institutions.

In the world of the UN, hours, if not days, are spent negotiating single words or phrases. Significant meanings are buried in subtle parsing and semantics. Thus final UN texts of the UNCED, Habitat II and WSSD period repeatedly used the phrase 'local communities' to replace the proposed use of 'local authorities' from early drafts, elevating an entirely different concept of how localities are represented.[4] In this way, the UN's persistent use of the phrase 'local Agenda 21' instead of the term 'Local Agenda 21', which has been consistently advanced by the local government community since before the UNCED, has a defined purpose: this thing we call LA21 is accepted to the extent that it localizes the global agenda, Agenda 21. An LA21 with a capital 'L', to imply that it is something distinctly local, has never been accepted or used in a negotiated UN text.

Why has the development community so frequently dismissed or marginalized LA21 activities in their programmes and budgets at the same time that they have constructed and financed their own, extensive 'local initiatives' programmes as a modality for international development assistance?

One of the primary arguments used in the international development community during the 1992–2002 period is that LA21 planning is restricted to, and addresses the concerns of, developed country communities. However, as shown in Table 13.2, the LA21 survey data undermine these claims. In fact, the LA21 movement is presently growing fastest in middle gross national product (GNP) (US$756–$9265 per capita) developing countries.

Another mischaracterization of the LA21 movement, often used to marginalize if not also to de-legitimate its status in the development community, is that LA21 is primarily a part of the global 'green' movement and its ecology-focused agenda. As such, the argument goes, LA21 imposes a global (green) agenda rather than uplifting local agendas, as the LA21 movement claims. It thereby risks misdirecting the attentions of developing country communities from their more urgent 'brown' needs.

This characterization of LA21 is also contradicted by the ICLEI survey data. The African survey respondents list the top priorities in their LA21s as: capacity-building, community development, economic development, employment, health, land use, poverty alleviation, water resource management

Table 13.2 *LA21 activity by national GNP*

	1996	2001	% change 1996–2001
Low GNP (<US$755)	63	183	190
Mid GNP (US$756–$9265)	118	883	606
High GNP (>US$9265)	1631	5400	231
*Totals minus Germany	(1601)	(3348)	(109)

Source: ICLEI/UNDPCSD (1997); ICLEI/UNDESA (2002)

and women's issues. In Latin America, we find a unique regional emphasis on culture and tourism, education and literacy, and natural resource management, along with many of the same concerns as in Africa. The Asia-Pacific respondents (most from Japan, South Korea and Australia) give unique emphasis to consumption patterns, along with air quality, biodiversity, climate change, community development, energy, land use, transportation, and water resource management. The European respondents share many of the same 'green' concerns as their Asia-Pacific counterparts, but also list community development, education, and health among the top 10 issues addressed.

The sheer number of issues reported to be addressed would appear to indicate either very extensive and comprehensive planning processes or an approach that considers these different concerns in an integrated fashion during the design of specific solutions, as exemplified in the earlier examples of Betim and Mwanza. On the basis of my work with LA21 communities, and the considerable body of ICLEI case studies of LA21 practice, the latter explanation better represents the practice in the developing world, where there are neither resources nor time for extensive, comprehensive LA21 planning as undertaken in some European cities.

Hence, an explanation for the international development community's tepid response to LA21 – which persisted and even deepened as the LA21 movement grew – must be found in other issues. One could claim that poor information-sharing about emerging LA21 practice in developing countries, or the competition for declining ODA funds, or the simple desire of each UN programme to have and control its own campaign – rather than to support such a 'public domain' endeavour – provide explanations. Such explanations might suffice, were it not for the fact that the development community was simultaneously promoting an alternative model of urban governance and development planning.

'Private interest governance' as an alternative global reform agenda

For more than a decade, seemingly in response to the neoliberal policies of the time, international development institutions and development INGOs have steadily constructed a model of what can be called 'private interest governance', which fundamentally competes with traditional notions of public governance through the local state. As the central focus of the LA21 movement was to renew and strengthen forms of public governance, the LA21 approach clashed fundamentally with the mainstream of development community thinking as it emerged in the post-UNCED years.

Responding to the policies of their national government benefactors, many international development organizations emerged in the post-UNCED years as active agents of the neoliberal project to re-structure local political economy in support of market liberalization, privatization and diminished public sector roles and obligations. Emphasizing private investment as an alternative to public sector strategies, whether at the scale of micro-credit or of corporate foreign direct investment (FDI), the development community enlisted politically willing local governments into this alternative development and governance paradigm. They were joined by development NGOs, whose capabilities and techniques for providing stop-gap support to marginalized communities were well honed during decades of local state intransigence and ineffectiveness towards their growing local informal areas.

In this context, the development community developed a new international discourse on 'good governance' and 'local initiatives' in parallel to LA21. 'Governance', to paraphrase Jones, serves as code for 'alternative state projects' advancing 'new forms of representation' that 'support the ideological and material effects of policy aimed at ameliorating crisis' in capital accumulation (Jones, 1998). The neoliberal, private interest governance project supported and legitimized new *localized* (*not 'local'*) mechanisms for planning and resource allocation to replace national corporatist economic management. These were frequently defined by centrally mandated local government reforms, steered by 'public-private partnerships' and programmed according to the priorities and terms of international development assistance organizations (Peck, 1995). The new governance discourse employed the same friendly, albeit vague, terms as the LA21 effort, such as 'bottom-up', 'grassroots', 'responsiveness', 'participation', 'partnerships' and 'accountability', but represented quite distinct and even incompatible strategies, consistent with the neoliberal vision of an 'era of entrepreneurial governance and hollowed out states' (MacLeod and Goodwin, 1999).

As stated by the Governance Cooperative, a Canadian coalition of NGOs, trade unions and business associations, supported by the Canadian International Development Agency:

> *The need for the concept of governance derives from the fact that today, government is widely perceived as an organisation. In its early form government was seen as a process whereby citizens came together to deal with public business... Today, government is viewed as one of several institutional players, like business or labour, with its own interests.* (Martin, 1998)

This deterioration of consensus that public policy and government process is or ought to be the locus of social decisional authority is a point celebrated by an increasing chorus of development studies experts and many INGO and other mission-oriented private organizations that themselves seek a stronger, more direct and more legitimated role in development. 'Instead of seeing weak institutions of the state, which require strong measures designed to strengthen', writes McCarney (1999), 'weak states are instead regarded as an opening for alternative understandings, if not nurturings, of altered power concentrations in state-society relations'. Through the 'governance' discourse, a transfer of decisional authority is promoted from mechanisms of the state (including citizenship) to evolving groupings of 'stakeholders' representing their diverse group interests.

The post-UNCED urban sector programmes of the World Bank exemplify this new governance approach and its primary purposes. The aforementioned CDS initiative,[5] which quickly grew to involve hundreds of cities throughout the developing world, provided a highly effective mechanism for engaging local public and private sector institutions in joint development planning. While, as in the case of LA21, participating cities made diverse application of the CDS process, the dominant urban development paradigm that was promoted, reinforced and rewarded by the CDS programme was the 'competitive cities' development paradigm. To quote the World Bank's country director for the Philippines, in his December 2003 speech to leaders of 31 participating Philippines cities, the CDS aims to advance a particular global agenda:

> *Under the global context where capital, talent and jobs move more freely and quickly to locations offering the best business environment, Philippine cities will have to achieve fast and continued progress in governance, infrastructure and environment for the national economy to remain competitive.* (World Bank, 2003)

The official CDS vision statement for the city of Olongapo, Philippines, illustrated how the programme's central emphasis on growth and competitiveness in synch with a globalized economy is frequently localized:

> *Olongapo seeks to be the first full-fledged free port city in the Philippines within the next decade. It must grow into a dynamic entrepot for trade, commerce and tourism. It must be a hub, a warehouse, a marketplace, a transshipment area, a center for the exchange of goods and services and a window of the country directly linked to the world... The reduction of government intervention in trade matters as well as a liberalized economy will be its foundation. Trade incentive packages should be offered within the area to encourage business and services to relocate to the City...* (City of Olongapo, 1999)

In this agenda for development, where cities are primarily viewed as 'locations' for global business, localized strategies for city competitiveness are implementing globally homogenous urban infrastructure, services and environmental management norms, thus creating recognizable new urban environmental transition patterns.

As documented by Sassen and her collaborators (Sassen, 2002) this 'global city' transitional pattern is characterized by a central business district (CBD) that receives priority investment for a high standard of environmental infrastructure and services, connected through transportation and communications infrastructure to the global market. These central business hubs are supported by residential quarters and educational and research facilities for professional workers, where public-private partnerships provide a similarly high standard of urban infrastructure and services. Environmental problems (e.g. air quality, flooding) affecting these areas receive priority government attention.

In the metrics of this paradigm, certain urban investments are raised to obligatory, urgent status to secure and position the competitive, global, liveable, tourist-friendly city. Hence, in the aforementioned example of Mwanza, Tanzania (where the LA21 process prioritized municipal expansion of water and sanitation services to informal settlements and protection of the local/regional fishery), a simultaneous, parallel World Bank urban infrastructure programme allocated its resources instead to improving services in the CBD and to upgrading the regional primary road network to facilitate (largely fisheries-related) trade shipments in and out of the city. Only 2 per cent of the total project budget was applied to low-cost sanitation facilities (UNDESA, 1998).

Of course, this kind of resource allocation (which in a case like Mwanza ignores the prioritized needs of more than half the local population) leaves large development gaps unfilled. In the expanding urban and commercial districts where informality reigns (often to the wage structure benefit of the desired global trade) an entirely different standard for development is applied, requiring its own parallel institutional strategy. Under the new project and its governance model, the hollowed-out state abandons all pretenses of pursuing historic public obligations and of developing the fiscal, legal or administrative capacity to pursue equitable city-wide development according to locally defined priorities. Here also, the transnational private sector, in spite of the promises of FDI, shows no interest to invest. The resulting institutional gap is fixed through the local establishment of a quasi-private sector in low-income communities that, to use the jargon of the development community, starts with small 'upgrading' activities that, over time, might be 'brought up to scale'. Here, a variety of private, non-state 'local initiatives' are welcomed to fill the gap: self-help programmes, service-providing community-based organizations, NGOs, collectives and micro-enterprises. A global infrastructure of best practices recognition is mobilized to celebrate these gap-filling local initiatives, thereby providing anecdotal doses of encouragement as the objective trends of declining access to urban services, declining public health, and declining environmental justice become increasingly stark.

In this gap-filling 'local initiatives' strategy, NGOs provide a strategic global mechanism to promote and coordinate the new market-based, local collective action sector. The volunteering NGOs thus advance from their traditional roles as chroniclers of base local realities and as advocates for state intervention into the role as alternative to the state, often acting without reference to local state projects, such as LA21. Through this perhaps unwitting alignment with the broader neoliberal project, the NGOs and CBOs are engaged, and now celebrated, in the official development assistance community, as the primary institutions responsible for development of basic services to the poor. Co-opted in this way, the participating NGO retires much of its historic strategic position as coordinator of social movements.

Indeed, as de Azevedo (1998) observes, the multiplying community-based initiatives are not generally aligned with social movements, focused on broader public or social claims, but rather are 'demand-driven movements' focused on specific private, group claims. 'Their goals are therefore negotiable', writes de Azevedo, and are therefore suited to the 'governance' paradigm of interest-group negotiation, 'as there are no questions of principle at stake ... they do not challenge the broader political and social system'.

de Azevedo further observes the limitations of the 'participation' techniques advanced by this governance paradigm, noting its restricted and 'instrumental' nature. 'Participation', he writes, 'is restricted to communities that are to benefit directly from a specific project or local programme...'

> Community organizations participate in these programmes for pragmatic reasons, to obtain extra resources from the authorities. Such programmes, which have their virtues ... are at best palliative measures, such that their attractions to the low-income population would evaporate if government agencies fulfilled their legal objectives by providing a basic minimum of social services. (emphasis added) (de Azevedo, 1998)

Azevedo's last observation is just the point. The private interest governance model presumes a government that will fail and that should be dismantled, devolved and de-centred. It moves us away from the compelling question of how we provide services *as a public obligation and norm* and of how we build the capacity to respond democratically and legitimately to local developmental values and aspirations. It redirects our attention towards the ongoing negotiation of incremental, palliative gains through 'partnerships' in which international aid agencies, INGOs and/or trans-national companies make time-bound local project investments with the 'participation' of the 'targeted' poor and their free labour. These partnerships replace commitment to the development of local public institutions capable of long-term investment in and maintenance of public infrastructure and services, not to mention regulation of private practices. Traditional inequalities in North–South relations and associated aid and trade conditionality thus are augmented by a more trenchant reshaping of local social relations as private arrangements multiply and erode public ownership, control, procedures and cultures of choice making.

Thus, in the private interest governance model, stakeholder participation is used as a social process to orchestrate a new concept of 'inclusiveness', based not on effective enfranchisement in a functioning state, but on one's ability to project one's private claims. By providing a project-based solution to a stakeholder's immediate private interest, the neoliberal project is re-legitimated. 'Local initiatives' provide a short-term institutional fix by isolating smaller private claims (e.g. for increased household water supply) from abandoned and burgeoning public needs (e.g. for public health). This fix can be described as a hybrid of market-based and collective action models of urban services provision (see McGranahan et al, 2001, pp84–111). But the model builds no *local* institutional capacity to address broader public interests or to advance a strategic local developmental project – a distinctive local agenda tailored to local specificities. Strategic control often remains with the international partners, which reflexively assert universalizing, global institutional norms, such as 'cost recovery', 'bankability', 'soundness', 'scalability', or 'replicability'. The strategic leadership – INGOs and international agencies – have no direct local legitimacy; that is, they are supra-national and report to inherently non-local boards of directors. To address this problem, the strategic agents depend upon enlistment of the weak, 'facilitating' local state and willing national and international municipal associations, which often merely lend their names and logos.

Through this form of 'governance' we continue the development tradition of advancing uniform norms of development – and uniform environmental transitions. Diluted are the broader, integrated and non-instrumental public claims, among them local conceptions of equity and sustainability, whose 'stakeholders' may have weak or – in the case of future generations or ecosystems – no voice.[6] These are claims that can only be secured by strong, public norms, legitimated through political process, and enforced by strong public institutions.

Conclusions: A public interest movement in search of an agenda

LA21 proposed to address the failures of the modern state *vis-à-vis* key unmet local public interests and to build its capacity to advance locally defined forms of development. It did so by promoting strengthened capacities for the local state; more inclusive, accountable and integrated local planning and resource allocation processes; and greater local state capabilities as a strategic agent for development. It envisioned a role for cities not as uniform host locations for corporate headquarters and call centres, but as generators of culturally distinctive development pathways in the tradition of cities like Curitiba and Porto Alegre (Brazil), Bologna (Italy) and Barcelona (Spain). Each of these widely recognized cities illustrate the vision of the Local Agenda: places with distinctive local models of *public* interest governance, led by strong, interventionist local governments, which oversee the coherent implementation of a distinctive local developmental strategy over a period of decades. But today the LA21 vision and movement often competes with the neoliberal development project for the commitment of local authorities, and the latter has exponentially greater resources and institutional means to localize its agenda than the LA21 movement ever had to globalize its own.

The focus of LA21 on planning practice, adaptable in a low-cost fashion to specific places, gave it practical merit that enabled its remarkable spread. Yet the primary shortcoming of LA21 in the face of a neoliberal political tide may have been its lack of a clearer, more explicitly communicated and advocated governance or political project.

In the early stages of methodological formulation and negotiation of LA21 movement protocols between local, national, regional and global LA21 actors, the Local Agenda itself – that is the global political project – was left loosely defined. In LA21, the Local Agenda was little more than an ethos and set of planning principles; it rarely produced explicit political proposals. Thus LA21 failed to ally with and support compatible social or political movements.

Table 13.3 *Substantive focus of LA21 processes, by region and income level*

Responses to survey question: 'Which of the following statements best describes the focus of your LA21 process (please check only one)?'

Responses by region	Global (%)	Africa (%)	Asia-Pacific (%)	Europe (%)	Latin America & Caribbean (%)	North America (%)
Focus area						
Economic development	13	26	11	9	25	23
Environmental protection	45	16	54	40	14	34
Social Issues	5	13	6	1	7	0
Focus on economic, social and environmental issues equally	35	45	27	50	50	36
No answer	2	0	2	0	4	7

Responses by Income Level (GNP/capita)	Low income (%)	Middle income (%)	High income (%)
Economic development	34	14	10
Environmental protection	17	43	50
Social issues	9	5	4
Focus on economic, social and environmental issues equally	40	35	35
No answer	0%	2%	2%

The Local Agenda's most compelling proposals were left poorly defined: the centrality of local agendas in a period of globalization; the need for a strong local state to advance compelling, heterogeneous public interests in a rapidly urbanizing world; and requirement for significant local state reform to make the local state a functional, viable, and strategically capable institution for a new 'public' that would include long-excluded majorities and voices.

Lacking a fully explicit political-institutional agenda, LA21 also failed to develop effective institutional strategies. Proposals were made to key development organizations, such as UNDP, to create new finance 'windows' for LA21 plan implementation or to better interface with established project cycles, but these were poorly pursued. Key questions, such as how to engage the private sector *as agents* of local state strategy (as opposed to parallel and competing actors), were not addressed in the vast majority of LA21 experiments. LA21 developed expertise in environmental and development planning, but remained weak and failed to align with innovations in local finance, such as micro-finance and the development of municipal bond markets. The resulting lack of innovation in the spheres of finance and enterprise left LA21 with no substantial capacity to deliver local strategies.

Failing viable delivery mechanisms in a period of market-based opportunity, many cities with active LA21 processes eventually dedicated greater political and institutional resources to alternative private interest governance initiatives, which offered options for finance and private sector engagement. Lacking effective and unified leadership in the wider local government community, with a clear and compelling political and institutional programme, most cities saw few alternatives to neoliberal reform.[7]

However, this competing reform project has largely failed its public promises. Structural adjustment, privatization, FDI and increased trade, deregulation and voluntary industry codes, and the notion of scaled 'local initiatives' together have failed to narrow the gap between poor and affluent, to extend environmental justice or to increase sustainability. The failures of 1990s neoliberal experiments have created new openings for a Local Agenda. While neoliberal experiments have re-engineered the urban world, the reforming and recasting of the local state has not been finished. In fact, neoliberal programmes in many countries may have cleared the way for the rebuilding effort, having curtailed historically problematic urban governance and management practices, rooted in colonialism, centralized planning and resource allocation, and legalistic or elitist approaches. Like autopoietic systems, many cities that have simultaneously experimented with LA21, CDS and other private or community action initiatives are creating their own models, which renew or define new functions for the state and public processes.

In a time of renewed calls for state-building, even in the bastions of neoliberalism (e.g. to address public issues like security), the timing for a Local Agenda may never be better.

Acknowledgements

I would like to thank Peter Marcotullio, Gordon McGranahan, Laila Smith and Wayne Wescott for their insightful comments on this chapter.

Notes

1 See the ICLEI Cities for Climate Protection Campaign website www.iclei.org/co2

2 I would like to specifically thank Karen Alebon, Txema Castiella, Sam Chimbuya, Stefan Kuhn, Sean Southey and Wayne Wescott for clarifications about practice in their country or region.
3 I was the principal representative for local government in the drafting committees for the Habitat Agenda.
4 From 1991 to 2000 I served as a principal technical representative of local government in major UN negotiations on urban development and sustainable development, and witnessed increased substitution of 'local communities' for the proposed references to 'local authorities' as the decade went on.
5 For more information see www.citiesalliance.org.
6 Thus the 2nd World Water Forum in 2000 rejected the notion of an individual's right to a minimum quantity of potable water.
7 During the 1990s, most international associations of local government were fiscally weak, lacked coherent programmes (even on central themes like decentralization), and invested most of their political resources in internecine competition and symbolic positioning. The International Union of Local Authorities (IULA), United Towns Organisation/Cities Unis, and World Association of Major Metropolises had marginal, if any, LA21 involvement. They participated in the major international policy fora of the decade, but provided little if any specific policy proposals to these fora, other than general calls for recognition of local government as an order of government.

References

Brugmann, J. (1994) 'Who can deliver sustainability? Municipal reform and the sustainable development mandate', *Third World Planning Review*, vol 16, no 2, pp129–146
City of Olongapo (1999) 'City development strategies of Olongapo: Presented to the World Bank, Volume 1, July 1999', available at www.citiesalliance.org
de Azevedo, S. (1998) 'Law and the future of urban management in the third world metropolis', in E. Fernandes, and A. Varley (eds), *Illegal Cities: Law and Urban Change in Developing Countries*, Zed Books, London, pp258–273
Desai, N. (2002) 'Opening address to the World Summit on Sustainable Development', Johannesburg, 26 August, www.johannesburgsummit.com
Grossman, E. (1995) 'Nile perch and Lake Victoria infestation problem', *TED Case Studies*, vol 4, no 2, American University, Washington, DC, www.american.edu/TED/PERCH.HTM
Guha, R. and Martinez-Alier, J. (1997) *Varieties of Environmentalism: Essays North and South*, Earthscan, London
Hom, L. (2002) 'The making of Local Agenda 21: How local authorities got involved in the 1992 Rio Earth Summit', *Local Environment*, vol 7, no 3, pp 251–256
ICLEI (1995) *Case Study 30 – Participatory Regional Development Planning, The Provincial Municipality of Cajamarca, Peru*, International Council for Local Environmental Initiatives, Toronto
ICLEI (1996) *The Local Agenda 21 Planning Guide*, International Council for Local Environmental Initiatives/International Development Research Centre/United Nations Environment Programme, Toronto
ICLEI (1998) *Local Agenda 21 Model Communities Programme: Case Studies*, International Council for Local Environmental Initiatives, Toronto

ICLEI (2000) *Case Study 62 – New Goiabinha Project – Betim, Brazil*, International Council for Local Environmental Initiatives, Toronto

ICLEI, CAG Consultants and United Nations Department of Economic and Social Affairs (UNDESA) Division for Sustainable Development (1998) *Barriers to the Implementation of Local Agenda 21*, International Council for Local Environmental Initiatives, Toronto

ICLEI/UNDESA (2002) *Second Local Agenda 21 Survey, Background Paper No. 15*, United Nations, New York, www.iclei.org

ICLEI/UNDPCSD (1997) *Local Agenda 21 Survey*, United Nations, New York, www.iclei.org

Jessop, B. (2001) 'The social embeddedness of the economy and its implications for economic governance', in F. Adaman and P. Devine (eds), *Economy and Society: Money, Capitalism and Transition*, Black Rose Books, Montreal

Jones, M. (1998) 'Restructuring the local state: Economic governance or social regulation?' *Political Geography*, vol 17, no 8, pp959–988

Lafferty, W. (ed) (2001) *Sustainable Communities in Europe*, Earthscan, London

Lo, F.-C. and Marcotullio, P. J. (2000) 'Globalisation and urban transformations in the Asia-Pacific Region: A review', *Urban Studies*, vol 37, no 1, pp77–111

Luhmann, N. (1995) *Social Systems*, Stanford University Press, Stanford

MacLeod, G. and Goodwin, M. (1999) 'Reconstructing an urban and regional political economy: On the state, politics, scale and explanation', *Political Geography*, no 18, pp697–730

Martin, I. (1998) *Building a Learning Network on Governance: The Experience of the Governance Cooperative*, Institute on Governance, Ottawa, p1, www.iog.ca

McCarney, P. (1999) 'Considerations on governance in global and local perspective: Towards a framework for addressing critical disjunctures in urban policy', Paper presented to the 1999 Commonwealth Association for Public Administration and Management (CAPAM) on Educating the Next Generation of Public Administrators, Oxford

McGranahan, G., Jacobi, P., Songsore, J., Surjadi, C. and Kjellen, M. (2001) *The Citizens At Risk: From Urban Sanitation to Sustainable Cities*, Earthscan, London

OECD (2001) *The DAC Guidelines Strategies for Sustainable Development: Guidance for Development Co-operation*, Organisation for Economic Co-operation and Development, Paris

Peck, J. A. (1995) 'Moving and shaking: Business elites, state localism and urban privatism', *Progress in Human Geography*, vol 19, pp16–46

Sassen, S. (ed) (2002) *Global Networks/Linked Cities*, Routledge, New York and London

Satterthwaite, D. (2001) 'Reducing urban poverty: Constraints on the effectiveness of aid agencies and development banks and some suggestions for change', *Environment & Urbanization*, vol 13, no 1, pp137–157

Shuman, M. (1986) 'Local foreign policies', *Foreign Policy*, Winter 1986–87, pp154–174

UN (1992) *Agenda 21*, United Nations, New York

UN (2002) *World Summit on Sustainable Development – Plan of Action*, advanced unedited version, 4 September, United Nations, www.johannesburgsummit.com

UNDESA Division for Sustainable Development (1998) *Barriers to the Implementation of Local Agenda 21, Background Paper No 31*, DESA/DSD, New York

World Bank (2003) *City Development Strategies Conclude Workshop Grant Agreement to Promote Sustainable City Development Signed by RP, WB*, Press Release No. 04/14, 16 December, World Bank, Washington, DC

Index

Page references in *italics* refer to figures, tables and boxes

Accra
  air pollution 139, 149–50
  children's risk factors 6, 134, 142–5, *146–7, 149*
  disease prevalence 134, 142–8, 144, *145, 146–7, 149, 150*
  ecological footprint 151, 152
  economic development 134, 135
  environmental health issues 6, 11, 134, 139, *140,* 142–8, *149, 150*
  housing issues 134, 139, *141*
  hygiene issues 6, 138, 139, 142, *146*
  overview 133–4
  poverty issues 6, 135–6, 139
  resource consumption 151
  sanitation provision problems 6, 136, *137,* 138, 139, 142, *144, 147*
  socio-economic issues 142, *143,* 147–8
  structural adjustment policies 11–12, 134, 136, 139
  urban environmental transition 134–5
  waste problems 138–9
  water supply problems 6, 136–8, 139, 142, *144, 145, 147*
  women's risk factors 6, 11, 134, 135, 142, 143, *149*
Africa
  air pollution 150–1
  child mortality rates *84–5,* 87
  disease prevalence 83, 86, 144–5
  environmental health issues 133

  and LA21 *338,* 340, 344, 347–8, *351*
  local government 92–3
  sanitation provision 2, *80,* 93, 113, 133
  transport *188,* 220
  water supply problems *80,* 90, 93, 113, *114,* 116, 124, 133
  see also Accra; sub-Saharan Africa
air pollution
  ambient 19, 21, *22,* 26, 37, 38–9, 133, 187
  carbon emissions 19, *22,* 30, 322
  as a city-regional scale burden 19, 21, *22,* 37, *190,* 235–6
  and disease 26, 229
  economic development relationship 25, 180, *181,* 227–8
  economic issues 56, 194
  and environmental health 55, 149–50, *190,* 191, 194, 199, 248–9
  extent of exceedance *192*
  as a global scale burden 19, *190,* 191, 195, 200, 284
  government policies 31, 201, 254–61, 262–4
  in high-income cities 58, 60, 222, 228, 284, 322
  indoor 19, 21, *22,* 26
  as a local scale burden 19, 21, *22,* 29, *190*
  in low-income cities 5, 55, 82, 222–3, 227, 230
  in middle-income cities 5, 56, 133, 189, 201, 222–3, 227, 235–6

motorization *see* vehicle emissions below
smog 38, 39, 189, 198, 237, 254, 255, 302
transport systems *see* vehicle emissions below
and urbanization 180, 255
vehicle emissions 6, 180, 187, 189–94, *213,* 222–3, 227, 257–8
see also cities and regions in case studies

Asia
  air pollution 191, *193,* 194, 222–3, 227, 230
  income classification *208,* 209
  local government 92, 94
  motorization 182, *183,* 184, *185,* 186–7, *188,* 191, *193,* 214–15
  sanitation provision 2, *80,* 89, 90, 96, 113
  time-space telescoping 51–2, 180
  transport systems 211, *212–13,* 214–17, 219–20, 221–5, 226–8
  urban environmental transition 4, 179–80
  water supply problems *80,* 89, 90, 96, 124
  see also Asia-Pacific region; China

Asia-Pacific region
  air pollution 55
  child mortality rates *85,* 87
  economic development 52–5
  environmental burdens 47, 52, 53–60
  and LA21 *338,* 344–5, *351*
  motorization 53, 54, 56, 58, 182, *183*
  sanitation provision 53, 54, 57
  time-space telescoping 47–61
  urban environmental transitions 48–61
  waste disposal problems 53, 54, 57, 59
  water supply problems 53, 54, 56, 60
  see also China

Australia
  and LA21 345

motorization 184, *185*
transport systems *208,* 209, 211, *212–13,* 216, 217, 219, 221, 226–7

Bangalore 80, *81,* 93–4, 98
Bangkok *186, 188,* 194, *208,* 209, 214–15, 216
Beijing *30,* 189, *193,* 197, 198, *208*
Brazil 69–70, 91, 92, 96, 333–4, 350
Britain 35–7, 38, 41, 122 *see also* Manchester; UK

Cairo 194, *208*
Canada
  income classification *208*
  and LA21 336, 341
  transport systems 209, 211, *212–13,* 217, 219, 226
children
  education and play 70, 79, 82
  environmental health issues 75, 109, 118, 124, 127, 142
  ill-health problems 6, 75, 86, 108, 134, 142–5, *146–7, 149*
  mortality rates 75, 83, *84–5,* 86
China
  air pollution 56, 194, 195, 197–9, 198
  automobile industry 12, 196–7
  economic development 195, 196
  environmental health issues 195, 199
  government policies 196–7, 199
  income classification *208*
  motorization 12, 181, 182, 194, 195–200, 197–9, 215
  transport systems 217, 220
cities
  boundary problems 5–6, 117–20
  defining 292–3
  environmental burdens and size 31–2, 61, *181, 192,* 242, 246, 278
  high-income *see* high-income cities
  low-income *see* low-income cities
  middle-income *see* middle-income cities
  quantitative comparisons 207–10

resource flows 33, 57, 286, *297,* 308, 313, 315–23
rural areas compared 1, 2–3, 28, 74, 76–7, 277–8
socio-economic issues of size 54, 70, 77, 209–10, 242, 278, 293, 335
urban environment framework 292–3, *294*
city-regional scale burdens
  air pollution 19, 21, *22,* 37, *190,* 235–6
  defining 3
  economic development relationship 13, 14, 19, 21, 52, *310,* 323
  global dimension 14, 15, 40, 295
  of high-income cities *22,* 311–13
  sanitation provision 34
  sustainability issues 301–2, 325–6
  and urban environmental transition 235–6, 325–6
  in urban metabolism 314–15
  waste problems 21, *22,* 37
  water pollution 21, *22,* 37, 57
  see also Accra; Manchester
civil society 31–2, 90, 241–2, 247–8, 333, 334, *338,* 340, 343

Delhi
  air pollution 189, *193*
  degenerated peripheralization 10, 157, 162, 167, 173, 176
  economic development 156–7, 160, 174–5
  electricity provision 171, *172–3*
  employment issues 156–7, 160–2, 169, 175, 176
  environmental burdens 6, 10, 156, 157, 176
  environmental health issues 173
  income classification 209
  industry 10, 156, 157, 162–7, 174–5, 176
  migrant population 156–8, 159, 161, 162, 165, 167, 169, 174
  motorization *193,* 215
  overview 157–60
  rural areas 157, 158, 159, 160, 161–2

sanitation provision 171, *172–3*
slums 10, 157, 161, 163, 165, 167–70, 173–4, 175
socio-economic segmentation 6, 10, 156, 166, 170, 175–6
water pollution 171
water supply problems 170–1, *172–3*
women in employment 160, *161,* 162, 169
developing countries
  developed countries compared 45–6, 47, *55,* 61–2, 191
  disease prevalence 133, 147
  economic development 46–7, 327
  environmental health issues 133, 191
  income classification 209
  and LA21 341
  motorization 180, 182, 200, 229
  time-space telescoping 46, 236
  transport systems 214, 215–16, 230
  urban environmental transition 46, 48–9, 133
  see also cities and regions in case studies
disease prevalence
  and air pollution 26, 229
  Bradley-Feachem classification 107–9
  in developing countries 133, 147
  diarrhoeal 54, 70, 83, 86, 95, 108, 115, 124
  environmental burdens relationship 26–7, 229
  F-diagram model 109–11, 126
  faecal-oral disease 107–8, *109*
  in high-income cities 70, 86, 97–8
  hygiene issues 26, 107–8, 109, 110–11, 115, 126, 142, *146*
  in low-income cities 54, 70, 82–3, 86, 97–8
  malaria 26, 70, 98, 108, *109,* 144
  in middle-income cities 70, 82–3, 97–8
  and overcrowding 2, 142, *145*
  poverty relationship 2, 86, 142, *147,* 148
  and sanitation 2, 21, 26, 72, 82, 108–10, 115

in slums 2, 86, 145, 147
socio-economic issues 142, *143*, 147–8
waste disposal issues 82
water-related 5, 106, 107–9, 110–11, 115, 123–4, 254
see also under cities and regions in case studies

ecological footprint approach
  comparisons 323, *324*
  criticized 286, 316
  defining 27, 28, 274, 285, 315–16
  environmental burdens relationship 8, 20, 274, 286
  high-income cities 5, 20, 286, 319, *322*, 323, *324*
  see also under cities and regions in case studies
economic development
  air pollution relationship 25, 180, *181*, 227–8
  in developing countries 46–7, 327
  environmental burdens relationship 2, 4, 13–14, 18–20, 24–6, 48–9, 179, 302
    city-regional scale 13, 14, 19, 21, 52, *310*, 323
    global scale 13, 19, 179, *310*
    local scale 13, 14, 19, 21, 170, *310*
  environmental health relationship 4, 19–20, 133
  equity issues 1–2, 18–19, 34, 47, 323
  and government policies 20, 30–1
  long waves theory 49–50
  motorization relationship 180, 182, 184, *185*, 187–9, 195, 196, *228*
  'staged-type environmental evolution' (STEE) model 179–80
  and sustainability 35, 325
  time-space telescoping 46, 47, 48–9, 50–2, 180
  urban environmental transition (UET) theory 14, 48–9, 179, 226–9, 235–6, 302, *310*, 323–6
  urbanization relationship 1, 35–6

see also under cities and regions in case studies
economic issues
  of air pollution 56, 194
  externalities 27–8, 29, 71, 121, 122, 127, 283, 316
  globalization 9, 40–1, 50–1, 53, 162, 170, 284
  of sanitation 72, 91–2, 96
  transport systems *212*, *213*, 214, 224–5, 226, 300
  of water supply 57, 91–2, 96
  see also socio-economic issues
Egypt 76–7, 194
electricity provision 38, 88, *141*, 171, *172–3*, 322
environmental agendas 39–41, 52, 179, 275, 279–80, 292, 303–5, 326–7
  see also under Mexico City
environmental burdens
  city-regional scale *see* city-regional burdens
  and city size 31–2, 61, *181*, *192*, 242, 246, 278
  civil society involvement 31–2, 241–2, 247–8
  defining 21, 28, 236–7, 275
  disease relationship 26–7, 229
  displacement issues 13, 19, 36–8, 157, 176, 274–5, 281–8, 331
  dualist approach 13, 276–81
  and ecological footprint approach 8, 20, 274, 286
  economic development relationship 2, 4, 13–14, 18–20, 24–6, 48–9, 179, 302
    city-regional scale 13, 14, 19, 21, 52, *310*, 323
    global scale 13, 19, 179, *310*
    local scale 13, 14, 19, 21, 170, *310*
  environmental health relationship 19–20, 26–7
  equity issues 6, 9, 18–19, 31, 34, 228–9, 236
  externalities 13, 28, 29, 121, 228
  globalization 2, 7, 9, 20, 40, 57, 284
  see also global scale burdens

and governance 8, 9–10, 14, 30–5, 41, 236
government policies 10–11, 14, 20, 29–35, 236, 237–8, 275–6
of high-income cities 58–60
city-regional scale *22*, 311–13
global scale 13, *22*, 39–40, 229, 236, 275, 281–2, 284–5
local scale *22*, 28–9, 52–3
and international agencies 10, 41
local government policies 10, 41, 248, 285, 335
local scale *see* local scale burdens
of low-income cities 2, 52, 53–5, 70, 76–7, 276
of middle-income cities 13, 52, 55–7, 70, 76–7, 229
of poverty 5–6, 29, 228–9, 276, 279
socio-economic issues 45, 51–2, 235, 277–8, 280
spatial dimension 34–5, 41, 235, *312*
time-space telescoping 46, 50, 51, 236, 237
transport systems 6–7, 12, 221–4, 229, 230, 300
of urbanization 2, 35–6
see also under cities and regions in case studies
environmental health
access to services 112, 115–17, 118, 119–20, 127, 128
and air pollution 55, 149–50, *190*, 191, 194, 199, 248–9
boundary problems 5–6, 118, 120, 122, 123–4, 127
of children 75, 109, 118, 124, 127, 142
cost transference 69, 71, 99, 281–5
in developing countries 133, 191
and disease *see* disease
displacement issues 98–9
economic development relationship 4, 19–20, 133
environmental burdens relationship 19–20, 26–7
environmental risk transition 26–7, 133
and governance 11, 88, 91, 96–7

hazard exposure issues 70–1, 82, 98, 139, *141, 149*
in high-income cities 19–20, 23, 71, 97
and housing 71, 75, 79, 82
infrastructure relationship 71, 82, 88, 89, 90
international agencies' policies 11, 74–6, 77
local government responsibility 5, 11, 88, 97
at local scale 2, 6, 13, 21, *310*
in low-income cities 11, 13, 70–2, 75–6, 78–80, 82–7, 97–9, 132–3
in middle-income cities 70–2, 75–6, 78–80, 82–3, 97–9
motorization affects 26, 180–1, 201
overcrowding relationship 70, 71, 142, *145*
pest exposure issues 6, 82, 139, *140*
poverty relationship 5–6, 148, 229, 276
in rural areas 83, *84–5*, 133, 143–4
socio-economic issues 134, 142, *143*, 147–8
urban bias 11, 87–8
and women 6, 11, 135, 142
see also under cities and regions in case studies
environmental Kuznets Curve (EKC) 13–14, 19, 23, 24–5, 28, 31, 179, 302
environmental movements 20, 31, 37–9, 41, 132
equity issues
of economic development 1–2, 18–19, 34, 47, 323
and environmental burdens 6, 9, 18–19, 31, 34, 228–9, 236
of urbanization 1, 2
Europe
air pollution 187, 189, 191, *193*, 223
child mortality rates *85*
income classification *208*, 209
and LA21 *338*, 339, 340, 341, 345, *351*
motorization 182, *183*, 184, *185*, 189, 191, *193*, 211

transport systems 206, 211, *212–13*, 214, 216, 217, 219, 221, 224

global scale burdens
   air pollution 19, *190,* 191, 195, 200, 284
   biodiversity loss 21, 39
   climate change 7, 14–15, 23, 26, 39, 40, 41, 133, 284–5
   defining 3
   economic development relationship 13, 19, 179, *310*
   greenhouse gas emissions 14–15, 19, 21, *22,* 32–3, 40, 58, 236, 284
   of high-income cities 13, *22,* 39–40, 229, 236, 275, 281–2, 284–5
   of motorization 181–7, 200, 221, 284
   ozone depletion 21, 32, 39, 133, 284–5
   political issues 32–3
   resource consumption 21, 23, *310, 314*
   sanitation provision 34
   and sustainability 40, 48, 55, 62
   in urban metabolism *314,* 318–19
   waste problems *22*
globalization
   economic issues 9, 40–1, 50–1, 53, 162, 170, 284
   of environmental burdens 2, 7, 9, 20, 40, 57, 284 *see also* global scale burdens
   localization compared 179, 278–9
governance
   and environmental burdens 8, 9–10, 14, 30–5, 41, 236
   and environmental health 11, 88, 91, 96–7
   'glocalization' 9, 278
   international agencies strategies 345–50
   and LA 21 335–6, 350
   localization 9–10, 278–9, 346, 347–50
   private sector involvement 345–50, 352
   scale issues
      global scale 8, 10, 14–15, 32–3, 41, 279, 334
      local scale 8, 9–10, 14, 41, 91, 96–7, 332, 334 *see also* local government
   and urban agendas 10, 41, 334–5
government policies
   air pollution 31, 201, 254–61, 262–4
   and economic development 20, 30–1
   and environmental burdens 10–11, 14, 20, 29–35, 236, 237–8, 275–6
   global agendas 332, 334
   in high-income cities 10–11, 59–60
   housing 73, 74
   industry 162, 164, 165, 166, 167, 175, 200
   and LA21 8, 10, 236, 333, 336, *338,* 341–2
   motorization 12, 196–7, 199–200, 201
   poverty 74–5
   sanitation 31, 88–9
   slums 73, 167–70, 174, 175
   transport systems 215, 216–17, 230
   water supply 88–90, 252–4
   see also local government

high-income cities
   air pollution 58, 60, 222, 228, 284, 322
   disease 70, 86, 97–8
   displacement issues 19, 281–5, 287–8, 331
   ecological footprints 5, 20, 286, 319, *322,* 323, *324*
   environmental agendas 39–41, 236, 279–80, 292, 326–7
   environmental burdens 58–60
      city-regional scale *22,* 311–13
      global scale 13, *22,* 39–40, 229, 236, 275, 281–2, 284–5
      local scale *22,* 28–9, 52–3
   environmental health issues 19–20, 23, 71, 97
   government policies 10–11, 59–60
   income classification *208,* 209
   and LA21 *344, 351*

motorization 58, 60, 215
pollution control 59, 133
poverty issues 279, 280
quality of life issues 59, 60, 283, *307,* 308
resource consumption 8, 33, 34, 37, 286, 325
roadside populations *194*
sanitation 93-4
sustainability issues 7-8, 39-41, 280, 301-2
transport systems 211-14, 215, 216, 217, 219-20, 221, 223-5, 226-8
urban environmental transition 7, 228, 281-2, 331
urban metabolism 286-7, 292, *312,* 313-15, 319-22
waste management 59, 61, 319
water pollution 58, 60-1
water supply 60, 93-4
see also cities and regions in case studies
Hong Kong *186, 193, 208,* 209, 216, 228
housing
 and environmental health 71, 75, 79, 82
 government policies 73, 74
 infrastructure deficient 71, 72
 international agencies' involvement 75
 in low-income cities 71, 72-3, 79
 in middle-income cities 71, 72-3
 overcrowding problems 2, 70, 71, 79, 133
 private sector involvement 166
 see also slums; squatter settlements
hygiene issues
 and access to services 5, 107, 110, 111-12, 113, 118, 126, 127
 and disease 26, 107-8, 109, 110-11, 115, 126, 142, *146*
 low-income cities 5, 6, 111
 see also under cities and regions in case studies

India 94, 97, 98, 215 *see also* Delhi
industry

automobile industry 12, 196-7, 200
 government policies 162, 164, 165, 166, 167, 175, 200
 relocating 10, 33, 58, 157, 163, 165-6, 175, 176
 spatial structure 157, 163-7
informal or illegal settlements 73, 75, 79, 83, 90, 93, 94, 97 *see also* slums; squatter settlements
international agencies
 and environmental burdens 10, 41
 environmental health policies 11, 74-6, 77
 global agendas 332
 governance strategies 345-50
 housing involvement 75
 and LA21 332, 340, 341-5
 poverty alleviation 74-5, 277-8
 and sanitation 94-6, 98
 on urban bias 5, 11, 74, 95, 277-8
 and water supply 94-6, 98
 see also World Bank

Jakarta 135, *186, 188, 194, 208*
Japan 211, 216, 217, 345

Kuala Lumpur *186, 188, 193,* 194, *208,* 215

Latin America and the Caribbean
 air pollution 191, *193*
 child mortality rates *85,* 87
 and LA21 336, *338,* 340, 342, 344
 motorization 182, *183,* 191, *193*
 poverty issues 70
 sanitation provision 2, *80,* 91, 96, 98
 transport systems 220
 water supply problems *80,* 89-91, 96, 98
 see also Brazil; Mexico City
Local Agenda 21 (LA21) 8, 10, 236, 332, 333-6, 337-45, 350-2
local government
 on air pollution 201
 climate change policies 14-15
 environmental health responsibility 5, 11, 88, 97

environmental policies 10, 41, 248, 285, 335
and LA21 10, 236, 335–6, *338,* 339–41, 343–4, 350–1, 352
services responsibility 91, 92–4, 96, 335
slum policies 174
weaknesses 92–3
see also under cities and regions in case studies
local scale burdens
air pollution 19, 21, *22,* 29, *190*
defining 3
economic development relationship 13, 14, 19, 21, 170, *310*
environmental health issues 2, 6, 13, 21, *310*
global dimension 14, 15, 40
of high-income cities *22,* 28–9, 52–3
and LA21 8, 333
of motorization 220
sanitation provision 6, 21, 29, 34, 235
waste problems 21, *22*
water supply problems 6, *22,* 235
low-income cities
air pollution 5, 55, 82, 222–3, 227, 230
child mortality rates 83, *84–5,* 86
defining 69
disease 54, 70, 82–3, 86, 97–8
displacement issues 19, 34–5, 98–9
environmental burdens 2, 52, 53–5, 70, 76–7, 276
environmental health issues 11, 13, 70–2, 75–6, 78–80, 82–7, 97–9, 132–3
housing issues 71, 72–3, 79
hygiene issues 5, 6, 111
income classification *208,* 209
and LA21 *344, 351*
motorization 54, 55, 229–30
overcrowding problems 5, 70
poverty issues 77–9, 228–9
resource consumption 76, 151
roadside populations *194*
sanitation provision 5, 53, 54, 70, 72, 79, 80, *81,* 89–91

slums 10, 79
transport systems 211–14, 215, 216–17, 219, 220–1, 224, 225, 226–8
urban environmental transition 226–9, 331
urbanization 76–7
waste problems 53, 54, 55, 80, 82
water supply 5, 53, 54, 70, 72, 79, 80, *81,* 89–91
see also cities and regions in case studies

Manchester
airport *296,* 299–301
ecological footprint 8, 323, *324*
environmental burdens 8, *296,* 298–301
motorization 8, *296*
overview 293–7
socio-economic issues 297–301
sustainability issues 301–2, 325–6
transport systems *296,* 299–301, 302
urban environmental transition 298–9, 308, 309–11
Mexico City
air pollution 12–13, 236, 238, 243, 246, 248–9, 254–64 *see also* vehicle emissions below
civil society 241–2, 247–8
economic development 235, 236, 239, 242, 247
environmental agendas 12–13, 237, 239–46, 248–9, 250–64
environmental burdens 7, 12–13, 236–40, 246, 248–9, 254–5
government policies 236, 237–8, 243–5, 248, 252–61, 262–4
housing issues 70, 73, 241, 242–3, 246, 264
motorization *188, 193,* 255, 259, 263, 264
overview 235
sanitation provision 236, 246
time-space telescoping 236
transport systems 248, 255, 256, 259

urban environmental transition 236
vehicle emissions 189, 191, *193,* 194
water supply problems 13, 236, 238, 239–40, 248, 249–54
middle-income cities
  air pollution 5, 56, 133, 189, 201, 222–3, 227, 235–6
  child mortality rates 83, *84–5*
  defining 69
  disease 70, 82–3, 97–8
  environmental burdens 13, 52, 55–7, 70, 76–7, 229
  environmental health issues 70–2, 75–6, 78–80, 82–3, 97–9
  housing issues 71, 72–3
  income classification *208,* 209
  and LA21 *344, 351*
  motorization 6–7, 52, 56, 180, 189, 195–201, 215, 229–30
  overcrowding problems 5, 70
  poverty issues 77–9, 228–9
  resource consumption 76
  roadside populations *194*
  sanitation provision 5, 57, 70, 72, 79–80, 93–4
  transport systems 6–7, 210–25
  urban environmental transition 6–7, 133, 226–9
  waste problems 57, 80, 82
  water pollution 5, 56–7, 133
  water supply 5, 56, 70, 72, 79–80, 93–4
  see also cities and regions in case studies
motorization
  air pollution results 6, 12, 180, 187, 189–94, 197–9
  in developing countries 180, 182, 200, 229
  economic development relationship 180, 182, 184, *185,* 187–9, 195, 196, *228*
  and environmental health 26, 180–1, 201
  as a global scale burden 181–7, 200, 221, 284
  government policies 12, 196–7, 199–200, 201

growth in 6, 180, 181–2, *183,* 187, *189,* 194, 200
  in high-income cities 58, 60, 215
  as a local scale burden 220
  in low-income cities 54, 55, 229–30
  in middle-income cities 6–7, 52, 56, 180, 189, 195–201, 215, 229–30
  see also cities and regions in case studies
Mumbai 93–4, *188, 193, 208,* 216

Nairobi 83, 86, 93, *188*
New Zealand *208,* 209, 211, *212–13,* 216, 217, 219, 225, 226–7
NGOs 73, 240–2, 282, 301, *338,* 346, 348
Niger 113, *114*
North America 182, *183,* 222, *338, 351 see also* Canada; US

pollution 2–3, 26, 59, 133, 199, 229, 299, 304 *see also* air pollution; water pollution
poverty
  and access to services 5, 115, 119–20, 126, 229
  defining 77–8
  disease relationship 2, 86, 142, *147,* 148
  environmental burdens 5–6, 29, 228–9, 276, 279
  environmental health relationship 5–6, 148, 229, 276
  government policies 74–5
  in high-income cities 279, 280
  international agencies on 74–5, 277–8
  in low-income cities 77–9, 228–9
  in middle-income cities 77–9, 228–9
  overcrowding relationship 5
  pollution issues 229
  in rural areas 5, 73, 78, 133
  see also under cities and regions in case studies
private sector
  governance involvement 345–50, 352
  in housing provision 166

and LA21 340, 341, 346, 351
in sanitation provision 11, 95–6
in water supply 11, 95–6

resource consumption
  displacement issues 19, 28, 33–4, 106, 274–5, 286, 287, 299
  as a global scale burden 21, 23, *310, 314*
  in high-income cities 8, 33, 34, 37, 286, 325
  in low-income cities 76, 151
  in middle-income cities 76
  and resource flows 33, 57, 286, *297,* 308, 313, 315–23
rural areas
  access to services 94, 113
  cities compared 1, 2–3, 28, 74, 76–7, 277–8
  employment issues 161–2
  environmental health issues 83, *84–5,* 133, 143–4
  migration 5, 61, 157, 158, 159, 161, 165
  poverty issues 5, 73, 78, 133

sanitation provision
  access issues 5, 111, 112–13, 115, 116–17, 118, 119–20, 126, 127
  boundary problems 5–6, 121, 122–3, 127–8
  as a city-regional scale burden 34
  and disease 2, 21, 26, 72, 82, 108–10, 115
  displacement issues 34–5, 37, 38
  economic issues 72, 91–2, 96
  as a global scale burden 34
  government policies 31, 88–9
  in high-income cities 93–4
  in informal or illegal settlements 90, 93, 94
  infrastructure problems 5, 88, 89, *118,* 119–20
  and international agencies 94–6, 98
  local government responsibility 91, 92–4, 96, 335
  as a local scale burden 6, 21, 29, 34, 235

in low-income cities 5, 53, 54, 70, 72, 79, 80, *81,* 89–91
in middle-income cities 5, 57, 70, 72, 79–80, 93–4
overcrowding relationship 5, *145*
private sector participation 11, 95–6
the sanitary revolution 3, 5, 9, 20, 31, 35–7, 41
sewerage systems 2, 37–8, 72
see also under cities and regions in case studies
São Paulo 19, 135, *188, 193, 208,* 220, 246
Seoul *186, 188, 193, 208*
Singapore *186, 188, 208,* 216, 228
slums
  access to services 2, *81,* 97, 157, 174
  child mortality rates 83
  and disease 2, 86, 145, 147
  government policies 73, 167–70, 174, 175
  in low-income cities 10, 79
  migrant population 157, 165
  relocating 10, 163, 167–8, 169–70, 174, 175
  upgrading 73–4, 161, 168–9
socio-economic issues
  and city size 54, 70, 77, 209–10, 242, 278, 293, 335
  of consumption 306, 307–8
  of degenerated peripheralization 10, 157, 162, 167, 173, 176
  of employment 156–7, 160–2, 169, 175, 176
  of environmental burdens 45, 51–2, 235, 277–8, 280
  of environmental health 134, 142, *143,* 147–8
  segmentation issues 6, 10, 156, 166, 170, 175–6, 297–8, *310*
  and urban environmental transition 306, 307–8, *310,* 311, 325
  see also under cities and regions in case studies
squatter settlements 73–4, 145, 167–8, 170, 229, 240, 327, 335
sub-Saharan Africa 79, 83, *84,* 87, 89, 90, 92–3, 94

sustainability
  at city-regional scale 301–2, 325–6
  and economic development 35, 325
  as a global agenda issue 332, 333, 335
  and global scale burdens 40, 48, 55, 62
  of high-income cities 7–8, 39–41, 280, 301–2
  and LA21 333, 335, 343
  transport systems 6, 201, 228, 300–1

Tokyo *30, 186,* 187, *188, 208,* 215, 228
transport systems
  air pollution 180, 189–94, *213,* 222–3, 227, *228*
  and city size 209–10, 214
  in developing countries 214, 215–16, 230
  economic issues *212, 213,* 214, 224–5, 226, 300
  energy use *213,* 221–2, 227, *228,* 322
  environmental burdens 6–7, 12, 221–4, 229, 230, 300
  government policies 215, 216–17, 230
  in high-income cities 211–14, 215, 216, 217, 219–20, 221, 223–5, 226–8
  infrastructure *212,* 214, 216–17, 226, *228*
  and land use *212,* 214, *218*
  in low-income cities 211–14, 215, 216–17, 219, 220–1, 224, 225, 226–8
  in middle-income cities 6–7, 210–25
  private vehicle ownership/use 210–11, *212–13,* 214, 216–17, 221, 226, *228*
  public transport *212, 213,* 217–21, 221–2, 230
  sustainability issues 6, 201, 228, 300–1
  transport deaths 26, *213,* 223–4, 227, *228*

  and urban environmental transition 207, 210, 226–9

UK *55,* 299, 306, 307, 310, 316, 319–22, 341
urban environmental transition
  at city-regional scale 235–6, 325–6
  and consumption 28, 33–4, 285–6, 305–8, *310,* 314–22, 324–5
  defining 3–4, 207, 226, 235, 302
  in developing countries 46, 48–9, 133
  economic perspective 20, 27–30, 48–9, 132, 179, 302–8, *310*
  empirical analysis 13–14, 19, 23, 24–6
  and environmental justice 18–19, 235
  environmental perspective 20, 27–30, 48–9, 132, 179, 311–13
  health perspective 19–20, 26–7, 132, 179, *310*
  in high-income cities 7, 228, 281–2, 331
  and information 296, *297, 303, 310,* 324
  and LA21 8, 10, 236, 332
  in low-income cities 226–9, 331
  in middle-income cities 6–7, 133, 226–9
  physical dimension 6–8
  political perspective 13, 30–5, 282–5, 331–2
  and production 28, 34, 285–6, 302–5, 306–7, *310,* 314–22, 324–5
  socio-economic issues 306, 307–8, *310,* 311, 325
  spatial dimension 4–6, 14, 227, *303,* 309–11, *312*
  stylized perspective 19, 21–3
  and time-space telescoping 46, 47–52, 180, 236, 237
  and transport systems 207, 210, 226–9
  UET theory 14, 48–9, 179, 226–9, 235–6, 302, *310,* 323–6

urban metabolism model 274–5, 285, 286–7, 292, *312*, 313–15, 319–23
see also under cities and regions in case studies
urbanization 1–3, 35–6, 76–7, 180, 255, 310, *312*
US
  air pollution 38–9
  economic development 49–50
  environmental movements 37–9, 41
  income classification *208*, 209
  motorization 184, *185*, 186, *188*, 211
  sanitation provision 37–8
  transport systems 211, *212–13*, 216, 217, 219, 221–2, 225, 226–8
  urban environmental transition 50
  water pollution 39

waste
  accumulation problems *22*, 37, 319
  as a city-regional scale burden 21, *22*, 37
  disease issues 82
  disposal problems 21, *22*, 53, 54, 55, 57, 118
  as a global scale burden *22*
  in high-income cities 59, 61, 319
  infrastructure problems *118*
  as a local scale burden 21, *22*
  in low-income cities 53, 54, 55, 80, 82
  in middle-income cities 57, 80, 82
  see also under cities and regions in case studies
water pollution
  as a city-regional scale burden 21, *22*, 37, 57
  displacement issues 106, 107
  in high-income cities 58, 60–1

  in middle-income cities 5, 56–7, 133
water supply
  access issues 5, 106, 111, 112–17, 118, 119–20, 124–6, 127–8
  boundary problems 5–6, 121–6, 127–8
  economic issues 57, 91–2, 96
  Falkenmark indicator 124, 128
  government policies 88–90, 252–4
  in high-income cities 60, 93–4
  in informal or illegal settlements 90, 93, 94
  infrastructure problems 5, 88, 89, 90, 118–20
  and international agencies 94–6, 98
  local government responsibility 91, 92–4, 96
  as a local scale burden 6, *22*, 235
  in low-income cities 5, 53, 54, 70, 72, 79, 80, *81*, 89–91
  in middle-income cities 5, 56, 70, 72, 79–80, 93–4
  private sector participation 11, 95–6
  water quality issues 88, 108, 110–12, 118, 124, 127, 138
  water-related disease 5, 106, 107–9, 110–11, 115, 123–4, 254
  water stress 124, *125*, 127–8, 275
  see also under cities and regions in case studies
women
  employment issues 160, *161*, 162, 169
  environmental health issues 6, 11, 135, 142
  ill-health problems 134, 142, 143, *149*
World Bank 11, 12, 73, 95–6, 342, 346–7, 348
World Health Organization (WHO) 75–6, 113, 148